ACS SYMPOSIUM SERIES **640**

Designing Safer Chemicals

Green Chemistry for Pollution Prevention

Stephen C. DeVito, EDITOR
U.S. Environmental Protection Agency

Roger L. Garrett, EDITOR
U.S. Environmental Protection Agency

Developed from a symposium sponsored
by the Division of Environmental Chemistry, Inc.

American Chemical Society, Washington, DC

Library of Congress Cataloging-in-Publication Data

Designing safer chemicals: green chemistry for pollution prevention / Stephen C. DeVito, editor, Roger L. Garrett, editor.

    p. cm.—(ACS symposium series, ISSN 0097–6156; 640)

    "Developed from a symposium sponsored by the Division of Environmental Chemistry, Inc., at the 208th National Meeting of the American Chemical Society [held in Washington, D.C., in 1994]."

    Includes bibliographical references and indexes.

    ISBN 0–8412–3443–4

    1. Environmental chemistry—Industrial applications—Congresses. 2. Environmental management—Congresses.

    I. DeVito, Stephen C., 1955– . II. Garrett, Roger L., 1935– III. American Chemical Society. Division of Environmental Chemistry. IV. American Chemical Society. Meeting (208th: 1994: Washington, D.C.) V. Series.

TP155.D43 1996
660—dc20   96–24785
                                                                                             CIP

This book is printed on acid-free, recycled paper.

Copyright © 1996

American Chemical Society

All Rights Reserved. The appearance of the code at the bottom of the first page of each chapter in this volume indicates the copyright owner's consent that reprographic copies of the chapter may be made for personal or internal use or for the personal or internal use of specific clients. This consent is given on the condition, however, that the copier pay the stated per-copy fee through the Copyright Clearance Center, Inc., 222 Rosewood Drive, Danvers, MA 01923, for copying beyond that permitted by Sections 107 or 108 of the U.S. Copyright Law. This consent does not extend to copying or transmission by any means—graphic or electronic—for any other purpose, such as for general distribution, for advertising or promotional purposes, for creating a new collective work, for resale, or for information storage and retrieval systems. The copying fee for each chapter is indicated in the code at the bottom of the first page of the chapter.

The citation of trade names and/or names of manufacturers in this publication is not to be construed as an endorsement or as approval by ACS of the commercial products or services referenced herein; nor should the mere reference herein to any drawing, specification, chemical process, or other data be regarded as a license or as a conveyance of any right or permission to the holder, reader, or any other person or corporation, to manufacture, reproduce, use, or sell any patented invention or copyrighted work that may in any way be related thereto. Registered names, trademarks, etc., used in this publication, even without specific indication thereof, are not to be considered unprotected by law.

PRINTED IN THE UNITED STATES OF AMERICA

Advisory Board

ACS Symposium Series

Robert J. Alaimo
Procter & Gamble Pharmaceuticals

Mark Arnold
University of Iowa

David Baker
University of Tennessee

Arindam Bose
Pfizer Central Research

Robert F. Brady, Jr.
Naval Research Laboratory

Mary E. Castellion
ChemEdit Company

Margaret A. Cavanaugh
National Science Foundation

Arthur B. Ellis
University of Wisconsin at Madison

Gunda I. Georg
University of Kansas

Madeleine M. Joullie
University of Pennsylvania

Lawrence P. Klemann
Nabisco Foods Group

Douglas R. Lloyd
The University of Texas at Austin

Cynthia A. Maryanoff
R. W. Johnson Pharmaceutical Research Institute

Roger A. Minear
University of Illinois at Urbana–Champaign

Omkaram Nalamasu
AT&T Bell Laboratories

Vincent Pecoraro
University of Michigan

George W. Roberts
North Carolina State University

John R. Shapley
University of Illinois at Urbana–Champaign

Douglas A. Smith
Concurrent Technologies Corporation

L. Somasundaram
DuPont

Michael D. Taylor
Parke-Davis Pharmaceutical Research

William C. Walker
DuPont

Peter Willett
University of Sheffield (England)

Foreword

THE ACS SYMPOSIUM SERIES was first published in 1974 to provide a mechanism for publishing symposia quickly in book form. The purpose of this series is to publish comprehensive books developed from symposia, which are usually "snapshots in time" of the current research being done on a topic, plus some review material on the topic. For this reason, it is necessary that the papers be published as quickly as possible.

Before a symposium-based book is put under contract, the proposed table of contents is reviewed for appropriateness to the topic and for comprehensiveness of the collection. Some papers are excluded at this point, and others are added to round out the scope of the volume. In addition, a draft of each paper is peer-reviewed prior to final acceptance or rejection. This anonymous review process is supervised by the organizer(s) of the symposium, who become the editor(s) of the book. The authors then revise their papers according to the recommendations of both the reviewers and the editors, prepare camera-ready copy, and submit the final papers to the editors, who check that all necessary revisions have been made.

As a rule, only original research papers and original review papers are included in the volumes. Verbatim reproductions of previously published papers are not accepted.

ACS BOOKS DEPARTMENT

Contents

Preface vii

OVERVIEW

1. Pollution Prevention, Green Chemistry, and the Design
   of Safer Chemicals 2
   Roger L. Garrett

2. General Principles for the Design of Safer Chemicals:
   Toxicological Considerations for Chemists 16
   Stephen C. DeVito

METHODS

3. Cancer Risk Reduction Through Mechanism-Based Molecular
   Design of Chemicals 62
   David Y. Lai, Yin-tak Woo, Mary F. Argus,
   and Joseph C. Arcos

4. Isosteric Replacement of Carbon with Silicon in the Design
   of Safer Chemicals 74
   Scott McN. Sieburth

5. Design of Biologically Safer Chemicals Based
   on Retrometabolic Concepts 84
   Nicholas Bodor

6. Predicting Rates of Cytochrome-P450-Mediated Bioactivation
   and Its Application to the Design of Safer Chemicals 116
   Jeffrey P. Jones

7. Use of Computers in Toxicology and Chemical Design 138
   G. W. A. Milne, S. Wang, and V. Fung

8. **Designing Biodegradable Chemicals** 156
   R. S. Boethling

9. **Designing Aquatically Safer Chemicals** 172
   Larry D. Newsome, J. Vincent Nabholz, and Anne Kim

   APPLICATIONS AND EXAMPLES

10. **Designing Safer Nitriles** 194
    Stephen C. DeVito

11. **Designing an Environmentally Safe Marine Antifoulant** 224
    G. L. Willingham and A. H. Jacobson

12. **Imine–Isocyanate Chemistry: New Technology
    for Environmentally Friendly, High-Solids Coatings** 234
    Douglas A. Wicks and Philip E. Yeske

Author Index 247

Affiliation Index 247

Subject Index 247

Preface

WHEN CONGRESS PASSED THE POLLUTION PREVENTION ACT of 1990, a new era in the philosophy and policy of controlling the risks of toxic chemicals was born. Experience of the previous two decades clearly demonstrated that exposure-based solutions to the problems of toxic chemicals are both limited in their effectiveness and increasingly expensive for private industry and society as a whole. The Pollution Prevention Act is nonregulatory in nature but serves as guidance by providing a hierarchical series of approaches to pollution prevention. At the top of this series is "source prevention". In other words, the ultimate approach to preventing problems with toxic chemical substances is not to produce such substances in the first place.

The U.S Environmental Protection Agency's (EPA's) principal initiative to carry out the congressional mandate of source prevention has been the establishment of the Green Chemistry Program. The initial concept developed and implemented under Green Chemistry was the "Alternative Synthetic Pathways" concept presented in symposia at the 1993 and 1994 national meetings of the American Chemical Society (ACS). This concept has evolved into a highly successful program that will undoubtedly lead to the replacement of many of the time-honored, yet polluting, chemical synthesis methodologies employed by academia and private industry today. The sister concept under the Green Chemistry Program is that of "Designing Safer Chemicals". This concept relates to the systematic design or redesign of industrial and commercial chemicals with the specific intent of making them safe or safer for humans and the environment without substantial changes in their efficacy. The concept also encompasses the means by which molecular manipulation for both safety and efficacy can be integrated into every facet of the development, manufacture, and use of industrial-type chemicals. It is this concept, designing safer chemicals, that is the subject of this book.

To many, designing safer industrial chemicals as a fundamental approach in preventing pollution at its source may appear to be somewhat naive or idealistic. There exists an underlying assumption that most chemicals are inherently toxic and that the risks associated with their production and use are inevitable. However, on closer examination, the idea of producing nontoxic or less toxic chemicals does not only seem reasonable, we believe it is achievable. Our belief is rooted in the roles of the medicinal chemist and the pesticide chemist who strive to develop safe drug products and pesticides, respectively. Through careful forethought and strategic molecular manipulation, these chemists minimize unwanted toxicity while maintaining desired efficacy. Our belief is further supported by the fact that commercially viable, nontoxic or less toxic industrial chemicals have already been developed by private industry. Although the idea of designing safer chemicals is not new, the conceptual framework of how this idea can be developed and implemented was first initiated by us in 1992 and culminated with its presentation in the day-long session "Designing Safer Chemicals" at the 208th ACS National Meeting in Washington, DC, in August 1994. This publication presents virtually all of the important information presented at that session and includes additional information that is highly relevant to the concept of designing safer chemicals.

This book is intended primarily for synthetic chemists in industry, academia, and government. These individuals represent the principal architects in the design, redesign, or evaluation of chemicals that ultimately fulfill the needs of an ever-advancing technological society. This group and their expertise represent the nucleus of the multidisciplinary team and knowledge base that will be required to implement the concept of designing safer chemicals. The book is intended also for a broader audience of other scientists and technical managers in private industry and academia; in the fields of chemistry, biochemistry, pharmacology, and toxicology; as well as those in organizations that fund research in these areas. The implementation of the concept will require a concerted effort on the part of all of these individuals, organizations, and institutions.

Turning the concept of designing safer chemicals into a reality will also require the introduction of new paradigms and significant changes in many existing paradigms. These will include changes in our approach to chemical education, new emphasis on mechanistic research in toxicology, and the introduction of greater multidisciplinary collaboration in the research and development of industrial chemicals. As a part of the paradigm changes, we envision the evolution of a new hybrid chemist who acquires not only the knowledge and experience of the industrial chemist in terms of structure–efficacy relationships, but also the expertise of the medicinal and pesticide chemists with respect to structure–biological-activity relationships at the molecular level. This new multidisciplinary scientist, or more specifically, this "toxicological chemist", will serve as a new professional whose role is to design safer chemicals, a new subspecialty in the field of synthetic chemistry.

The goals of this publication are to create an awareness of and an interest in the concept of designing safer industrial chemicals for all those engaged in the research, development, and manufacture of industrial chemicals. To accomplish these goals we have attempted to explain the concept, present a framework of important methodologies associated therewith, and provide examples of the technical and commercial feasibility of the concept when applied to large-volume industrial chemicals. Our efforts really begin with an illustration on the front cover. Here the structure of benzidine is shown in red. Although this substance contains chemical properties that are highly desirable for use in dyestuffs, its commercial usefulness is greatly limited because it is highly carcinogenic. On the other hand, 2,2'-diethylbenzidine, the structure shown in green, represents an analog of benzidine that retains the properties that make it useful in dyestuffs, *but* it is considerably less carcinogenic. The design of the latter substance was based on an understanding of the mechanism of carcinogenicity of benzidene at the molecular level and knowledge of the chemistry required for commercial usefulness. We have also attempted to provide insight into the major steps that must be taken to fully implement the concept and establish the design of safer chemicals as an essential new paradigm for the future development or modification of industrial chemicals. To our knowledge, this is the first book of its kind.

The future of the Green Chemistry Program and, more specifically, the concept of designing safer chemicals, looks promising. In March 1995, President Clinton announced the creation of the Green Chemistry Challenge Program. This new national agenda initiated an ongoing partnership among the EPA, private industry, and academia to recognize and promote fundamental breakthroughs in chemistry that achieve the goals of pollution prevention. The president specifically mentioned the design of safer chemicals as one of the major components of the Green Chemistry Challenge Program. Further, this initiative established a grants program for academic institutions to pursue fundamental research in Green Chemistry, including the design of safer chemicals. The EPA and the National Science Foundation have recently announced the availability of funds for research in this area, and further funding by these and other federal institutions is expected to continue and indeed grow.

As we move into the 21st century, it is clear that we have the tools and the resources to further unlock the secrets of molecular toxicology and to integrate this knowledge with our understanding of the relationships between chemical structure and industrial application. What is needed now is the collective resolve of individuals and organizations in both the public and private sectors to build the proper infrastructure and make the necessary changes to effectively implement the concept of designing safer chemicals. It is hoped that this book will jump-start the process of understanding and acceptance and will serve as the initial catalyst for its implementation.

Acknowledgments

We are particularly pleased with ACS's interest in publishing a book on the subject of designing safer chemicals. We greatly appreciate their support in promoting the concept to chemists and relevant scientists and technical managers in other fields of study. The editors express their sincerest gratitude to the authors for careful preparation of their chapters, to Carol A. Farris for her encouragement and willingness to assist us with this undertaking, and to Barbara Pralle and Vanessa Johnson-Evans of ACS Books for their patience and assistance.

Disclaimer

This book was edited by Stephen C. DeVito and Roger L. Garrett in their private capacities. No official support or endorsement of or from the U.S. EPA is intended or should be inferred.

STEPHEN C. DEVITO
ROGER L. GARRETT
Office of Pollution Prevention and Toxics
U.S. Environmental Protection Agency
Mail Code 7403
401 M Street, SW
Washington, DC 20460

May 3, 1996

OVERVIEW

Chapter 1

Pollution Prevention, Green Chemistry, and the Design of Safer Chemicals

Roger L. Garrett

Office of Pollution Prevention and Toxics, U.S. Environmental Protection Agency, Mail Code 7403, 401 M Street, SW, Washington, DC 20460

There has been an ever growing concern over the adverse effects of industrial chemicals on human health and the environment. Although the U.S. Environmental Protection Agency (EPA) has strived to reduce the impact of toxic chemicals, both the statutory authority and the EPA efforts have focused primarily on exposure-based initiatives. The Pollution Prevention Act of 1990 introduced a new era in regulatory philosophy and policy. Since then the emphasis has been on prevention of toxic substances at their source. One major EPA program implemented to meet the goals of pollution prevention is the Green Chemistry Program, and one of its important initiatives is "Designing Safer Chemicals." This initiative represents a new approach to designing chemicals that emphasizes safety to human health and the environment as well as efficacy of use. Implementing the concept will require major changes in the current practices of both academia and industry. To accomplish these changes, the concept must be understood, accepted and practiced by all those associated with the development, manufacture, and use of industrial chemicals.

As stated in the Preface, this book introduces a new approach to address the adverse effects to human health and the environment of chemicals developed and used by an ever advancing industrialized society. This new approach is the concept and practice of designing safer chemicals. The concept involves the structural design of chemicals to meet the dual criteria of safety and efficacy. Further, the concept relates to the total integration of designing safer chemicals into our thinking, planning, development and implementation of all phases of the design, production and use of chemicals. As we shall see throughout the pages of this book, the design of safer chemicals is a key program of the broader concept of "green chemistry" which represents the ultimate interpretation of of "pollution prevention" as introduced by an act of Congress in 1990 *(1)* and subsequently embraced by the Environmental Protection Agency (EPA) during the past several years.

This new approach can be considered a new paradigm that will challenge the way we have historically viewed and designed chemicals to meet the demands of the continually expanding domestic and world markets. More specifically, it will challenge the way we have dealt with environmental and human health problems caused by industrial and commercial chemicals. In order to meet the aforementioned challenges, it will necessitate changes in the current emphasis, beliefs, policies, procedures and infrastructure in both academia and industry.

Academia must take the lead in advancing our present knowledge of the basic mechanisms of toxicological actions and their relationship to the specific structures of various chemicals and chemical classes. Academia must also re-evaluate its undergraduate and graduate training programs for synthetic chemists to better align them with the concept of designing safer chemicals. Industry must meet the challenge by studying the concept and initiating paradigm shifts across the broad spectrum of its development and use of chemicals. The emphasis on designing safer chemicals must rank with the time-honored considerations of efficacy, cost and market share of chemical products. Finally, government must serve as the catalyst for the introduction and acceptance of the new concept and the necessary paradigm shifts by both industry and academia. The government's role must be the disseminator of information regarding the concept, the initiator or promotor of funding by the appropriate institutions for new toxicological research in this area and the overseer of appropriate regulatory and non-regulatory programs to encourage its implementation by the private sector.

This book captures much of the information presented at the full-day session "Designing Safer Chemicals" held as part of the week-long Symposium "Design For The Environment" presented at the 208th annual ACS meeting in 1994. The publication of this book represents one approach taken by the EPA to disseminate information on designing safer chemicals to industry and academia.

The Historical Perspective

Synthetic chemistry, involving both organic and inorganic chemistry, has been a prominent discipline among the physical sciences for over two centuries. The basic scientific literature on chemical synthesis has been a virtual watershed of knowledge that has fueled the development of large numbers of new industrial and commercial products and intermediates by the chemical industry.

Since the early introduction of chemicals for use in commerce and industry, the emphasis has been primarily on efficacy. Motivated by highly competitive markets, chemists focused principally on how well a chemical functioned in its intended use. Chemists have continuously explored new structural designs and configurations in their attempts to develop better-performing, lower cost products or intermediates. As our chemical knowledge has expanded, many more fundamental principles of the relationships between structure and industrial function have emerged. With all of these efforts, only limited consideration was generally given to the effect that parent molecular structures and functional groups might have on human health or the environment. In most instances, toxicity studies were performed to evaluate the more obvious acute toxicity effects as they related to high exposure of workers during manufacture and processing or consumers during use. Indeed, most of the industrial synthetic chemists, the principal architects and designers of the chemicals, had (and continue to have) little or no training in, or understanding of, toxicology or the relationship of chemical structure to adverse biological effects.

Prior to the establishment of the U.S. Environmental Protection Agency (EPA) in 1970, no federal agency had the direct responsibility or authority to regulate chemicals in commerce for the purpose of protecting the health of the general public and the environment. From EPA's inception until recently, both the statutes providing regulatory authority and the agency's emphasis have been primarily on the reduction of exposure to toxic chemicals *(2-9)*.

During the past several decades academia has made progress in formalizing the curricula and training within the field of toxicology. However, it has remained largely a field of study where chemicals are investigated in terms of their toxic effect endpoints as measured by clinical signs and symptoms, histopathological changes and dose-response relationships. Although pharmacokinetics or toxicokinetics has been an integral part of the field of toxicology, only limited progress has been made in the elucidation of basic toxicological mechanisms at the molecular level on a chemical-specific basis for industrial-type chemicals, particularly the newer ones. This is especially true in the case of non-cancer toxicological effects.

Although structure activity relationships (SAR) have been used by EPA for nearly two decades to evaluate the risk of chemicals with little or no toxicity data, similar SAR considerations do not appear to be a major factor in the *design* of industrial (commercial) chemicals. On the other hand, considerations and investigations of the relationships between chemical structure and biological effects have flourished in the field of medicinal chemistry. Here individual scientists and teams of scientists appropriately trained have been systematically and laboriously investigating the specific impacts of molecular structure on pharmacological, biochemical and toxicological effects as the basis for product development in the pharmaceutical industry. Also, pesticide chemists continuously address the relationships between chemical structure and the toxicological effects on unwanted pests, humans and the environment. These areas of scientific effort represent excellent examples of the collaboration and cross-fertilization of several basic disciplines; namely chemistry, pharmacology, biochemistry, physiology, and toxicology, in pursuit of a common goal.

The New Perspective

During the last half of the 1980s there was a new mood in the Congress and at EPA. The effort to control potential risks through the traditional regulation of exposure was seen as a highly resource intensive approach with no real advantages for the EPA, private industry or society as a whole. There have been continually rising costs to implement new exposure-control procedures, major cross-media considerations, controversy over safe exposure levels and uncertainty regarding the potential effects of low concentrations of a large number of chemicals in the aggregate. With the arrival of William Riley as the EPA Administrator in 1988, a new spirit of cooperation was spawned between EPA and industry. During the next several years the Agency transferred both a sense of responsibility to achieve risk reduction goals and the freedom of how to achieve these goals directly to private sector companies. The approach was to reduce risks by initiating new, primarily non-regulatory programs that established general goals and monitoring mechanisms for risk reduction, and implemented the concepts of right-to-know for the general public. The methods and, in certain cases, the time frames were left to private industry to consider and develop. This approach has been initiated through programs such as the Toxic Release Inventory (TRI) Program, the 33-50 Program and the Design for the Environment (DFE) Program. Each of these programs has placed the initiative, the burden and the rewards directly on private industry.

Six years ago Congress passed the Pollution Prevention Act of 1990. Although the Act is of a non-regulatory nature, the bill introduced the concept of pollution prevention and clearly defined a hierarchy of pollution prevention activities. These included waste disposal, waste treatment, waste recycling, source reduction and source prevention. The latter activity, source prevention, represents the ultimate goal of the pollution prevention concept. It clearly articulates the guidance of not generating toxic chemicals in the first place. The overarching program for source prevention is the Green Chemistry Program. This program, which is non-regulatory in nature, is a composite of several programs which address the reduction or elimination of toxic chemicals through fundamental changes in the field of chemistry.

One major program related to source prevention through Green Chemistry has been the development of the concept of "alternative synthetic pathways." This program was presented as a symposium at the 206th and 208th annual ACS meetings and is designed to have chemists rethink the chemical synthetic steps and pathways used to manufacture industrial and commercial chemicals. The intention has been to reduce or eliminate the use of toxic feedstocks, solvents, catalysts, impurities and byproducts during the manufacture of chemicals. This requires basic research and an extension of our fundamental knowledge in synthetic chemistry. In the six years since its inception, the program has been highly successful in stimulating new research on environmentally benign chemical syntheses by academia and private industry. As of 1996, funding of research in this area has exceeded $10 million annually and promises to continue its rapid rate of expansion.

Within the Green Chemistry program, the sister concept to "alternative synthetic pathways" is the "design of safer chemicals." That is, in addition to finding safer routes of synthesis, it is essential that we design the molecular structure of the final product so that it is safe or safer than the chemical it replaces and yet is effective with respect to its intended use. Although the design of safer chemicals may require more basic research in chemistry, the larger requirement for the full implementation of the concept will be more basic research in toxicology and an effective integration of this information into the area of molecular design.

The Concept of Designing Safer Chemicals

The concept of designing safer chemicals can be defined as:

> "The employment of structure - activity relationships (SAR) and molecular manipulation to achieve the optimum relationship between toxicological effects and the efficacy of intended use."

The key word in this definition is "optimum" since it is often not possible to achieve zero toxicity or to achieve a maximum level of efficacy, but rather some optimum combination of the two goals. Further, the concept of designing safer chemicals encompasses both human health and the environment, including all other animal, aquatic and plant life. It applies also to the structural modification of existing chemicals as well as the structural design of new chemicals as shown in Figure 1.

The idea of designing safer chemicals is not new. It was the subject of a symposium held in Washington, D.C. in 1983 sponsored jointly by the EPA, Oak Ridge National Laboratory and the Society of Toxicology. In a paper presented at this symposium *(10)* the design of safer chemicals was referred to as the "domestication of chemistry" by Dr. E.J. Ariens, the noted medicinal chemist with exceptional expertise in organic chemistry, molecular toxicology and pharmacology. The use of the term "domestication" is ideal. As defined in Webster's Collegiate Dictionary, domestication is:

> "to adapt to life in intimate association with,
> and to the advantage of, man."

As Dr. Ariens suggests, we have only to extend this definition to chemicals to understand why the term domestication of chemistry is an apt description of the designing safer chemicals concept.

Although the idea of designing safer chemicals is not new, the *concept* of how this idea can be developed, introduced and integrated into the real world of commercially viable industrial chemicals is new and is the subject of this chapter and indeed this book. Throughout history chemists have continually designed new chemicals or redesigned existing chemicals to achieve greater effectiveness of a chemical intermediate or chemical product. The knowledge and skill of the chemist has been developed and focused primarily on the physical and chemical properties of chemical classes and chemical substituents and the molecular manipulation of these factors to attain the desired industrial properties of the end product. As a result, chemists have been the principal architects of industrial and commercial chemicals and it is their structural designs that dictate the biological ramifications of these substances in humans and the environment.

The concept of designing safer chemicals brings a new dimension to molecular design; namely, a far greater consideration of the impact of a new or existing chemical on human health and the environment throughout the chemical's life-cycle of manufacture, use and disposal. This involves not only important considerations by chemists regarding the design of new chemicals, but, equally important, the reconsideration and redesign of a wide range of existing chemicals.

These considerations must include both direct and indirect impact of the chemicals on human health and the environment. Also, they will involve a basic knowledge and understanding of the routes of exposure and mechanisms of access or uptake as well as *in vivo* mechanisms of toxicity to humans, animals, aquatic organisms and plant life.

Considerations in Designing Safer Chemicals

In general, the considerations in designing safer chemicals may appear rather straightforward. Design a chemical that cannot gain access to an organism, or if access cannot be denied, design a chemical that will not adversely affect the normal biochemical and physiological process of any organism. However, considering the complex, diverse and dynamic nature of living organisms, in practice this becomes a formidable challenge. As the principal architect, the chemist must be knowledgeable of the many potential approaches to designing safer chemicals and fully understand the details of the relationships between chemical structure and biological effects. They must then carefully evaluate the numerous approaches to structural design to avoid adverse biological effects at the molecular level. Additionally, the chemist must consider chemical structure as it relates to the fate of chemicals released to the environment in terms of degradative structural changes, the potential for dispersion in air, water and soil and the potential adverse effects in the environment. This includes not only considerations of direct impact on living organisms but also indirect and delayed impacts such as acid rain, the ozone layer and global warning. With a sound understanding of these relationships at the molecular level, the chemist can then evaluate the numerous approaches to structural design that will maintain the efficacy of the chemical and at the same time avoid adverse biological effects in humans or the environment.

Table I represents a summary of general considerations related to accomplishing safer chemical design. A more comprehensive discussion of these considerations and specific approaches to designing safer chemicals is presented in Chapter 2 of this book. The considerations in designing safer chemicals can be divided into those that are "external" to the organism, including man, animals, aquatic organisms, and plant life and those that are "internal" and address approaches to chemicals that may gain access to these same organisms.

The external approaches relate to a reduction in exposure by designing chemicals that influence important physical and chemical properties related to environmental distribution and the uptake of the chemical by man and other living organisms. Structural designs or redesigns that will increase degradation rates and those that reduce volatility, persistence in the environment or conversion in the environment to biologically active substances represent examples of major external considerations for reducing exposure. Additionally, molecular designs that reduce or impede absorption by man, animals and aquatic life also represent important external considerations. The routes of absorption may include the skin, eyes, lungs, gastrointestinal tract, gills or other routes depending on the organism. Bioaccumulation and biomagnification (the increase in tissue concentration of a chemical as it progresses up the food chain) represent other important considerations related to the design of chemicals to reduce exposure. It is well-known that certain chemicals, for example chlorinated pesticides and other chlorinated hydrocarbons, will be stored in the tissues of a wide range of living organisms and may accumulate to toxic levels. This phenomenon is exacerbated by the fact that the lower forms of life or the organisms at lower trophic stages are subsequently consumed as food by fish, mammals and birds. These species in turn may be consumed by humans. Hence, the substances of concern may both bioaccumulate in lower life-forms and biomagnify or increase their concentration in higher life-forms by orders of magnitude as they accumulate and migrate up the food chain.

The reduction or elimination of impurities in a chemical intermediate or final product may be associated with the Green Chemistry program on alternative synthetic pathways. However, the approach also may be considered as an integral part of designing safer chemicals. This includes considerations of the presence of impurities

Current Status	Efficacy and Safety Considerations	Final Status
need for a new chemical	———structural design——→	new chemical that is safe and efficacious
existing ("toxic") chemical that is efficacious	———structural redesign——→	modified chemical that is safe and efficacious

Figure 1. Designing Safer Chemicals.

Table I. Summary of General Considerations in Designing Safer Chemicals

I. "External" Considerations - Reduction of exposure or accessibility
   A. Properties related to environmental distribution/dispersion
      1. volatility/density/melting point
      2. water solubility
      3. persistence/biodegradation
         a. oxidation
         b. hydrolysis
         c. photolysis
         d. microbial degradation
      4. conversion to biologically active substances.
      5. conversion to biologically inactive substances

   B. Properties related to uptake by organisms
      1. volatility
      2. lipophilicity
      3. molecular size
      4. degradation
         a. hydrolysis
         b. effect of pH
         c. susceptibility to digestive enzymes

   C. Consideration of routes of absorption by man, animals or aquatic life
      1. skin/eyes
      2. lungs
      3. gastrointestinal tract
      4. gills or other species-specific routes

   D. Reduction/elimination of impurities
      1. generation of impurities of different chemical classes
      2. presence of toxic homologs
      3. presence of toxic geometric, conformational or stereoisomers

Table I (Continued)

II. "Internal" Considerations - Prevention of toxic effects

    A. Facilitation of detoxication
        1. facilitation of excretion
            a. selection of hydrophilic compounds
            b. facilitation of conjugation/acetylation
            c. other considerations
        2. facilitation of biodegradation
            a. oxidation
            b. reduction
            c. hydrolysis

    B. Avoidance of direct toxication
        1. selection of chemical class or parent compound
        2. selection of functional groups
            a. avoidance of toxic groups
            b. planned biochemical elimination of toxic structure
            c. structural blocking of toxic groups
            d. alternative molecular sites for toxic groups

    C. Avoidance of indirect biotoxication (bioactivation)
        1. addressing bioactivation
            a. avoiding chemicals with known activation routes
                (1) highly electrophilic or nucleophilic groups
                (2) unsaturated bonds
                (3) other structural features
            b. structural blocking of bioactivation.

during synthesis such as more toxic homologs, geometric isomers, conformational isomers, stereoisomers, or structurally unrelated impurities.

The internal considerations in designing safer chemicals, are more complex than the external considerations. In general they include approaches using molecular manipulation to facilitate: biodetoxication; the avoidance of direct toxicity; or the avoidance of "indirect biotoxication" or bioactivation. (These approaches are discussed in great detail in Chapter 2). Facilitating biodetoxication involves the design of chemicals that may be hydrophilic per se or that are readily conjugated with glucuronic acid, sulfate or amino acids to accelerate urinary or biliary excretion. The avoidance of direct toxicity involves the selection of nontoxic chemical classes or nontoxic parent compounds as well as the use of functional groups that are nontoxic. However, toxic functional groups may be used to satisfy the efficacy of the chemical if the molecular design includes a plan for the biochemical removal of the toxic groups through the normal metabolism of the organism or the design includes the strategic molecular relocation of the toxic group, or the structural blocking of the toxic group to prevent adverse biological effects.

Finally, internal considerations may involve molecular designs that avoid indirect biotoxication. The term "indirect biotoxication" or bioactivation describes the circumstances where a chemical is not toxic in its original structural form but becomes toxic after *in vivo* transformation to a toxic metabolite. Bioactivation represents a characteristic mechanism for the toxicity of many carcinogenic, mutagenic and teratogenic chemicals. Important approaches used here are: incorporation of structural modifications that prevent bioactivation; or the avoidance of the use of chemicals or chemical classes with known activation routes to produce substances that are electrophilic, and react irreversibly with endogenous nucleophiles (e.g. DNA).

It is clear that both the external and internal considerations provide a wide range of opportunities and approaches to the synthetic chemist for designing chemical structures that reduce or eliminate the toxicity of industrial and commercial chemicals. The opportunities and approaches are expanded further by the possibility of factoring more than one approach into the molecular design. For example, chemical design considerations may include both properties that reduce exposure and one or more properties that facilitate excretion or metabolic deactivation.

It is acknowledged that the effective harmonization of the safety considerations of complex living organisms with the efficacy considerations of chemical structures for industrial and commercial purposes is a difficult goal to achieve. However, this harmonization is a critical prerequisite for success. The design of safer chemicals will require the ready availability of the data and information on the relationship between chemical structure and industrial/commerical function. More importantly, it will require the development of more data and information on the structure-biological activity relationships of these same chemicals at the molecular level. Most importantly, it will require the careful integration of these two sources of data and information to achieve the delicate balance between safety and efficacy. To reach this level of scientific competency a new foundation must be built that will require major changes in current toxicological research and the training of chemists, and more collaboration and cross-fertilization of the relevant scientific disciplines.

Building the Foundation for Designing Safer Chemicals

To bring about a universal practice of the design of safer chemicals, substantial changes must take place both in academia and in private industry. The major changes are summarized in Table II.

Awareness of The Concept. As Table II indicates, the first step is an increased awareness in the concept of designing safer chemicals. As part of its overall program on pollution prevention, and the Green Chemistry Program in particular, the EPA has taken steps to increase the awareness of the concept to chemists and other relevant members of the scientific community. The 1994 symposium and this book are examples of the meaningful approaches to introducing the concept. EPA hopes to further this awareness through other scientific publications and meetings, and to solicit the aid of that sector of the news media that specifically focuses on the scientific and industrial communities. It is hoped that EPA's efforts to familiarize chemists and their respective managements in academia and industry will be further supplemented by individuals and organizations already familiar with the concept.

Scientific and Economic Credibility. The next steps are establishing the scientific credibility of the concept with respect to academia and the funding institutions and demonstrating the technical and economic feasibility from the standpoint of private industry. This book serves this purpose by providing excellent examples of the design and redesign of chemicals destined for practical commercial use. Chapter 3 demonstrates how the carcinogenic properties of many commercial aromatic amines (e.g. benzidines, anilines) can be greatly reduced by molecular modifications that facilitate excretion in the urine or prevent bioactivation. Chapter 4 demonstrates how chemical toxicity may be reduced by isosteric replacement of a carbon atom with a silicon atom. Chapter 5 provides examples of how toxic substances (e.g. DDT) can be redesigned such that they will retain their commercial efficacy but will decompose rapidly under physiological conditions to innocuous, readily excretable products. Chapter 8 discusses how commercial chemical substances that are toxic because of their ability to persist in the environment can be redesign such that they biodegrade readily. Chapter 10 describes how an understanding of the mechanisms of toxicity of nitriles and their structure-activity relationships can be used to design commercially useful but less toxic nitriles. Chapters 11 and 12 were written by chemists working in industry, and demonstrate the technical, economic and commercial feasibility of designing safer biocides and paint constituents, respectively. It is believed that there are other examples of similar successful efforts undertaken by private industry and it is hoped that these approaches will be shared with others in future workshops, symposia and the published literature as the program gains momentum.

Focus on Chemicals of Concern. There must be a sharper focus on, and the establishment of priorities for, those chemicals and chemical classes of greatest concern to human health and the environment. Both industry and academia should focus their attention on those commercial chemicals and chemical classes that have the greatest potential for adverse effects. This involves not only an assessment of the toxicological properties per se, but also a consideration of the extent of the potential exposure to humans and the environment. Factors such as production volume, use and physicochemical properties should be considered since it is the combination of the toxic effects and the extent of the exposure that determines the risk. Using these criteria, academia and industry can establish priorities for purposes of toxicological research and the evaluation of industrial chemical products, respectively.

Mechanistic Toxicological Research. Research in toxicology must shift its emphasis to mechanistic research, or the basic understanding of how a specific chemical or chemical class exerts its toxicological effect on living organisms at the molecular level. It is only with the accumulation of substantial data and information of this nature that the existing principles and concepts of structure - activity relationships (SAR) can be developed further. The acronym SAR is the short hand routinely used to describe the relationship between the specific structural features of a chemical and its biological activity or biochemical/toxicological impact on a living organism. The SAR of a chemical may involve one or more functional groups, the parent compound or a combination of functional groups and the parent chemical or chemical class. The elucidation of toxicological mechanisms on a chemical-specific or class-specific basis and the systematic compilation of this data will provide the necessary foundation and guidance for the molecular manipulation by synthetic chemists to develop safer chemicals. To stimulate interest and provide academia with the means to undertake more basic research in toxicology, the appropriate funding institutions must accept the concept and actively participate by making funds available in this specific area of research. Without financial support for the conduct of more basic mechanistic research and the subsequent expansion of the SAR database, the opportunities for new, creative molecular structures that are both efficatious and safe will be severely limited.

Revision of Chemical Education. For synthetic chemists to make effective use of these data and information, there will be a need for the revision of the existing concepts and practices of chemical education at both the undergraduate and graduate level. Traditionally, synthetic chemists have been trained to design and build chemical structures and to understand the relationship between these structures and the chemical's intended function. Only limited emphasis has been placed on the biological significance of the chemical structure in most instances. Exceptions to this include the medicinal chemists and pesticide chemists. However, in these cases the training and experience in functionality is focused primarily on biological effects as related to pharmaceuticals or pesticides with little knowledge or interest in the relationship of structure and industrial/commercial applications of other types of chemicals.

Although the function of designing safer chemicals can be accomplished through multi-disciplinary collaboration among chemists, toxicologists, pharmacologists, biochemists and others, it is believed that individuals with a combined knowledge of chemical structure, industrial application and biological activity at the molecular level will perform more efficiently and effectively. It is believed that this would be true even if such an individual functioned as part of a multidisciplinary team. This is born out by experience with medicinal chemists and pesticide chemists and their effectiveness in developing new pharmaceuticals and pesticides. The former utilize their knowledge and experience of pharmacology and biochemistry as an integral part of their efforts to structurally design new or improved pharmaceutical chemicals with minimal or no toxic effects. Pesticide chemists also employ their expertise on the impact of chemical structure on insects, rodents and other unwanted pests as well as their toxic effects on humans and animals in the design of pesticide chemicals.

To provide adequate training of synthetic chemists interested in designing safer chemicals and destined for careers in both academia and private industry, it is believed that new curricula should be developed to provide firm grounding in biochemistry, pharmacology and toxicology. At the graduate level this may be best accomplished through joint appointments and multi-disciplinary graduate committees comprised of the appropriate fields of study to oversee curricula and graduate research efforts directly related to the chemistry/biology relationships involved in designing safer chemicals. Although extra time and effort may be required for the completion of such a program, the rewards to the individual student, the chemical industry and society will justify the program. As the emphasis on harmful chemicals intensifies and EPA's pollution prevention programs grow, individuals with this background and training will be highly sought after by universities and private companies alike.

The introduction of these new multi-disciplinary curricula and training will result in the emergence of a new "hybrid chemist" or perhaps, the "toxicological chemist." As depicted in Figure 2, this hybrid chemist will evolve from the current subspecialties in synthetic chemistry. The classical industrial synthetic chemist focuses primarily on chemical structure as it relates to the intended industrial chemical function of the substance as an intermediate or final product. The medicinal chemist and pesticide chemist focus on molecular design as it relates to pharmacological effects in humans or animals and toxicological effects in unwanted pests, respectively. The new hybrid chemist or the "toxicological chemist" must consider both the function of the chemical in its industrial or commercial application and its toxicological effects in humans and the environment. In most respects achieving the delicate balance between safety and efficacy will undoubtedly prove to be the most difficult and challenging effort in the history of synthetic chemistry. However, with the appropriate resolve and focus, the development of such chemicals can be achieved.

Chemical Industry Involvement. As shown in Table II, major support and participation by the chemical industry is essential to fully realize the practical potential of the concept of designing safer chemicals. Further, industry must take steps to increase the awareness of the concept among its scientists and its management. Industry must encourage its people to approach the concept with open minds and to carefully evaluate its potential in terms of economic and technical feasibility. Both basic and applied research programs should be initiated by industry to address the design of safer chemicals both in-house and with other research organizations. This undoubtedly will require substantial investments in time, money and human resources. It will also require shifts in existing paradigms, the opening of new lines of communication and coordination with academia and the establishment of new relationships with the EPA.

To implement the concept of designing safer chemicals it is essential that all of the steps summarized in Table II be taken. Further, they cannot be accomplished without a concentrated and cooperative effort on the part of all relevant organizations in both the private and public sectors. EPA, with the help of the American Chemical Society (ACS), has introduced the concept. The ACS has also recognized the importance of the cross-disciplinary approach by establishing the new Division of Chemical Toxicology as the

Table II. Building the Foundation for Change

- Increased awareness of the concept of designing safer chemicals
- Establishing the scientific, technical and economic credibility of the concept
- Effecting a sharper focus on chemicals of concern
- Greater emphasis on mechanistic and SAR research in toxicology
- Revision in the concepts and practices in chemical education
- Major participation by the chemical industry

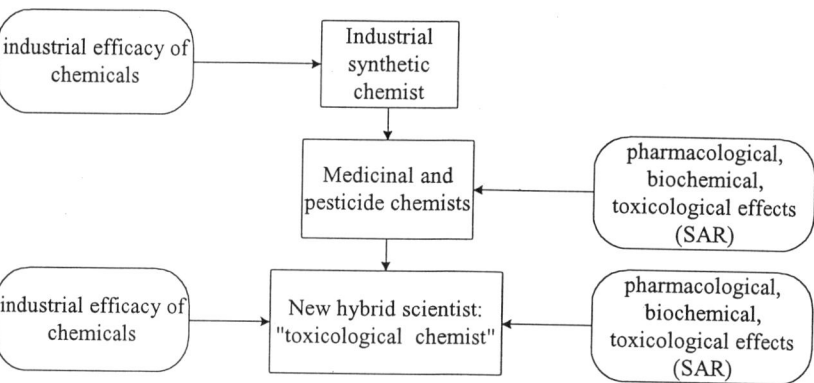

Figure 2. Evolution of a New Scientist Hybrid: the "Toxicological Chemist".

most recent addition to its organization. It is incumbent upon private industry, academia and the funding institutions to provide their support and begin participating in the implementation of the concept.

The subsequent chapters in this book provide scientific and technical insight into the concept and present useful, real-world examples of various approaches to designing safer chemicals. It is anticipated that this small but important group of pioneers will show the way to all others that industrial/commercial chemicals can be designed or redesigned to provide useful, cost-effective alternatives that reduce the present risks to human health and the environment.

Conclusion

We have entered a new era in the control of toxic substances in our environment and in our society; namely, that of prevention. One of the principal concepts of the pollution prevention/green chemistry program is that of designing safer chemicals used in industry and commerce. Although the concept is not new, it has never been fully developed or promoted. Since 1994, however, the EPA has taken steps to increase the awareness, the understanding, the acceptance, and the implementation of this concept as an integral part of the field of synthetic chemistry, both academically and industrially. The Agency's goals are to integrate this concept into every aspect of the science and technology employed in the design, manufacture and use of chemicals in our society. To reach this goal, there must be changes in existing paradigms, infrastructures, research in toxicology, and approaches to chemical education as well as changes in emphasis, attitudes, and practices by individuals and organizations throughout the public and private sectors. If this goal is achieved, we will have taken one of the major steps towards the establishment of a technological society that is more friendly to the health and well being of its people and their environment.

Disclaimer

This chapter was prepared by Dr. Roger L. Garrett. The opinions and views expressed in this chapter are those of the author, and do not necessarily reflect the views, policies or regulatory positions of the U.S. Environmental Protection Agency.

Literature Cited

1. The Pollution Prevention Act of 1990, codified as 42 U.S.C. 13101-13109
2. The Clean Water Act, codified as 33 U.S.C. 1251-1387
3. The Clean Air Act, codified as 42 U.S.C. 7401-7671
4. The Ocean Dumping Act, codified as 33 U.S.C. 1401-1445
5. The Safe Drinking Water Act, codified as 42 U.S. C.300f-300j-11
6. The Solid Water Disposal Act/Resource Conservation and Recovery Act 1976, codified as 42 U.S.C. 6901-6991k
7. The Superfund and Amendments Act, codified as 42 U.S.C. 9601-9675
8. The Toxic Substances Control Act, codified as 15 U.S.C. 2601-2671
9. The Federal Insecticide, Fungicide, and Rodenticide Act and Amendments Act, codified as 7 U.S.C. 136-136y
10. Ariens, E.J. Drug Metab. Rev. **1984,** *15,* pp. 425-504.

Chapter 2

General Principles for the Design of Safer Chemicals: Toxicological Considerations for Chemists

Stephen C. DeVito

Office of Pollution Prevention and Toxics, U.S. Environmental Protection Agency, Mail Code 7406, 401 M Street, SW, Washington, DC 20460

> The evolution of pollution prevention and green chemistry has given a new challenge to chemists. In addition to the traditional aspects of chemical design, which focus largely on commercial usefulness and ease of synthesis, modern day chemists are now expected to develop commercial chemical substances that are safe to human health and the environment as well. The most effective way for chemists to meet this challenge is to design chemicals such that they will have minimal toxicity. This paper discusses several approaches that can be used by chemists to design safer chemical substances. Theses approaches include: reducing absorption; understanding toxic mechanisms; using structure-activity relationships, retrometabolism, and isosterism; eliminating the need for associated toxic substances; and identifying equally useful, less toxic substitutes of another chemical class. Specific examples of the successful commercial application of these approaches are presented. This paper provides chemists with a strategic framework for the rational design of safer chemicals from the standpoint of pollution prevention and green chemistry.

Pollution prevention has become the preferred strategy for reducing the risks posed by the manufacture, use and disposal of commercial chemical substances. The strong emphasis on pollution prevention as a means of controlling the toxics that enter our environment and jeopardize our health has given new challenges to commercial chemists and has led to the concept of "green" chemistry. Green chemistry is based on the premise that the most desirable and efficient way of preventing pollution is to: design chemicals such that they are commercially useful but have minimal or no toxicity; and design alternative synthetic pathways for new and existing chemicals such that they neither utilize toxic reagents or solvents, nor produce toxic by-products or co-products. Green chemistry strives to encourage the development of safer commercial substances and non-polluting commercial syntheses for purposes of reducing the risks posed by existing substances and traditional chemical syntheses. Thus, traditional considerations of synthesis yield and commercial usefulness are no longer the only ones for a chemist during the design of a chemical. In addition to these traditional considerations, the modern chemist is also expected to consider toxicity as part of the design strategy of a new chemical substance, and to design chemicals such that they are commercially useful, have minimal toxicity to human health and the environment, and can be made by commercially feasible syntheses that do not contribute to pollution.

To reduce the toxicity of a chemical substance or to make a chemical safer than a similar chemical substance requires an understanding of the basis of toxicity. Once toxicity is understood, strategic structural modifications can be made that directly or indirectly attenuate toxicity but do not reduce the commercial usefulness of the chemical. There are several general approaches that provide the framework for molecular modification needed for the rational design of safer chemicals. These general approaches are:

- Reducing Absorption
- Use of Toxic Mechanisms
- Use of Structure-Activity (Toxicity) Relationships
- Use of Isosteric Replacements
- Use of Retrometabolic ("Soft" Chemical) Design
- Identification of Equally Efficacious, Less Toxic Chemical Substitutes of Another Class
- Elimination of the Need for Associated Toxic Substances

The approaches of toxic mechanisms, structure-activity (toxicity) relationships, isosteric replacement and retrometabolism are used extensively by medicinal chemists for the design of safe, clinically efficacious drug substances. These approaches, as well as molecular modifications intended to reduce exposure and absorption or eliminate the need for associated toxic substances (e.g., toxic solvents) can also be used by chemists to design safer, nonmedicinal commercial substances.

This chapter demonstrates how each of the seven approaches above, either alone or in combination, can be used for the rational design of safer chemicals. For the benefit of the reader, the next section of this chapter provides a brief discussion of the important aspects of chemical toxicity. The remaining sections discuss each of the above approaches and provide examples of how they have been or can be applied in the design of safer chemicals. The subsequent chapters in this book provide more detailed discussions and examples of these approaches. As with most scientific writings, this chapter draws on the perspectives of several previous authors. Earlier reviews of particular note that address the concept of designing safer chemicals include those by Ariens *(1-3)*, Finch *(4)*, Bodor *(5)*, Baumel *(6)*, and DiCarlo *(7)*.

Aspects of Chemical Toxicity

There are three fundamental requirements for chemical toxicity. First, there must be exposure to the chemical substance. Second, the substance must be bioavailable and third, the substance must be intrinsically toxic. Exposure to a chemical refers to the contact of that substance with the skin, mouth or nostrils. Bioavailability is the ability of a substance to be absorbed into and distributed within a living system (e.g., humans, fish) to areas where the toxic effects are exerted and is a function of the toxicokinetics (i.e., the interrelationship of absorption, distribution, metabolism and excretion) of the substance. Intrinsic toxicity is the ability of a substance to cause an alteration in normal cellular biochemistry and physiology following absorption, and is usually attributable to a particular structural portion, the toxicophore, of the substance.

The relative seriousness of the toxic effect depends upon the extent of exposure to the substance, its bioavailability, and the importance of the physiologic process that the substance has disrupted. Some substances contain structural features that are not directly toxic but undergo metabolic conversion (bioactivation) to yield a toxicophore. These structural features are toxicogenic, in that they yield a toxicophore subsequent to metabolism. In either case, the toxicity results from the interaction of the toxicophore with biomolecular sites of action within cells. The toxicophore-biomolecular interaction is known as the toxicodynamic phase of chemical toxicity. Intrinsic toxicity is an important aspect of chemical toxicity, as it distinguishes chemical substances that are toxic from those that are not toxic. The general aspects of chemical toxicity described above are illustrated in Figure 1.

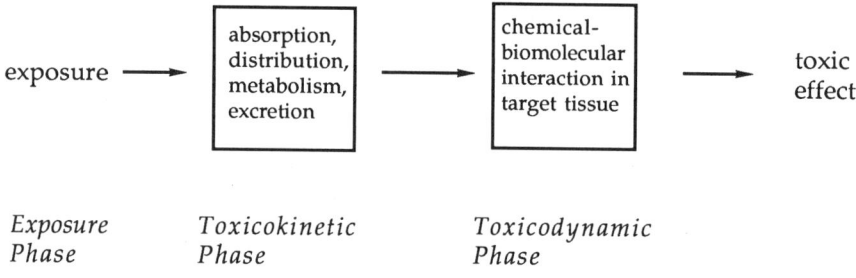

Exposure Phase   *Toxicokinetic Phase*   *Toxicodynamic Phase*

Figure 1. Aspects of Chemical Toxicity (adapted from ref. 2).

From the preceding paragraph one may begin to envision how safer chemicals may be designed. Structurally modifying a substance such that less of it is absorbed, avoiding the incorporation of substituents that are known to be toxicophoric or toxicogenic, or making structural modifications that direct metabolism to yield innocuous products are just some design considerations that can be used to make a chemical less toxic. But the utilization of these and the other approaches to design safer chemicals requires that the chemical designer has a basic understanding of the processes involved in chemical toxicokinetics and toxicodynamics. These processes are described briefly below. Excellent comprehensive discussions of toxicokinetics and toxicodynamics are available; these are highly recommended to the reader *(8-15)*.

Absorption. Absorption of a chemical substance refers to the entrance of the substance into the bloodstream from the site of exposure. Commercial chemical substances can gain access to the body as a result of absorption through the gastrointestinal tract, lung or skin, following oral ingestion, inhalation or dermal exposure, respectively. For a substance to be absorbed and become bioavailable, the *molecules* of the substance must pass through numerous cellular membranes and enter the bloodstream (which is mostly aqueous) where they are circulated throughout the body, and again cross many cellular membranes to gain entrance into the cells of organs and tissues. This means that the substance must have the necessary physicochemical properties that enable the molecules comprising the substance to reach their free molecular form, cross biological membranes and enter the blood. The membranes of essentially all cells of the body, particularly those of the skin, the epithelial lining of the lung, the gastrointestinal tract, capillaries, and organs, are composed chiefly of lipids. Therefore, absorption of a chemical substance into the body and its ability to travel throughout the bloodstream (distribution) to the area of the body where the toxic response is elicited requires that the substance has a certain amount of both lipid solubility (lipophilicity) and water solubility. In fact, all aspects of toxicokinetics of a substance are dependent, either directly or indirectly, on transport of the substance through cellular or organelle membranes, which in turn is based upon the physicochemical properties of the substance *(8,11)*.

Anatomical and biological factors are also important in absorption. These include surface area, thickness of the membrane barrier, and blood flow. The physicochemical, anatomical and biological properties that most significantly influence absorption and membrane diffusion are listed in Table I. A brief discussion of these factors with respect to route of exposure is provided below. Detailed discussions are available elsewhere *(8,11-16)*.

Gastrointestinal Tract. The gastrointestinal tract is a major site from which chemical substances are absorbed. Many environmental toxicants enter the food chain and are absorbed together with food from the gastrointestinal tract. In occupational settings, for example, airborne toxic substances enter the mouth from breathing and, if not inhaled, can be swallowed and absorbed from the gastrointestinal tract.

The major physiological factors governing absorption from the gastrointestinal tract are surface area and blood flow. From Table I, it can be seen that of the three possible sites of absorption, the gastrointestinal tract has the largest surface area and the second greatest blood flow. The majority of absorption from the gastrointestinal tract occurs from the small intestines. The pH of the gastrointestinal tract ranges from about 1-2 in the small intestines, and gradually increases to about 8 in the large intestines. This variation in pH influences the extent to which acidic or basic chemical substances are ionized, which influences the extent of their absorption. Acidic substances are absorbed much more readily from the small intestines than the large intestines, because they are less dissociated in the small intestines. The opposite is true for basic substances.

The physicochemical properties that most significantly affect the extent to which a substance is absorbed from the gastrointestinal tract include: physical state; particle size (for solids); the relative lipid to water partitioning; dissociation constant; molecular weight; and molecular size. A substance must be sufficiently water soluble such that it can undergo requisite dissolution to its free molecular form. Substances that are liquid in their neat form, or are already dissolved in a solvent, are generally absorbed more quickly from the gastrointestinal tract than a substance that is a solid. Generally, substances that are in the form of salts (e.g., hydrochloride salt, sodium salt) undergo dissolution more quickly than their un-ionized (neutral) forms, and are absorbed more quickly. For solids, particle size also affects the rate of dissolution and, thus, overall absorption. The smaller the particle size, the larger the surface area and the faster the dissolution and absorption of the substance. Larger particle size means less surface area and therefore a slower dissolution in the gastric fluids, and slower or even less absorption.

Lipid solubility is more important than water solubility in regard to absorption from the gastrointestinal tract. The more lipid soluble a substance is, the better it is absorbed. Highly lipophilic substances (log P > 5), however, are usually very poorly water soluble and generally are not well absorbed because of their poor dissolution in the gastric juices. On the other hand chemical substances with extreme water solubility and very low lipid solubility are also not readily absorbed *(16)*.

The higher the molecular weight the less a substance is absorbed from the gastrointestinal tract. Assuming sufficient aqueous and lipid solubility, a general guide is that substances with molecular weights less than 300 daltons are typically well absorbed; those with molecular weights ranging from 300 to 500 are not as readily absorbed, and substances with molecular weights in the thousands are sparingly absorbed, at best *(16)*.

Table I. Physicochemical and Biological Factors Influencing Membrane Permeation and Absorption[a]

Physicochemical Factors

molecular size; molecular weight; dissociation constant; aqueous solubility; lipophilicity (octanol/water partition coefficient, i.e., log P); physical state (solid, liquid, gas); and particle size.

Biological Factors

Route of Exposure	Surface Area (m^2)	Thickness of Absorption Barrier (μm)	Blood Flow (L/min.)
skin	1.8	100-1000	0.5
g.i.[b]	200	8-12	1.4
lung	140	0.2-0.4	5.8

[a] Adapted from ref. *12*.   [b] Gastrointestinal tract.

Lung. The function of the lung is to exchange oxygen for carbon dioxide. The continuous, repetitive branching of the airways from the trachea to the terminal alveoli (where gas exchange takes place) creates an enormous surface area. The lungs also receive 100% of the blood pumped from the heart. The thickness of the alveola cellular membranes (the absorption barrier of the lung) is only 0.2-0.4 μm. These anatomical and physiological characteristics of the lung enable the rapid and efficient absorption of oxygen and favor the absorption of other substances as well *(8,11-13)*.

Absorption of chemical substances from the lung differs from intestinal and dermal absorption in that lipid solubility is less important. In fact, water solubility, rather than lipid solubility, is the more important factor. This is because the cellular membranes of the alveoli are very thin (0.2-0.4 μm), so that the distance a substance has to traverse the alveolar membrane to enter the blood is very short. Chemicals absorbed through the lung can enter the blood within seconds *(8)*.

For solid substances, particles of 1 μm and smaller may be particularly well absorbed from the lung because they have a large surface area and can also penetrate deep into the narrow alveolar sacs of the lung. Particles of 2 to 5 μm are mainly deposited in the tracheobronchiolar regions of the lung, from where they are cleared by retrograde movement of the mucus layer in the ciliated portions of the respiratory tract.

These particles are eventually removed from the lung and may be swallowed and absorbed from the gastrointestinal tract. Particles of 5 μm or larger are usually deposited in the nasopharyngeal region and are too large for absorption from the lung, but also may be swallowed and absorbed from the gastrointestinal tract *(8)*.

Skin (Dermal). Unlike the lung and the gastrointestinal tract, the primary purpose of the skin is not for the absorption of substances essential to life, but rather protection against the external environment. Compared to the lung and gastrointestinal tract, the skin has much less surface area and blood flow, as well as a considerably thicker absorption barrier (Table I). Nonetheless, the skin represents a significant organ of exposure and absorption. For chemicals to be absorbed from the skin, they must pass through the seven cell layers of the epidermis before entering the blood and lymph capillaries in the dermis. This absorption barrier ranges from 100 to 1000 μm *(8)*. The rate determining step is diffusion through the stratum corneum (horny layer), which is the uppermost layer of the epidermis. Passage through the six other layers is much more rapid *(8,14)*.

Substances that are liquid in their neat form tend to be absorbed more readily than solids, because liquids cover more dermal surface area and are nearer to their free molecular state than are solids. Solids with higher melting points (> 125 °C) and substances (particularly solids) that are ionic or highly polar are generally not well absorbed from the skin. Substances with greater lipophilicity (higher log P) are absorbed more readily from the skin than are less lipophilic substances. Highly lipophilic substances (log P > 5), however, can pass through the stratum corneum but are generally too water insoluble to pass through the remaining layers and enter the bloodstream. These substances are poorly absorbed from the skin. Absorption of chemicals through the skin tends to decrease as molecular weight and size increase *(14)*.

Distribution. Distribution refers to the movement of a chemical throughout the living system from its site of entry into the bloodstream following absorption from the skin, gastrointestinal tract, or lung. Distribution usually occurs rapidly. The rate of distribution to organs or tissues is primarily determined by blood flow and the rate of diffusion out of the capillaries into the cells of a particular organ. Following absorption, many substances distribute to the heart, liver, kidney, brain, and other well-perfused organs. Where a substance is distributed, however, is largely dependent upon its physicochemical characteristics. For example, substances that are not very lipophilic generally do not enter the brain because they cannot penetrate the blood-brain barrier, whereas substances that are more lipophilic are more likely to enter the brain. Other factors, such as extent of binding to plasma proteins and accumulation in fatty tissues, are also dependent on physicochemical properties, and are discussed elsewhere *(8,10,17)*. Toxic substances generally are not toxic to all the organs to which they are distributed. Rather, the the toxicity of a substance is usually elicited in only one or two organs. These sites are referred to as the "target" organs of a particular substance. The target organ may not necessarily be the site of highest concentration of the substance.

Metabolism. The body has the ability to distinguish between non-food (foreign) chemical substances (e.g., industrial chemicals, drugs) that offer no nutritional value or may even be harmful, and substances that are nutritional (e.g., vitamins, carbohydrates, amino acids). Following absorption, the body will try to eliminate as quickly as possible any chemical substance that is non-nutritional. The major sites of elimination of most chemical substances are the urine and feces. Unlike absorption and distribution, which require greater lipid solubility over water solubility, elimination of a chemical substance in the urine or feces requires greater water solubility. Thus, the physicochemical properties that permit rapid passage across cellular membranes during absorption and distribution also impair subsequent excretion. The body, however, has enzyme-mediated mechanisms for converting substances into more water-soluble substances that are easier to excrete. These mechanisms are generally referred to as metabolism or biotransformation. The overall purpose of metabolism is detoxication: a defense mechanism to convert potentially toxic chemical substances to other substances (metabolites) that are readily excreted. The ability to metabolize chemical substances appears to have been an evolutionary necessity, because terrestrial animals (including humans) ingest many foreign substances with their food *(10)*.

The chemical reactions involved in the metabolism of chemical substances foreign to the body are classified as either phase-I or phase-II reactions. Phase-I reactions convert the chemical substance into a more polar metabolite by oxidation, reduction, or hydrolysis. Phase-II reactions involve coupling (conjugation) of the chemical substance or its polar (Phase-I) metabolite with an endogenous substrate such as glucuronate, sulfate, acetate, or an amino acid, which further increases water solubility and promotes rapid excretion. The enzyme systems responsible for Phase-I and Phase-II reactions are located predominately in the smooth endoplasmic reticulum of the liver. These enzymes are also present in other organs, including the kidney, lung, and gastrointestinal epithelium. Oxidation is the most common form of metabolism, as it generally leads to more water-soluble and, thus, readily excretable metabolites. Most oxidations are mediated by the cytochrome P450 enzymes and most commonly involve hydroxylations. Hydrolysis, reduction, and conjugation reactions are mediated by esterases, flavins, and transferases, respectively. Detailed discussions of metabolism *(9,10,17)* and recent reviews on the role of the cytochrome P450 enzymes in metabolism *(18, 19)* are available.

Metabolism of certain chemical substances does not result in detoxication. In fact, it is the metabolites of many toxic chemical substances that, ironically, are responsible for the toxicity. This phenomenon, the conversion of a chemical substance into a toxic metabolite, is known as bioactivation and is the basis of toxicity of many chemical substances *(3-5,9,18-25)*. A major element of designing safer chemicals is controlling bioactivation; that is, by adding molecular modifications to direct metabolism away from bioactivation and toward detoxication. This point is discussed in greater detail elsewhere in this chapter and is exemplified in other chapters in this book.

Toxicodynamics. The toxicodynamic phase comprises the processes involved in the molecular interaction between the toxic substance and its biomolecular sites of action (receptors for reversibly-acting agents or sites of induction of chemical lesions for irreversibly-acting agents) and the resultant sequence of biochemical and biophysical events that finally result in the observed toxic effect. In general, a toxic substance exerts its toxicity by the interaction of a particular portion of the molecule (the toxicophore) or a metabolite thereof with a cellular macromolecule, which disrupts normal biochemical function of the macromolecule and ultimately results in toxicity. The cellular macromolecules may be enzymes, nucleic acids, or proteins, to name just a few. The interaction may involve binding of the toxic substance to a particular receptor site of a macromolecule or to a nonreceptor portion of the molecule. The interaction may be irreversible, in which case covalent bonding usually occurs between the substance and the macromolecule, or reversible, in which case no covalent bonding occurs. In any case, the interaction defines the specific mechanism of toxic action (toxicodynamics) of a substance.

Excretion. Substances are eliminated from the body by excretion in the urine, feces, or breath, from the kidney, bile duct, and lung, respectively. The kidney and bile duct eliminate polar (more water-soluble) substances more efficiently than substances with high lipid solubility. The kidney is the most important organ for eliminating substances or their metabolites from the body. Substances excreted in the feces are typically the metabolites of absorbed substances, which enter the gastrointestinal tract through the bile duct. Excretion through the lung occurs mainly with volatile substances. Detailed discussions of excretion are available *(8, 10)*.

Designing Safer Chemicals: Molecular Modifications That Reduce Absorption

The toxicity of a substance can be mitigated if structural changes are incorporated that reduce or prevent absorption from the lung, skin, or gastrointestinal tract. To many people, properties such as physical state, melting point, boiling point, vapor pressure, water solubility, lipophilicity, molecular size, molecular weight, and others seem to have little to do with toxicity. Yet, the absorption of a substance and, hence, its potential to be toxic are quite dependent on these properties. Molecular modifications that alter these properties can likewise alter the potential for absorption. Absorption is also related to exposure. When designing a chemical to be absorbed less, one should consider the most likely route of exposure (i.e. dermal, inhalation, oral) expected to result from its use or environmental release and design the molecule accordingly. Guidance on how to modify substances such that they are absorbed less is provided below.

Reducing Absorption From the Gastrointestinal Tract. If oral exposure is expected to be significant, the chemical should be modified to reduce absorption from the gastrointestinal tract. Simple modifications, such as increasing particle size or keeping the substance in an un-ionized form (i.e., free base, free acid), should in many cases reduce oral absorption. Designing the substance such that its log P is greater than 5 should ensure that the substance is not sufficiently water soluble for oral absorption. Designing the substance to be greater than 500 daltons, to have a melting point above 150 ºC (for non-ionic substances) or to be a solid rather than a liquid should also reduce the likelihood of absorption from the oral route. The incorporation of several substituents (e.g., -SO_3-) that remain strongly ionized at a pH of 2 or below should make the substance so polar that it cannot easily penetrate the lipid membranes of the intestinal lining and other membranes, and should significantly reduce absorption. Even if some absorption takes place, substances containing sulfonate or equally ionizable moieties have great difficulty in penetrating the biological membranes of other tissues, and also should be excreted rapidly in the urine because of their extreme water solubility. This principle has been applied to reducing the toxicity of azo dyes *(2)*.

Reducing Absorption From the Lung. Designing chemicals to be less volatile (lower vapor pressure, higher boiling point), for example, should reduce inhalation exposure to such substances. Designing a chemical such that it has low water solubility (or high lipophilicity) or a high melting point (> 150 ºC), or has a particle size greater than 5 μm should also reduce the likelihood of absorption from the lung. Particle sizes less than 5 μm should be avoided when inhalation exposure to the substance is likely to occur.

Reducing Absorption From the Skin. Substances that are liquid in their neat form tend to be absorbed more readily than solid substances. Thus, designing a substance to be a solid rather than a liquid should reduce the likelihood of dermal absorption. Designing a substance to be polar or ionized (*e.g.,* sodium salt of an acid, hydrochloride salt of an amine), or to be water soluble or of low lipophilicity should also decrease its potential to be absorbed dermally. Increasing particle size or molecular weight and size should also reduce the likelihood of dermal absorption.

Designing Safer Chemicals from an Understanding of Toxic Mechanisms

Understanding *why* a chemical is toxic is extremely helpful in the design of a safer chemical. The chemist who is able to recognize toxicophoric or toxicogenic substituents, or obtain information pertaining to a toxic mechanism is in a much better position to design safer chemicals. From such knowledge one can often infer structural modifications needed to render a new substance much less toxic than the unmodified toxic substance. Structural changes intended to lessen toxicity should, of course, not interfere with the commercial usefulness of the substance. This section discusses a few well-known mechanisms of toxicity of commercial chemical substances and, using specific examples, demonstrates how knowledge of these mechanisms can be used by the chemist for the design of safer substances.

Toxic Mechanisms Involving Electrophiles. Chemical substances that are electrophilic or are metabolized to electrophilic species are capable of reacting covalently with nucleophilic substituents of cellular macromolecules such as DNA, RNA, enzymes, proteins, and others. These nucleophilic substituents include, for example: thiol groups of cysteinyl residues in protein; sulfur atoms of methionyl residues in protein; primary amino groups of arginine and lysine residues, or secondary amino groups (e.g., histidine) in protein; amino groups of purine bases in RNA and DNA; oxygen atoms of purines and pyrimidines; and, phosphate oxygens (P=O) of RNA and DNA *(25)*. These irreversible covalent interactions with important cellular macromolecules interfere with the function of the cellular macromolecules and, depending upon the type of electrophile and cellular macromolecule, can lead to a variety of toxic effects including cancer, hepatotoxicity, hematotoxicity, nephrotoxicity, reproductive toxicity, and developmental toxicity.

Fortunately, the mammalian body has several defense systems that offer "sacrificial" nucleophiles that can react with foreign electrophiles. These natural defense systems are located predominately in the liver and include, among others, the glutathione transferase system and the epoxide hydratase system *(2)*. Reactions of electrophiles with the nucleophiles of these systems form readily-excretable water soluble products, and allow the safe elimination of the electrophiles before they can react with the nucleophiles in the more biologically-critical cellular macromolecules. The nucleophiles of these defense systems may become depleted or overburdened, however, upon continuous or high exposure to electrophilic substances. Under these conditions toxicity ensues. The interaction of electrophiles with the natural defense systems and cellular macromolecules is shown in Figure 2.

Examples of electrophilic substituents commonly found in commercial chemical substances are shown in Table II. Also shown are the types of nucleophilic reactions these electrophiles undergo with cellular macromolecules, and the toxicity that may occur. Lists of specific electrophilic chemical substances are available *(24, 31, 32)*. Although the presence of an electrophilic substituent raises the possibility that the substance may be toxic, it should not be inferred that the presence of any of these substituents always means that the substance is necessarily toxic. Whether or not the substance is toxic is also dependent on factors such as its overall bioavailability, its metabolism and the presence of other substituents that may attenuate the reactivity of the electrophilic substituent.

Designing Safer Electrophilic Substances. Ideally, electrophilic substituents should never be incorporated into a substance. However the electrophilic group is often necessary for the intended commercial use of the substance. This poses

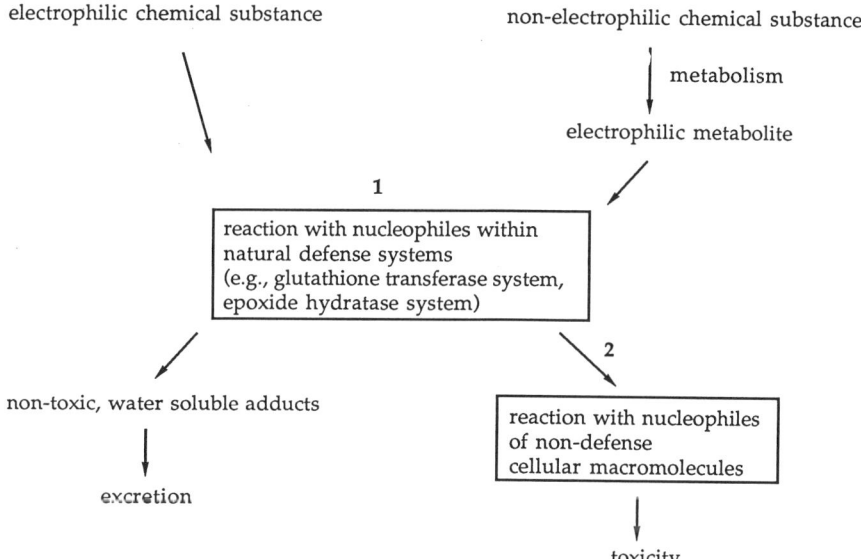

Figure 2. Detoxification of Electrophilic Substances or Electrophilic Metabolites. Electrophiles can be detoxified by reaction with defense nucleophiles (pathway **1**). If the defensive nucleophiles are depleted upon continuous exposure, the electrophiles will begin to react with important cellular nucleophiles (pathway **2**) which results in toxicity.

Table II. Examples of Electrophilic Substituents Commonly Encountered in Commercial Substances, the Reactions they Undergo with Biological Nucleophiles, and the Resulting Toxicity [a]

Electrophile	General Structure	Nucleophilic Reaction	Toxic Effect
alkyl halides	R-X x= Cl, Br, I, F	substitution	various; e.g., cancer, granulocytopenia (27)
α,β–unstaturated carbonyl and related groups	$C=C-C=O$ $C=C-\underset{\underset{O}{\|\|}}{\overset{O}{\|\|}}S-$ $C\equiv C-C=O$ $C=C-C\equiv N$	Michael Addition[b]	various; e.g., cancer, mutations, hepatotoxicity, nephrotoxicity, neurotoxicity, hematotoxicity (27,29-33)
γ - diketones	$R_1\overset{O}{\underset{}{\|\|}}C-CH_2CH_2-\overset{O}{\underset{}{\|\|}}C-R_2$	Schiff base formation	neurotoxicity (26,63)
epoxides (terminal)	$-CH-CH_2$ $\quad\ \backslash\ /$ $\quad\ \ O$ $-O-CH_2-CH-CH_2$ $\qquad\qquad\ \backslash\ /$ $\qquad\qquad\ \ O$	addition	mutagenicity, testicular lesions (6,27,31)
isocyanates	$-N=C=O$ $-N=C=S$	addition	cancer, mutagenicity, immunotoxicity (e.g., pulmonary sensitization, asthma) (27,28,31)

[a] The presence of any of these substituents in a substance does not automatically mean that the substance is or will be toxic. Other factors such as bioavailability, and the presence of other substituents that may reduce the reactivity of these electrophiles can influence toxicity as well. [b] The carbon-carbon double and triple bonds may also be oxidized to epoxides and other reactive species. See text for details.

a dilemma for the chemist who wishes to design an electrophile to react with a nucleophile necessary for intended commercial use but **not** with biological nucleophiles in individuals exposed to the substance. As impossible as this may seem, there are approaches that chemists can use to design safer, commercially-useful electrophilic substances.

Acrylates, for example, contain an α,β-unsaturated carbonyl system and as such undergo Michael addition reactions. This is believed to be the basis of the carcinogenic properties of acrylates *(31)*. Incorporation of a methyl (-CH_3) group onto the α-carbon (to provide a methacrylate) decreases the electrophilicity (i.e., reactivity) of the β-carbon *(34)* and, hence, methacrylates do not undergo 1,4-Michael addition reactions as readily. Methacrylates often have commercial efficacy similar to acrylates in many applications, but are less likely to cause cancer because they are less reactive. This point can be demonstrated by comparing ethyl acrylate, **1**, which causes cancer in experimental animals *(35)*, to methyl methacrylate, **2**, which does not cause cancer in a similar assay *(36)*. It seems logical that placement of a methyl group onto the α-carbon of similar α,β-unsaturated systems (Table II) may also decrease toxicity without sacrificing commercial utility.

1   **2**

carcinogenic   non-carcinogenic

Isocyanates (Table II) are a very important commercial class of electrophilic substances. They are widely used for a variety of purposes, including adhesive formulations, monomers, and synthetic intermediates. Following inhalation exposure many isocyanates are pneumotoxic, causing pulmonary sensitization and asthma *(28)*. Some isocyanates are believed to be mutagenic and carcinogenic as well *(27,31)*. The toxicity of isocyanates has been attributed to the reactivity of the isocyanate group with endogenous nucleophiles *(27-31)*. To reduce the likelihood of isocyanate-induced toxicity, some commercial companies that manufacture and use isocyanates in coating systems mask the electrophilic character of this functional group by converting it into a ketoxime derivative, as represented by **3**.

3

The isocyanate-ketoxime is usually made *in-situ* during the manufacture of the isocyanate, thereby minimizing environmental and worker exposure to the isocyanate. In ketoxime form, the isocyanate is no longer electrophilic and, thus, not reactive. When ready to be used in coatings, the ketoxime moiety is removed thermally during application, thereby regenerating the isocyanate and enabling it to react with the intended nucleophile (usually an amine) without the risk of human exposure.

A similar electrophilic-masking approach is used with vinyl sufone moieties (Table II). The vinyl sulfone group is highly electrophilic and frequently used in fiber reactive dyes. (Vinyl sulfones react covalently with hydroxy groups of cellulosic fibers.) They are made by reacting the appropriate ß-hydroxyethyl sulfone (Figure 3) with sulfuric acid, which provides the sulfate ester (**4**). The sulfate ester, which is not electrophilic and, thus, not reactive, is removed by treatment with strong base to give the vinyl sulfone (Figure 3). Many commercial manufacturers of fiber reactive dyes that involve vinyl sulfone moieties sell the dyes in the non-reactive sulfate ester form; the ultimate user of the dye then removes the sulfate ester moiety of **4** in the dye bath (when the free vinyl sulfone is needed) and not beforehand. In doing so, the manufacturer and user greatly reduce the potential for occupational and environmental exposure to the vinyl sulfone.

Toxic Mechanisms Involving Bioactivation to Electrophiles. The majority of biochemical reactions that lead to formation of electrophilic metabolites involve cytochrome P450-catalyzed oxidations *(22)*. In these reactions a particular portion of the molecule is bioactivated to become an electrophile. Bioactivation and its relation to toxicity have been extensively studied and reviewed *(3-5, 9, 18-25)*. This section gives examples of structural substituents commonly found in commercial chemical substances that are known to be bioactivated to toxic electrophiles, and provides insight into deducing structural modifications that partly or completely block their metabolic conversion to electrophiles. For additional examples and more comprehensive discussions of this topic, the reader may find references *3-5, 9, 18-25, 32* and *37* useful.

4-Alkyl-Phenols: Bioactivation to *p*-Quinone Methides. Phenolic substances substituted in the 4-position (i.e. *para*) with an alkyl substituent are used in a number of applications, including synthetic processes, photography, paints, antiseptics, and disinfectants. As shown in Figure 4, phenolic substances containing a 4-alkyl substituent (represented by general structure **5**) with at least one benzylic hydrogen may be oxidized in experimental animals to *p*-quinone methides (**6**) by cytochrome P450 enzymes *(38-45)*. Quinone methides are electrophilic: the exocyclic methide carbon (=C) is conjugated to the carbonyl moiety and is, therefore, capable of undergoing Michael addition reactions with cellular nucleophiles. Many 4-alkyl-phenols produce hepatotoxicity, pneumotoxicity, and promote tumor formation by metabolism to quinone methides and subsequent Michael addition reactions with cellular macromolecules *(38-42)*. Thompson and co-workers observed that increasing the length or branching of the 4-alkyl substituent increases the rate of quinone methide formation and toxicity *(45)*. For example, 4-isopropylphenol (**5**, R_1 and R_2 = CH_3) is more hepatotoxic than 4-ethylphenol, which is more toxic than 4-methylphenol *(45)*.

When designing alkyl-substituted phenolic substances, chemists should avoid, whenever possible, placing the alkyl group *para* to the phenolic OH group. This obviates *p*-quinone methide formation and eliminates toxicity that otherwise would be caused by this metabolite. For example, 2-methylphenol and 3-methylphenol (which cannot be metabolized to *p*-quinone methides) have 0.1 to 0.2 the hepatotoxicity of 4-methylphenol, which is oxidized to its corresponding *p*-quinone methide *(44)*. Alternatively, if the alkyl substituent must be *para* in a molecule, the substituent should not contain any benzylic hydrogens as these hydrogens are necessary for *p*-quinone methide formation *(40,41)*.

Allyl Alcohols. The allyl alcohol substituent (C=C-C-OH) is found in many commercial chemical substances, either as the free alcohol or as an ester or ether. Allyl alcohols (as well as their esters or ethers) that contain at least one hydrogen on

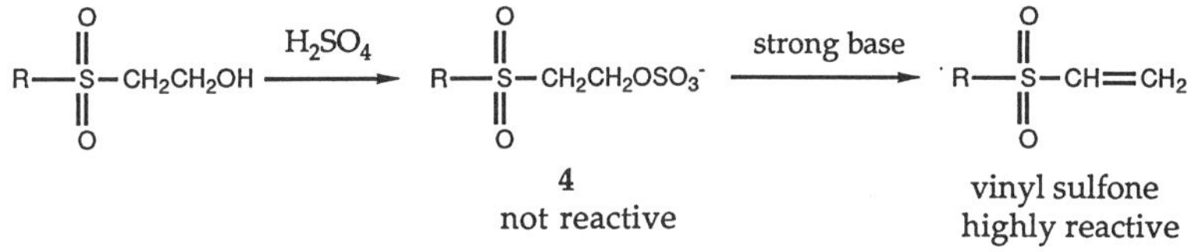

Figure 3. Synthesis of Vinyl Sulfones.

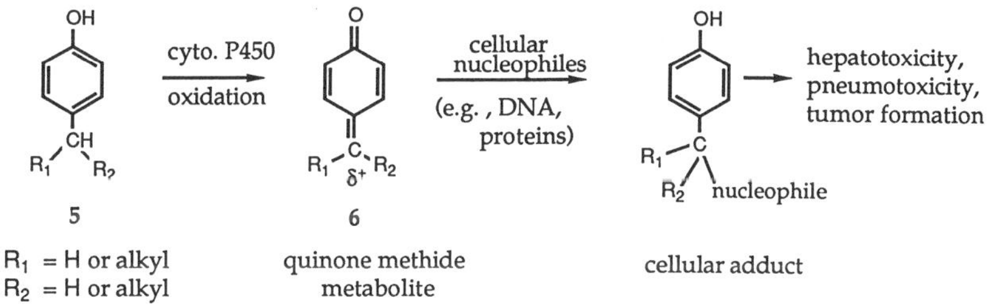

Figure 4. Bioactivation of 4-Alkyl-Phenols to Quinone Methides.

the alcoholic carbon can be oxidized in the liver by alcohol dehydrogenase (ALDH) to the corresponding α,ß-unsaturated carbonyl metabolite, which is toxic in many cases *(30,46,47)*. The hepatotoxicity of allyl alcohol **(7)**, for example, is due to its oxidation by ALDH to acrolein **(8)**, an α,ß-unsaturated aldehyde, which undergoes Michael addition with cellular nucleophiles in the liver *(30)* (Figure 5). Cyclic allyl alcohols (Figure 5) are expected to undergo similar enzymatic oxidation to yield α,ß-unsaturated carbonyl metabolites and are likely to be toxic as well.

In addition to the bioactivation pathway described above, allyl alcohols that contain an aromatic substituent on the alcoholic carbon (such as **9**, Figure 5) may undergo an alternative bioactivation that involves Phase II sulfate conjugation (discussed earlier) of the alcohol to form a highly electrophilic sulfate ester, as represented by **10** in Figure 5. The sulfate moiety of **10** is a particularly good leaving group because the resulting carbocation **(11)** is both benzylic and allylic and, thus, highly resonance stablized. Although sulfate conjugation usually represents a detoxification pathway, sulfate conjugation of 1-aryl-substituted allyl alcohols yields highly reactive electrophiles that undergo SN_1 reactions with biological nucleophiles and are often quite toxic *(4,21)*.

When designing alcohols, chemists should avoid placing C=C moieties on any carbon atom that contains both the -OH group and at least one hydrogen atom, as such allyl alcohols are likely to undergo enzymatic oxidation to α,ß-unsaturated carbonyl metabolites. Chemists should also avoid placing an aromatic substituent onto the alcoholic carbon of allyl alcohols. The metabolism of allyl alcohols to α,ß-unsaturated carbonyl substances can be prevented by replacing each of the hydrogen atoms on the alcoholic carbon atom with nonaromatic substituents that are not abstractable by ALDH (e.g., methyl, ethyl, etc.). If this structural modification is not feasible, then replacing the alkene hydrogens with bulky alkyl substituents should attentuate the reactivity of any α,ß-unsaturated carbonyl metabolite and thereby reduce toxicity.

Propargyl Alcohols. As with allyl alcoholic substituents, propargyl alcoholic substituents (C≡C-C-OH) that contain at least one hydrogen on the alcoholic carbon are also known to undergo enzymatic oxidation to the corresponding α,ß-unsaturated carbonyl metabolites and are quite toxic *(46-48)*. In addition, propargyl alcohols that contain an aromatic substituent on the alcoholic carbon may undergo Phase-II sulfate conjugation reactions to yield electrophilic species analogous to **10** *(49)*. It is advisable that chemists either avoid propargyl alcoholic moieties altogether during the design of a chemical substance or, if this moiety is necessary, incorporate the same structural modifications described above for allyl alcoholic moieties to prevent or reduce bioactivation to electrophilic metabolites.

Alkenes and Alkynes. Substances containing terminal carbon-carbon double- (olefinic) or triple-(acetylenic) bonded moieties can undergo cytochrome P450-mediated oxidations to yield metabolites that can cause a variety of toxic effects, including hepatotoxicity, mutagenicity, and carcinogenicity *(21, 50-52)*. As shown in Figure 6, substances containing terminal olefinic (e.g., 1-heptene, 4-ethyl-1-hexene, ethene) or acetylenic moieties (e.g., propyne, acetylene, octyne) can be metabolized to epoxide and ketene intermediates, respectively, which are highly electrophilic and capable of interacting with biological tissues *(50, 51)*. Terminal olefinic moieties that contain halogens can yield upon metabolism highly toxic α-halo carbonyl and acid halide metabolites, as shown in Figure 6 *(52)*.

Alternatively, substances containing terminal olefinic or acetylenic moieties may undergo metabolism to allyl or propargyl alcohols if allylic or propargylic

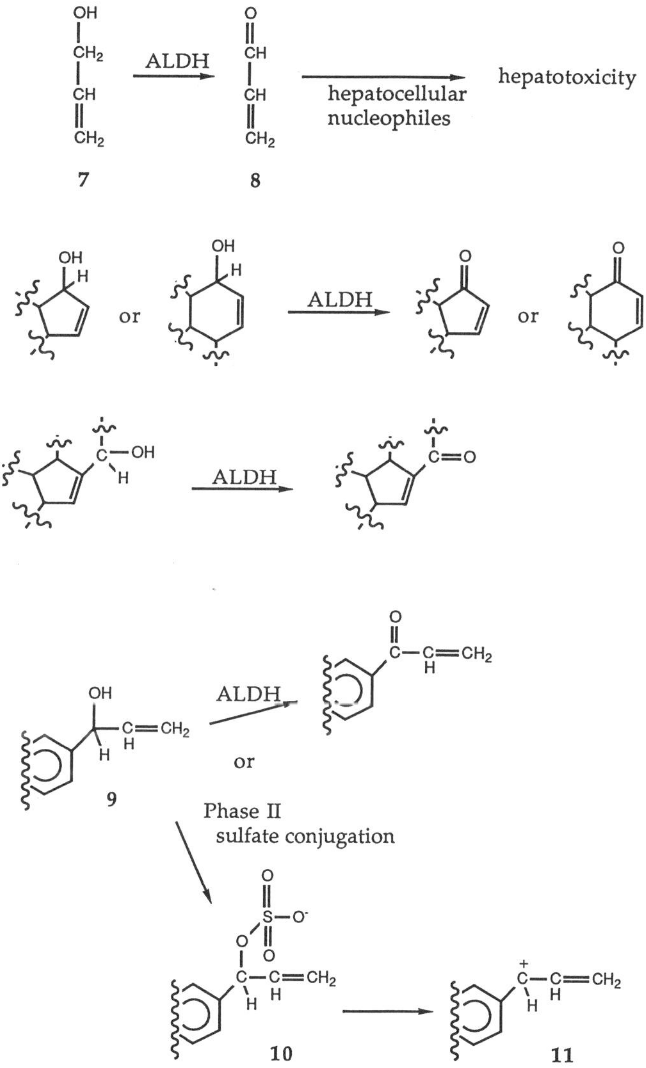

Figure 5. Bioactivation of Allyl Alcoholic Substances to α,β-Unsaturated Carbonyl-Containing Metabolites. (ALDH = alcohol dehydrogenase).

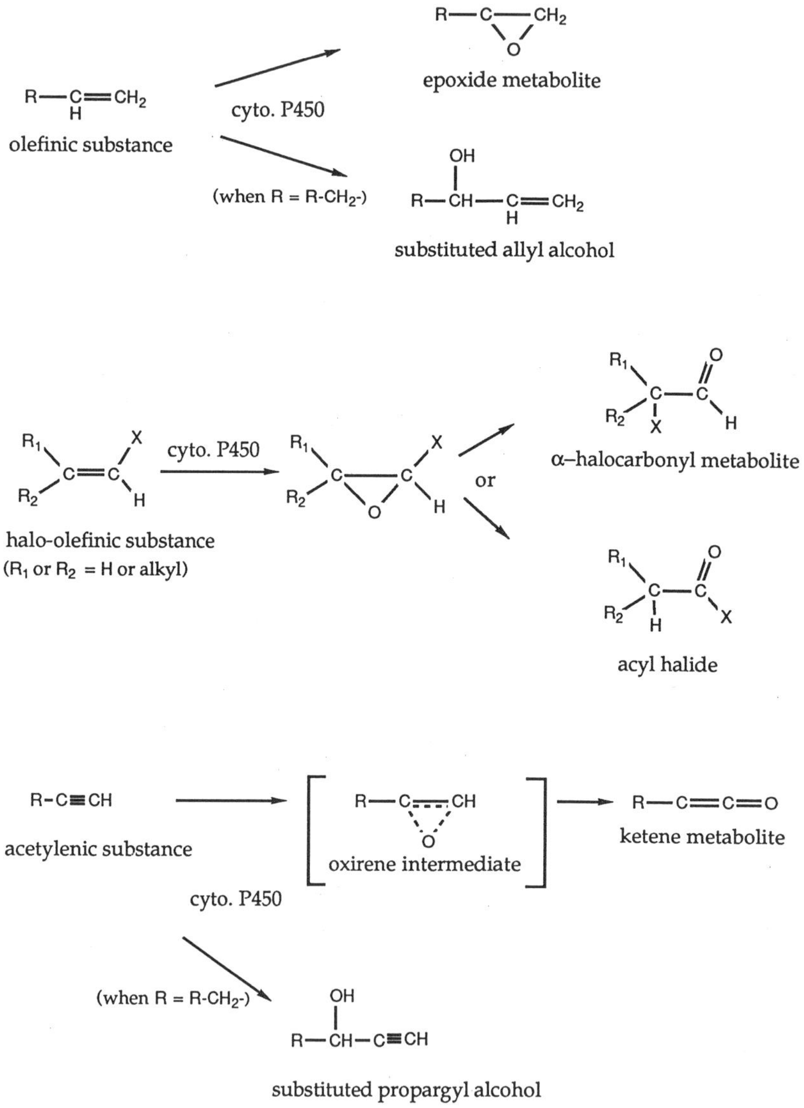

Figure 6. Bioactivation of Olefinic and Acetylenic Substances.

hydrogens are present (Figure 6). These allyl and propargyl alcohol metabolites may undergo further bioactivation, as discussed in the preceding paragraphs. In fact, the toxicity of allyl- and propynyl-substituted aromatic substances is believed to be due to their initial bioactivation to their respective allyl and propargyl alcohols, followed by Phase-II sulfate conjugation *(4,21,53)*.

The metabolism of olefinic or acetylenic substances to electrophilic metabolites is greatly reduced when the olefinic or acetylenic moieties are not terminal, or are terminal but contain alkyl substituents on the allylic and propargylic positions. 1-Heptene, for example, is bioactivated to electrophilic metabolites whereas 3-hexene, 2-methyl-1-heptene and 3,3-dimethyl-1-hexene are not *(50,51)*. Similarly, 1-decyne is metabolized to electrophilic metabolites whereas 3- and 5-decynes are essentially not *(50,51)*.

When designing substances that contain olefinic or acetylenic moieties, chemists should keep in mind the potential for these substituents to be bioactivated to electrophilic species and, whenever possible, incorporate other structural changes that lessen the likelihood for bioactivation. Terminal carbon-carbon double or triple bonds should be avoided or at least contain an alkyl substituent on the C-2 carbon (in the case of olefins) and alkyl substituents on the allylic or propargylic carbons. Halogens on terminal unsaturated carbons should be avoided. Aromatic substituents at the allylic or propargylic positions should also be avoided if allylic or propargylic hydrogens are present.

Mechanisms Involving Free Radicals. Free radicals are highly reactive species that have an unpaired electron. Many chemical substances form free radicals following their absorption and metabolism in the body. The formation of free radicals is a natural process of metabolism and cellular physiology, and many of the reactive intermediates formed during metabolism are free radicals or involve initial free radical formation. Free radical formation, for example, is a requisite step in cytochrome P450 hydroxylations. It is becoming increasingly apparent, however, that the toxicity of many chemical substances is due to formation of free radicals during their metabolism. An analysis of free radical-based toxicology is beyond the scope of this text, but it is important to stress here that chemical substances that easily generate free radicals often present a danger to human health because radicals are capable of interacting with and, subsequently, damaging cells and tissues if the body's natural defense systems against them are overpowered or depleted. Excellent reviews of free radical toxicology are available *(54-56)* and highly recommended to the reader.

Examples of Using Mechanisms of Toxicity to Design Safer Chemicals. The preceding paragraphs discuss several toxic mechanisms and emphasize the avoidance or molecular modification of well-known toxicophoric or toxicogenic substituents. There are many instances, however, in which the toxicity of a substance is known but the toxic mechanism is not apparent or predictable from its structure; no obvious or known toxicophoric or toxicogenic substituents may be present. In such cases the chemist who wishes to design a safer chemical will need to learn what the mechanism of toxicity is (assuming that it is known). The chemist will often need to conduct literature searches on the original toxic substance, to identify toxicity studies containing mechanistic information. From such information the environmentally conscious chemist can often design a new substance such that it will be both safe and commercially useful. The remainder of this section provides specific examples of the design of safer analogs of important commercial chemical substances whose toxic effects and mechanisms are well known but not apparent from their structures. The chapters entitled Cancer Risk Reduction Through Mechanism-Based Molecular Design of Chemicals (by Lai, et al) and Designing Safer Nitriles (by DeVito) provide more detailed examples on how an understanding of toxic mechanisms can be used to design safer commercial chemical substances.

Toluene as a Safer Substitute for Benzene. Benzene is known to cause hematoxicity and leukemia in humans *(37, 57)*. As shown in Figure 7, benzene (**12**) undergoes a series of oxidations in the liver to yield several highly electrophilic metabolites that include *(E,E)*-muconaldehyde (**14**); it is believed that **14** is at least partly responsible for the the toxicity of benzene *(58-60)*. Toluene (**13**), the methyl analog of benzene, is considerably less toxic than benzene. The relatively low toxicity of toluene is attributed to the fact that it is readily metabolized in the liver to benzoic acid (**15**) which is essentially nontoxic *(61)*. The methyl group of **13** is much easier to oxidize than is the aromatic ring, and is therefore oxidized preferentially to yield **15** rather than toxic metabolites analogous to those of **14**. The methyl group of **13** serves as "metabolic handle" for detoxication. Toluene is often used as a substitute for benzene because it is equally efficacious in most applications and is considerably less toxic. Although toluene was not developed as a safer substitute for benzene, this example demonstrates how a slight molecular modification can redirect metabolism from bioactivation to detoxication (i.e., drastically reduce toxicity) without necessarily affecting the commercial usefulness of the substance. In addition, this example suggests that the addition of a methyl group onto other toxic aromatic hydrocarbons may provide a metabolic handle for detoxication.

Design of Safer Glycol Ethers. Glycol ethers find extensive use in industry as solvents in the manufacture of lacquers, varnishes, resins, printing inks, textile dyes, and as additives in brake fluids and gasoline. They are also used extensively in consumer products such as latex paints and cleaners. Ethylene glycol monomethyl ether (**16**) and ethylene glycol monoethyl ether (**17**) (Figure 8) are known to cause reproductive and developmental toxicity in several species of laboratory animals *(62)*. Ethylene glycol monobutyl ether (**18**) is not as potent as **16** or **17** in causing reproductive or developmental toxicity, but it is known to destroy red blood cells when administered to rats *(62)*. The mechanism responsible for the toxicity of **16-18** is believed to include metabolism. Following absorption, the primary alcohol (-CH$_2$OH) moiety of these substances is known to be oxidized readily by hepatic alcohol dehydrogenase (ALDH) to yield corresponding alkoxyacetic acids represented by the general structure **19** (Figure 8). These alkoxyacetic acids are believed to be responsible for the toxic effects observed *(62)*. The precise mechanism (i.e., toxicodynamics) by which alkoxyacetic acids are toxic is under investigation.

Because the mechanism of toxicity of ethylene glycol ethers involves their bioactivation to alkoxyacetic acids, the most logical approach to designing less toxic glycol ethers should be to incorporate a structural modification that prevents their metabolism to alkoxyacetic acids. This is demonstrated in Figure 9 with 1-(alkoxy)-2-propanols (**20**). The alcoholic moiety of **20** is secondary, and therefore cannot be oxidized to a carboxylic acid. Not surprisingly, glycol ethers of type **20** are not oxidized by ALDH and do not cause reproductive, developmental, or hematoxicity *(62)*. In fact, this structural modification redirects metabolism to a detoxication pathway: it has been observed that 1-methoxy-2-propanol (**21**) is metabolized by hepatic cytochrome P450 enzymes to undergo O-demethylation to formaldehyde and propylene glycol, which are relatively non-toxic (Figure 9). From a commercial standpoint **21** is essentially equal to **16** in commercial efficacy as a solvent, and has replaced **16** in many commercial applications. This example demonstrates how a subtle molecular modification based on toxic mechanism can drastically reduce toxicity without affecting commercial efficacy. Other 1-(alkoxy)-2-propanols of type **20** are not expected to cause developmental, reproductive or hematoxicity since these cannot be metabolized to alkoxyacetic acids. These glycol ethers will presumably be metabolized similarly to that of **21**.

2-(Alkoxy)-1-propanols (Figure 9, **22**), on the other hand, are likely to be toxic because the alcohol moiety of these substances is primary and subject to oxidation to the carboxylic acid. 2-Methoxy-1-propanol (**22a**), for example, is

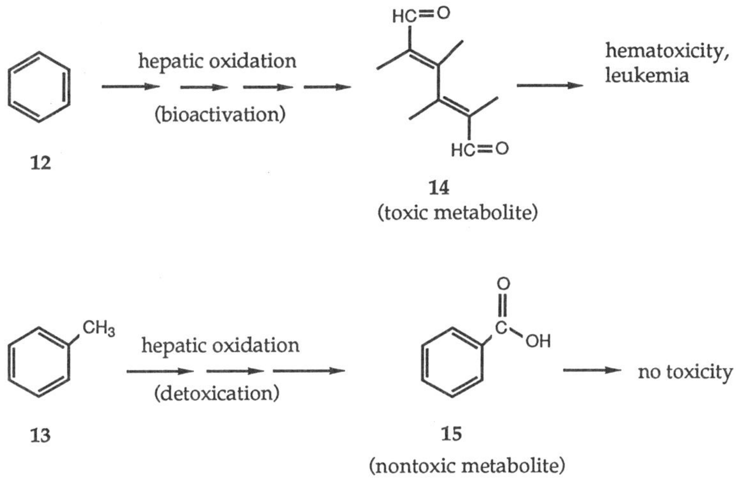

Figure 7. Metabolism of Benzene (**12**) and Toluene (**13**).

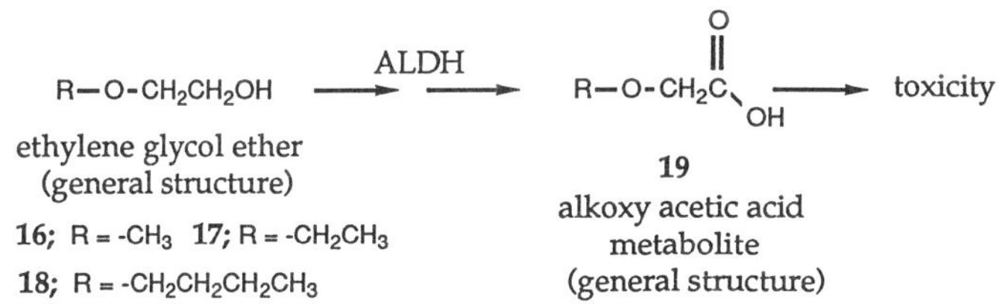

Figure 8. Mechanism of Toxicity of Ethylene Glycol Ethers: Bioactivation to Alkoxy Acetic Acid Metabolites. (ALDH = alcohol dehydrogenase.)

metabolized by ALDH to 2-methoxypropionic acid (**25**) which causes developmental toxicity (Figure 9) *(62)*. Thus, in designing a safer chemical, the chemist must exercise great care and judgment in selecting structural modifications that attenuate toxicity.

Design of Safer Substitutes of *n*-Hexane. *n*-Hexane (**26**) is used industrially as a solvent. Excessive exposure to *n*-hexane causes neurotoxicity *(63)*. The clinical manifestations of *n*-hexane-induced neurotoxicity include sensory numbness, loss of sensation of the extremities, and motor weakness. The mechanism by which *n*-hexane produces neurotoxicity involves oxidative metabolism of the 2 and 5 positions by cytochrome P450 enzymes. These positions are first oxidized to alcohols, which undergo further oxidation to 2,5-hexanedione (**27**) (Figure 10). The keto groups of **27** react with the epsilon amino groups of lysine residues within the proteins of axonal nerve fibers to form pyrrole adducts (Figure 10). These pyrrole adducts undergo protein crosslinking, which disrupts nerve function and ultimately leads to the neurotoxic effects. Formation of the gamma-diketone (**27**) is therefore essential for *n*-hexane neurotoxicity. Logically, modifying **26** such that it cannot be bioactivated to **27** should prevent the neurotoxicity. Methyl or other substituents at the 2 and 5 positions, for example, would permit metabolism to the corresponding alcohols but not to the 2,5-diketone. Possibly safer analogs of *n*-hexane include substances such as 2,5-dimethylhexane (**28**) (Figure 10). It has been shown in rats that the predominant metabolites of **28** are 2,5-dimethyl-1,2-hexanediol (**29**) and 2,5-dimethyl-1,5-hexanediol (**30**) (Figure 10) *(64)*. Unlike *n*-hexane, no neurotoxic effects of **28** have been reported (although no studies assesing the neurotoxic potential of **28** appear in the literature). Based on the mechanism of neurotoxicity for *n*-hexane and on the metabolic data for **28,** it seems logical to assume that **28** is not neurotoxic and therefore may be a useful substitute for *n*-hexane. However, the boiling point of **28** (108 ºC) is approximately 40 º higher than that of *n*-hexane, which may limit its application as a substitute for *n*-hexane.

Designing Safer Chemicals Using Structure-Activity (Toxicity) Relationships

As discussed earlier, substances that are capable of producing a biological effect (pharmacological or toxicological) contain a structural feature that bestows the intrinsic biological property. In the case of drugs, in which the biological response is desired, this structural feature is referred to generically as the pharmacophore. In the case of commercial chemical substances, in which the biological effect is undesired (toxic), the structural feature is referred to generically as the toxicophore. In either case, the structural feature elicits its biological effect through interaction with a specific biomolecular site of action to cause changes in cellular biochemistry. Substances that contain the same pharmacophore or toxicophore are therefore likely to exhibit the same pharmacological or toxicological properties. The relative potency amongst the substances in their ability to cause the biological effect may vary substantially. The relative potency is directly related to the specific or incremental structural differences between the substances and the influence these differences have on the ability of the toxicophore (or pharmacophore) to interact with its biomolecular site of action. The ability of substances belonging to the same chemical class to a cause a particular biological effect and the influence that their structural differences have on potency are referred to as structure-activity relationships. The relationship between structure and activity for a given group of substances becomes much clearer when the mechanism of biological action is known.

Structure-activity relationships are useful for several reasons. First, a series of structurally-similar chemicals with a measured pharmacological or toxicological response may allow one to infer similar pharmacological or toxic effects for an

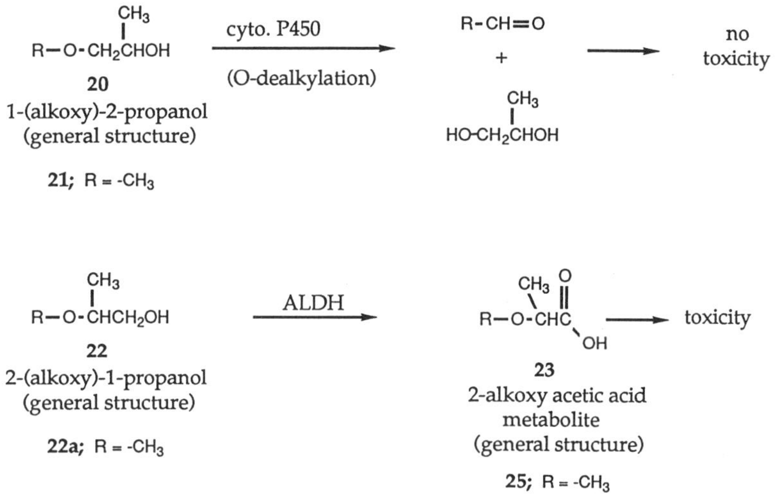
R—O-CH2CHOH (CH3)
20
1-(alkoxy)-2-propanol
(general structure)
21; R = -CH3
cyto. P450
(O-dealkylation)
R-CH=O
+
HO-CH2CHOH (CH3)
no toxicity
R—O-CHCH2OH (CH3)
22
2-(alkoxy)-1-propanol
(general structure)
22a; R = -CH3
ALDH
R—O-CHC(O)OH (CH3)
23
2-alkoxy acetic acid metabolite
(general structure)
25; R = -CH3
toxicity

Figure 9. Metabolism of 1-(Alkoxy)-2-Propanols (**20**) and 2-(Alkoxy)-1-Propanols (**22**). ALDH = alcohol dehydrogenase.

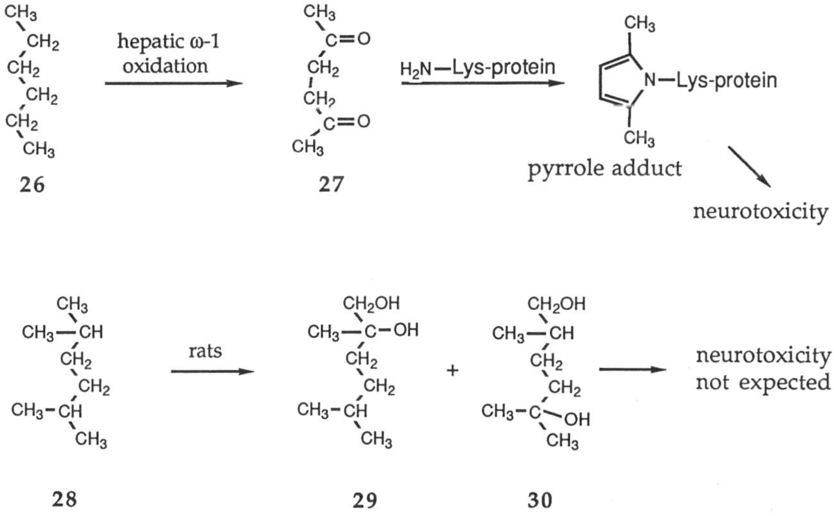
26
hepatic ω-1 oxidation
27
H2N—Lys-protein
N—Lys-protein
pyrrole adduct
neurotoxicity
28
rats
29
30
neurotoxicity not expected

Figure 10. Mechanism of Neurotoxicity of *n*-Hexane (**26**). Metabolism of 2,5-Dimethylhexane (**28**), a possible safer substitute for **26**.

analogous untested substance. Second, structure-activity relationships can be used to design new, analogous substances such that the biological activity is either maximized (in the case of drug substances) or minimized (in the case of commercial chemical substances). Structure-activity relationships have been used for decades by medicinal chemists in the design of highly efficacious drug substances *(65)* and by the U.S. Environmental Protection Agency for assessing the toxicity of new, untested commercial chemicals prior to commercialization *(66)*. Despite the structure-activity data available for many classes of commercial chemical substances, however, the use of structure-activity relationships has been given little attention by chemists as a rational approach for designing new, less toxic commercial chemical substances. This section will discuss how structure-activity relationship data can be used by chemists as a powerful tool for designing safer chemical substances. Unless otherwise indicated in this section, the word "activity" refers to toxicity. Subsequent chapters in this volume (i.e., Designing Safer Nitriles, by DeVito, and Cancer-Risk Reduction Through Mechanism-Based Design of Chemicals, by Lai, et. al.) provide more detailed discussions on the use of structure-activity relationships in combination with toxic mechanisms, and demonstrate the usefulness of this approach with respect to the design of safer nitriles and aromatic amines, respectively.

Qualitative Structure-Activity Relationships. With qualitative structure-activity relationships, the correlation of toxic effect with structure is made by visual comparison of the structures of the substances in the series and the corresponding effects on the toxicity. From qualitative examination of structure-activity data the chemist may be able to see a relationship between structure and toxicity, and identify the least toxic members of the class as possible commercial alternatives to the more toxic members. In addition the chemist may infer from the relationship the structural characteristics that reduce toxic potency, thereby providing a rational basis to design new, less toxic analogous substances. The larger the data set the more apparent the relationship between structure and activity becomes, but small data sets can nonetheless be quite useful. The application of qualitative structure-activity relationships for the design of safer chemicals is demonstrated below using several classes of important commercial chemical substances.

Polyethoxylated Nonylphenols. Polyethoxylated nonylphenols, represented by the general structure **31**, are used as emulsifiers/surfactants, predominately in detergents and inks. It has been observed that these substances cause an intense myocardial necrosis in dogs and guinea pigs within 5 days when administered orally at a dose of 40 mg/kg/day when the extent of ethyoxylation ranges between 14 to 29 ethoxy units *(67)*. However, when the extent of ethoxylation is less than 14 ethoxy units or greater than 29 ethoxy units, no myocardial effects are observed. Although the mechanism of myocardial toxicity is unknown, the structure-activity relationship data described above are nonetheless useful in designing safer polyethoxylated nonylphenols. Clearly, chemists should intentionally design and use polyethoxylated nonylphenols with fewer than 14 or more than 29 ethoxy subunits.

C_9H_{19}—⟨C₆H₄⟩—$O(CH_2CH_2O)_nCH_2CH_2OH$

31

$n = 14\text{-}19$; intense myocardial necrosis

$n < 14$ or > 29; no myocardial effects

Glycidyl Ethers. Glycidyl ethers, represented by the general structure **32**, are used as synthetic reagents for a variety of purposes. It has been shown that glycidyl ethers of the type represented by **32** are mutagenic and cause testicular lesions in rats and rabbits following oral and inhalation administration when the alkyl substituent is an n-octyl (n = 7), n-nonyl (n=8) or n-decyl (n=9) *(6)*. These toxic effects are not observed, however, when the alkyl substituent ranges from dodecyl (n=11) to tetradecyl (n=13). As in the case of polyethoxylated nonylphenols, the mechanism responsible for the toxicity of these glycidyl ethers is not known, although the epoxide moiety is almost certainly the toxicophore. Nonetheless, these structure-activity relationship data are useful for the design of safer glycidyl ethers. Chemists should avoid designing and using glycidyl ethers of the type represented by **32** in which the length of the alkyl moiety ranges from 8 to 10 carbons. Whenever possible, chemists should design and use glycidyl ethers in which the length of the alkyl moiety is at least 12 carbon atoms.

CH_2—CH—O—$(CH_2)_n CH_3$ n = 7-9; mutagenic, testicular lesions

32   n = 11-13; non-mutagenic, no testicular lesions

1,2,4-Triazole-3-thiones. The thiocarboxamide (-C-N) group is found in a variety of commercial substances. However, the thiocarboxamide group is often toxicophoric. Many thiocarboxamides are toxic to the thyroid gland (i.e., thyrotoxic) *(68-70)*. The thyrotoxicity is manifested by an inhibition in the thyroid's ability to synthesize thyroid hormone, which ultimately leads to hypothyroidism. In fact, some thiocarboxamides (e.g., propylthiouracil, methimazole) are used medically to treat hyperthyroidism *(70)*. The specific mechanism by which the thiocarboxamide moiety is thyrotoxic is unknown, but is believed to involve inhibition of thyroid peroxidase, the enzyme that catalyzes the incorporation of iodine into tyrosine residues during thyroid hormone synthesis *(69,70)*.

Chemists who design thiocarboxamide-containing substances need to be aware of the thyrotoxic potential of these substances and, whenever possible, should design these substances such that their potential for causing thyrotoxicity is minimal. The potential for thyroid toxicity of a particular thiocarboxamide may be kept to a minimum if structure-activity relationship data are available and are taken into account during chemical design. This point can be exemplified with 1,2,4-triazole-3-thiones, represented by the general structure **33**, in Table III. 1,2,4-Triazole-3-thiones are used commercially for a variety of purposes. Structure-activity relationship data pertaining to inhibition of thyroid hormone production are available for this class *(71)*, and are presented in Table III. The least thyrotoxic of the 1,2,4-triazole-3-thione congeners shown in Table III are **34** and **35**, which differ only in the position of the methyl group attached to nitrogen at position 2 or 4 of the triazole ring. Relocating the methyl group to the carbon at position 5 of the triazole ring (to give **36**) results in an over 200-fold increase in toxicity over **34** and **35**. Compound **36** is highly thyrotoxic, and presents a risk to any individuals exposed to this substance. Chemists who design 1,2,4-triazole-3-thiones can use the data in Table III to minimize the potential for thyrotoxicity in new compounds belonging to this class. Unless absolutely necessary for commercial usefulness, one should avoid placing a methyl group (and probably longer alkyl substituents) at position 5, especially if positions 2 and 4 contain hydrogens. If a methyl or other alkyl substituent must be present at position 5, the potential for thyrotoxicity can be reduced significantly if an additional methyl group (or perhaps another alkyl group) is added to the methyl ring at position 4, as apparent by comparing the thyrotoxicity of **36** to that of **37**.

Carboxylic Acids. Carboxylic acids are an important and widely used class of commercial chemical substances. The commercial uses of carboxylic acids include synthetic intermediates, plasticizers, catalysts, and preservatives, to name a few. Many commercial carboxylic acids have pharmacological and toxicological properties. Valproic acid (2-propylpentanoic acid), for example, is used medically as an anticonvulsant. Toxicological properties of carboxylic acids include liver toxicity and teratogenicity. Although carboxylic acid-induced liver toxicity can be severe, only a few carboxylic acids are known to cause it *(72)*.

The teratogenic properties of carboxylic acids, however, are known to occur for many carboxylic acids *(73-76)*. Teratogenicity (also known as developmental toxicity) is the ability of a substance to cause a toxic effect in the fetus or offspring, following maternal administration of the substance anytime prior to or during gestation. These toxic effects can vary widely, and frequently include: structural abnormalities; altered growth; mental retardation; blindness; deafness; cleft palate; functional deficits; and death. The teratogenic effects of carboxylic acids include skeletal defects (extra presacral vertebrae, fused ribs), exencephaly, reduced fetal weight, neural tube defects, and death *(74)*.

The structure-activity (teratogenicity) relationships have been discussed in two comprehensive reviews *(75,76)*. The teratogenic properties of carboxylic acids are highly structure-dependent. Table IV summarizes the structural requirements for high teratogenic potency of carboxylic acids. Chemists who develop carboxylic acids can use the structure-activity relationship information in Table IV to design carboxylic acids such that the likelihood for teratogenicity is minimized. Safer carboxylic acids are those that contain, for example, two hydrogens (i.e., no substitution) or no hydrogens (complete substitution) at the C-2 carbon. Other examples of safer carboxylic acids include ones that contain unsaturation between C-2 and C-3, or C-3 and C-4.

Quantitative Structure-Activity Relationships (QSARs). It is often possible to quantify structure-activity relationship data by correlating into a regression equation the biological property with one or more physicochemical properties of a set of analogous substances. In quantitative structure-activity relationships (QSARs) chemical structure is transformed into quantitative numerical values that describe physicochemical properties relevant to a given biological activity. Quantification of structure-activity relationships for a given series of substances depends, therefore, on the successful identification of one or more physicochemical properties correlating with the biological property. The physicochemical properties that correlate with the biological property are most likely related to the mechanism of biological activity, and are often referred to as "descriptors" of biological activity. An example of a general QSAR equation is illustrated by equation 1:

$$\log(1/C) = a(X)^2 + b(X) + c(Y) + d \qquad (1)$$

$$n \quad r \quad s$$

where biological activity is expressed as 1/C (C is a standard concentration or dose of a substance required to elicit the biological activity), X and Y are the physicochemical descriptors of the activity, a,b,c and d are coefficients, n is the number of substances comprising the dataset used in the regression, r is the correlation coefficient and s is standard deviation of the regression.

Table III. Structure-Activity (Thyrotoxicity) Relationships of 1,2,4-Triazole-3-thiones (34-40)[a]

Compound No.	R_1	R_2	R_3	Relative Toxic Potency [b]
34	CH_3	H	H	1.0
35	H	CH_3	H	1.2
36	H	H	CH_3	212.0
37	CH_3	H	CH_3	7.1
38	H	H	phenyl	5.7
39	CH_3	CH_3	H	4.7
40	H	H	H	3.6

[a] Data obtained from ref. 71. [b] Expressed as the ability to inhibit radioactive iodine uptake by the thyroid gland.

Table IV. Structural Requirements for High Teratogenic Potency of Carboxylic Acids[a]

i) a free carboxyl group[b]

ii) only one hydrogen atom at C-2

iii) an alkyl substituent larger than a methyl at C-2

iv) no double bonds between C-2 and C-3, or C-3 and C-4

[a] Teratogenic potency based on *in vivo* data. [b] Esters of carboxylic acids that have the requisite structural requirements for teratogenicity are also likely to be teratogenic because esters are often metabolized to the free carboxylic acid.

By quantifying structure-activity relationships, one is able to delineate the change in biological potency that is (or would be) accompanied by a given change in structure more precisely than is possible by intuition or qualitative structure-activity relationships alone. Using a QSAR correlation for acute lethality, for example, one can predict the median lethal dose (LD_{50}) of an untested substance directly from a physical property of that substance. Or, using a QSAR correlation for bacterial growth inhibition, one can predict the median effective dose (ED_{50}) of an untested potential antibiotic. Because there are methods for accurately estimating most physicochemical properties directly from structure, is not necessary to synthesize a substance in order to measure those physicochemical properties that need to be used as descriptors in a QSAR model. One can estimate those properties, incorporate them into the appropriate QSAR regression equation and predict the biological property of the substance even though the substance does not exist!

Medicinal chemists have, for many years, used QSAR as a tool for drug design. The U.S. Environmental Protection Agency (EPA) has used QSAR since 1981 to predict the aquatic toxicity of new, untested commercial chemical substances in the absence of test data. Chemists who are interested in designing safer chemicals will find QSAR very helpful because it enables one to rapidly assess the toxicity of substances *without* having to synthesize and test the substances. In fact, the EPA distributes a computer program called ECOSAR, which contains over 100 QSAR regression equations for at least 42 chemical classes *(77)*. The EPA encourages chemical manufacturers to use ECOSAR as a tool for designing new substances that are less toxic to aquatic species.

A discussion on how to derive a QSAR regression equation is beyond the scope of this chapter. The chapter in this book entitled Designing Safer Nitriles (DeVito) provides a relevant example and discusses how to derive and use a QSAR equation for purposes of predicting toxicity and designing safer nitriles. In addition, a number of books are available on QSAR, and most provide many QSAR regression equations that correlate the specific biological activities (e.g, acute toxicity) of various classes of substances with physicochemical properties. A few of these books *(78-83)* are listed in the Literature Cited section. The reader is encouraged to examine these books for detailed discussions on the development and applications of QSAR equations.

Designing Safer Chemicals Using Isosteric Replacements

Substances that have similar molecular and electronic characteristics but are not necessarily structurally related often have similar physical or other properties. Langmuir called this phenomenon isosterism, and coined the term isostere *(84)*. According to Langmuir's definition *(ref. 84 published in 1919)* isosteres are substances or substituents that have the same charge, caused by the same number and arrangement of electrons and the same number of atoms. As our understanding of molecular orbital theory evolved over the years, several variations of Langmuir's definition of isosterism were expressed by others; these are discussed in recent reviews *(85,86)*. Burger's definition *(85)* of isosterism encompasses the aspects of the previous definitions and states that isosteres are chemical substances, atoms or substituents that possess near equal or similar molecular shape and volume, approximately the same distribution of electrons, and which exhibit similar physicochemical properties. A few examples of isosteric atoms and substituents are provided in Figure 11. Many more examples are available in the literature *(85)*.

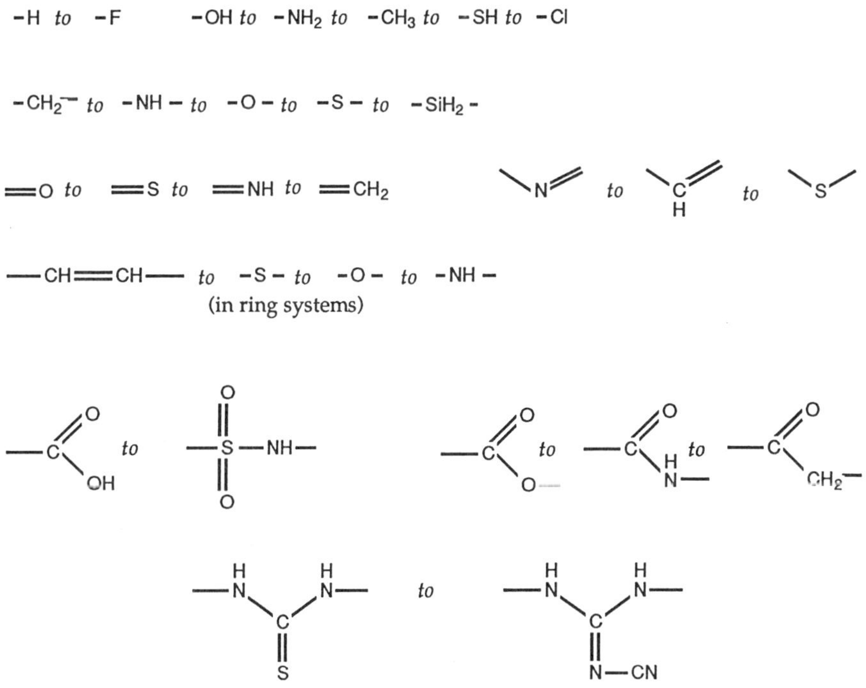

−H *to* −F   −OH *to* −NH₂ *to* −CH₃ *to* −SH *to* −Cl

−CH₂− *to* −NH− *to* −O− *to* −S− *to* −SiH₂−

=O *to* =S *to* =NH *to* =CH₂   \N= *to* \C(H)= *to* \S−

—CH=CH— *to* −S− *to* −O− *to* −NH−
(in ring systems)

−C(=O)OH *to* −S(=O)(=O)−NH−   −C(=O)O− *to* −C(=O)N(H)− *to* −C(=O)CH₂−

−N(H)−C(=S)−N(H)− *to* −N(H)−C(=N−CN)−N(H)−

Figure 11. Examples of Isosteric Atoms and Substituents.

Benzene (**12**) is isosteric with thiophene (**41**) and pyridine (**42**) because the -CH=CH- group is isosteric with -N= and -S- atoms (Figure 11). Although these substances are structurally different, some of their chemical properties are nonetheless similar. All of them are aromatic, all are liquid, and all are about equal in molecular size and volume. In fact, both **12** and **41** boil at about 81 ºC.

12    **41** 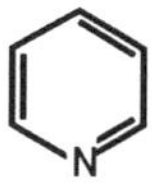   **42**

Substances that are isosteric equivalents of substances that are toxic or pharmacologically active may also possess these biological properties. It is also possible that biological properties may be bestowed, exacerbated or attenuated when isosteric modifications are made. This point is illustrated by the following examples. 7-Methyl-benzo[a]anthracene (**43**), is a known carcinogen, while 7-methyl-1-fluorobenzo[a]anthracene (**44**) is not *(1)*. The carcinogenicity of **43** is due to its bioactivation to an epoxide metabolite. The epoxidation occurs between the 1 and 2 positions. Isosteric replacement of the hydrogen atom at position 1 of **43** with a fluorine atom provides **44**. This substance (**44**) is not carcinogenic because the fluorine atom blocks bioactivation of the 1,2 position and, hence, the 1,2 epoxide metabolite is not formed *(1)*. Replacement of the hydrogen atom at position 2 of **43** with a fluorine atom also abolishes carcinogenicity *(1)*.

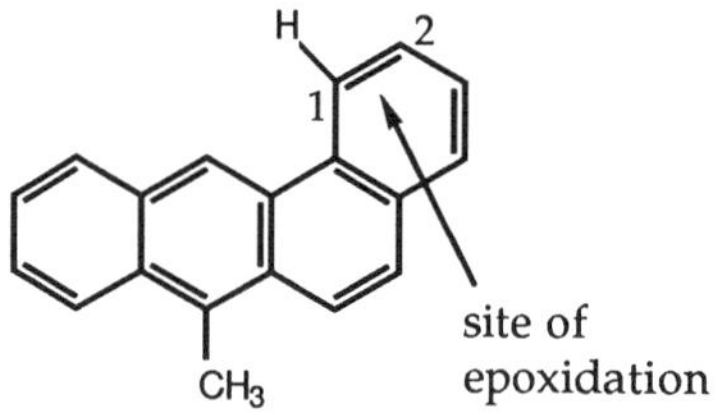

43

carcinogenic

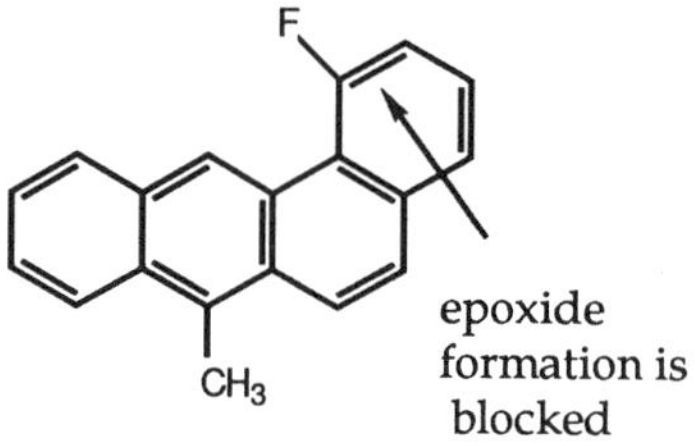

44

non-carcinogenic

Acetic acid (**45**), on the other hand, is essentially nontoxic but its fluoro-isostere, fluoroacetic acid (**46**), is highly toxic (human oral LD_{50} is estimated to be 2-5 mg/kg; *87*. In the body acetic acid reacts with coenzyme A (CoA) to for acetyl-CoA, which is an important precursor of the citric acid cycle (a biochemical cascade essential for energy production). Fluoroacetic acid is so sterically similar to acetic acid that it also reacts with CoA, and forms fluoroacetyl-CoA. Fluoroacetyl-CoA enters the citric acid cycle and forms fluorocitrate, which is a potent inhibitor of aconitase, a critical enzyme of the citric acid cycle *(87)*.

$$H-CH_2-C\begin{smallmatrix}O\\OH\end{smallmatrix}$$

45
non-toxic

$$F-CH_2-C\begin{smallmatrix}O\\OH\end{smallmatrix}$$

46
toxic

Medicinal chemists have used isosterism for the design of safe, effective drug substances for many years. During the development of anti-ulcer medications, for example, it was found that metiamide (**47**) greatly reduced acid secretion in the gastrointestinal tract by antagonizing H_2-receptor sites. Its potential as a useful anti-ulcer medication was lessened by the toxic effects caused by the thiourea moiety. This moiety is essential for H_2-receptor blockade, but bestows toxicity. Isosteric replacement of the thiourea moiety with the cyanoguanidine moiety gave cimetidine, (**48**), a potent H_2-receptor antagonist that lacks the toxicity of **47**. Cimetidine is one of the most widely used anti-ulcer medications in the world because of its effectiveness in treating ulcers and relative safety. It is noteworthy that in this example this isosteric modification selectively reduced toxicity without affecting pharmacological activity. This is a main reason why isosteric substitution is a common practice among medicinal chemists for the design of drug products. Many other examples of the use of isosterism in drug design are in the literature *(85,86)*.

47

48

The application of isosterism for the design of safer commercial chemicals is much less common than it is for the design of safer drugs. There are a few recent examples, however, of how isosterism has been used to design safer commercial chemical substances. One very successful application was recently reported for the design of metallized azo dyes *(88,89)*. Historically, chromium was a metal of choice in many metallized azo dyes because it imparts the desired color and fastness. Hexavalent chromium (Cr VI) was often used in making such dyes. Hexavalent chromium is a known human carcinogen, however, and its commercial use is strictly regulated and highly discouraged by environmental authorities. An alternative metal to chromium in premetallized azo dyes would have to have the same color and fastness properties as chromium but without the toxicity. It has been found that iron (Fe), which is essentially nontoxic, often imparts the same desirable qualities as chromium when used in azo dyes *(88,89)*. This is exemplified in comparing azo dyes **49** and **50**. Dyestuff **50** has the same color and fastness as **49**, but does not contain chromium. Other examples of dyestuffs that use iron rather than chromium are available *(88,89)*.

49; M = Cr
50; M = Fe

Another example of the use of isosterism in the design of safer chemicals is in the case of the insecticide MTI-800 (**51**) and its silane isostere **52**. MTI-800 is a potent insecticide that is also highly toxic to fish (its LC_{50} is 3 mg/liter), which limits its commercial usefulness. Isosteric substitution of the quaternary carbon of **51** with silicon resulted in a substance (**52**) that has moderately less insecticidal potency (0.2-0.6) but is considerably less toxic to fish (no fish mortality occurs at concentrations of 50 mg/liter) *(90,91)*. It is interesting to see in this example how the aquatic toxicity is reduced drastically while the insecticidal activity is only moderately affected. Substitution of silicon for carbon for the design of safer chemicals is discussed in greater detail in the chapter entitled Isosteric Replacement of Carbon with Silicon in the Design of Safer Chemicals (Sieburth).

51   fish LC_{50} = 3 mg/liter

52   No mortality to fish at 50 mg/liter

Based on the examples provided above and the many successful applications of isosterism in the design of safer drugs, it appears that isosteric replacement should have an expanded role in the design of safer commercial chemical substances in the future.

Designing Safer Chemicals Using Retrometabolic Design (i.e., "Soft" Chemical Design)

The concept of "soft" drug or retrometabolic design has emerged in the field of medicinal chemistry within the past 15 years *(5,92)*. Soft drugs are defined as biologically active, therapeutically useful drugs deliberately designed to be metabolized quickly to non-toxic substances after they accomplish their therapeutic purpose *(5)*. A soft drug is usually an analog of a pharmacologically active substance whose clinical utility is limited by toxicity or adverse effects. The soft drug retains the pharmacologic property but lacks the toxicity because of its rapid detoxication. The ideal soft drug, therefore, is one that possesses the desired pharmacologic property and is converted into non-toxic readily excretable substances in a single, non-oxidative, metabolic step (e.g., hydrolysis). Using a pharmacologically active but toxic drug substance as a guide, the design of a soft drug begins with deducing non-toxic metabolites that can be retrometabolically combined (hence the term "retrometabolic design") to form a single structure: the soft drug. This relatively new approach to drug design has led to the development of a number of non-toxic, highly useful drug substances; many examples are in the literature *(5,92)* and elsewhere in this book (Bodor).

An example of the concept can be seen in comparing cetylpyridinium chloride (**53**) with **54**, its "soft" analog (Figure 12). Cetylpyridinium chloride is an effective antiseptic but is regarded as being quite acutely toxic to mammals because it has a rat oral median lethal dose (LD$_{50}$) of 108 mg/kg *(92)*. Using **53** as a guide, the design of **54** captured the important structural elements that are necessary for antiseptic activity (i.e., the pyridinium and C-16 alkyl moieties), and structural modifications (the pyridinium methyl ester) that enable rapid breakdown of the substance to comparatively less toxic substances in mammals *(5,92)*. Pyridine, formaldehyde and tetradecanoic acid were chosen as the "metabolites" because they are relatively non-toxic and can be retrometabolically combined into a single easily-hydrolyzable substance (**54**) that is a soft analog of **53**. The side chains of **53** and **54** are essentially 16 carbons in length, and these substances share the same physicochemical and antiseptic properties. They differ greatly, however, in their mammalian toxicity: **54** is 40 times less toxic than **53** (the rat oral LD$_{50}$ of **54** is greater than 1000 mg/kg). Substance **54** is less toxic than **53** because the pyridinium methyl ester moiety undergoes facile hydrolytic cleavage in the blood to pyridine, formaldehyde, and tetradecanoic acid.

The concept of soft drug or retrometabolic design can be extended to commercial chemical design. A "soft" commercial substance could be defined as a substance deliberately designed such that it contains the structural features necessary to fulfill its commercial purpose but, if absorbed into exposed individuals, it will break down quickly and non-oxidatively to non-toxic, readily excretable substances. The concept has been applied to the design of safer alkylating agents and safer analogs of DDT *(5,92)*, and undoubtedly has the potential for broad application for a variety substances. But, like other approaches used by medicinal chemists for the design of drugs, the soft drug approach is seldom seen in commercial chemical design. An entire chapter of this book is devoted to the application of the principles of soft drug design to the design of safer commercial chemical substances (Bodor), where the reader will find additional discussion and examples of this approach.

Designing Safer Chemicals: Identification of Equally Useful, Less Toxic Chemical Substitutes of Another Class.

Another approach to designing safer chemicals is the identification of an equally useful less toxic substance that belongs to another chemical class. Unlike the approaches discussed thus far, the focus of this approach is on commercial use (not on molecular modification), and depends upon the successful identification of a less toxic substance of a different chemical class that can fulfill this use. This approach may be particularly attractive in situations in which molecular modification cannot eliminate or reduce the toxicity of a substance without having a negative affect on commercial usefulness. Some examples of the application of this approach are provided below.

Acetoacetates as Substitutes for Isocyanates in Sealants and Adhesives. Isocyanates are widely used in industrial sealants and adhesives. In these applications the sealant or adhesive effect results from reaction of an isocyanate with a nucleophile (such as an alcohol or amine) to yield a cross-linked adduct. Isocyanates are particularly useful in sealants and adhesives because of their fast cure, ability to adhere to most substrates, and relative low price. A major disadvantage of isocyanates, however, is their toxicity. Isocyanates cause cancer, mutations, pulmonary sensitization, and asthma (Table II) and, as such, pose serious health risks to manufacturing personnel. They also require special handling and storage, and have limited package stability and weatherability.

The Tremco Corporation (Beachwood, Ohio) has focused much attention on finding safer alternatives to isocyanates, and has recently developed a non-isocyanate based adhesive and sealant system that can replace isocyanate-based systems. This alternative sealant-adhesive system utilizes acetoacetate as a functional equivalent of isocyanate *(93)*. The chemistry of the new system (represented in Figure 13) involves the reaction of a polyol (**55**) with *t*-butyl acetoacetate (*t*-BAA) to produce an esterified product (**56**). A portion of **56** is reacted with a diamine (**57**) to provide **58**. The remaining portion of **56** is reacted with **58** to effect cure. The acetoacetate-based system has many of the advantages of the isocyanate-based system and is equally useful *(93)*. The primary advantage is that acetoacetates are essentially non-toxic; this new system eliminates the risks posed by the isocyanate-based system.

Isothiazolones as Substitutes for Organotin Antifoulants. The growth of marine organisms on submerged structures such as the hulls of ships can cause increased hydrodynamic drag, which is commonly referred to as fouling. Although seemingly harmless, fouling leads to increased fuel consumption, decreased ship speed, increased vessel servicing and cleaning costs, and increased dry dock time. It is estimated that the U.S. government spends over a billion dollars each year as a result of fouling of its military vessels (see chapter by Willingham and Jacobson).

Antifouling agents are often applied to hulls of ships to prevent fouling. Organotin substances are effective antifouling agents, but they are highly toxic to mussels, clams, and other non-fouling aquatic species. In addition, because many organotin substances are regarded as hazardous wastes, their removal from ships during cleaning operations must be performed carefully and is costly. Because of their ecotoxicity, the use of organotin antifoulants has been banned throughout the world. The Rohm and Haas Company (Spring House, PA) has devoted much effort to finding antifouling agents that are not toxic to non-fouling aquatic species. They have found that isothiazolones are effective marine antifoulants. 4,5-Dichloro-2-*n*-octyl-4-isothiazolin-3-one (**59**) is a particularly useful antifoulant. In addition to being an excellent biocide, it presents little risk to non-fouling aquatic organisms: it decomposes quickly in marine environments and the decomposition products bind strongly to sediment and are not available to aquatic species. This substance has recently been approved as an antifoulant by the Office of Pesticides of the U.S. Environmental Protection Agency. The use of **59** as an antifoulant involves its incorporation into a polymeric matrix which is applied to the hulls of ships. Fouling organisms are exposed to the substance as the polymeric matrix slowly wears from the hulls. Thus, fouling organisms are continuously exposed to the substance as the polymeric matrix wears away. If any of the substance (**59**) enters the water it quickly decomposes. See the chapter entitled Designing an Environmentally Safe Marine Antifoulant (Willingham and Jacobson) for a more detailed discussion of the development of **59**.

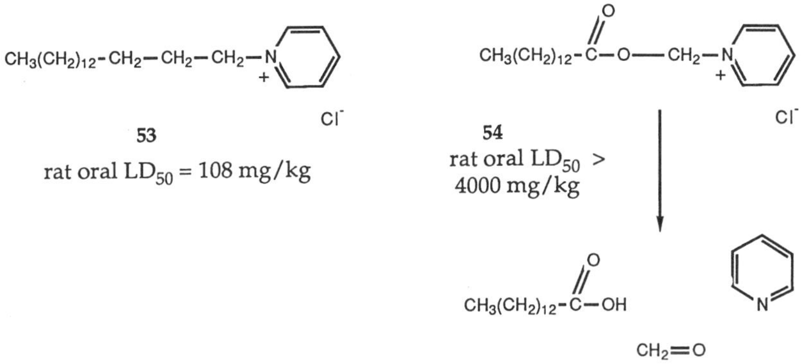

Figure 12. Cetylpyridium Chloride (53) and its Soft Analog, 54.

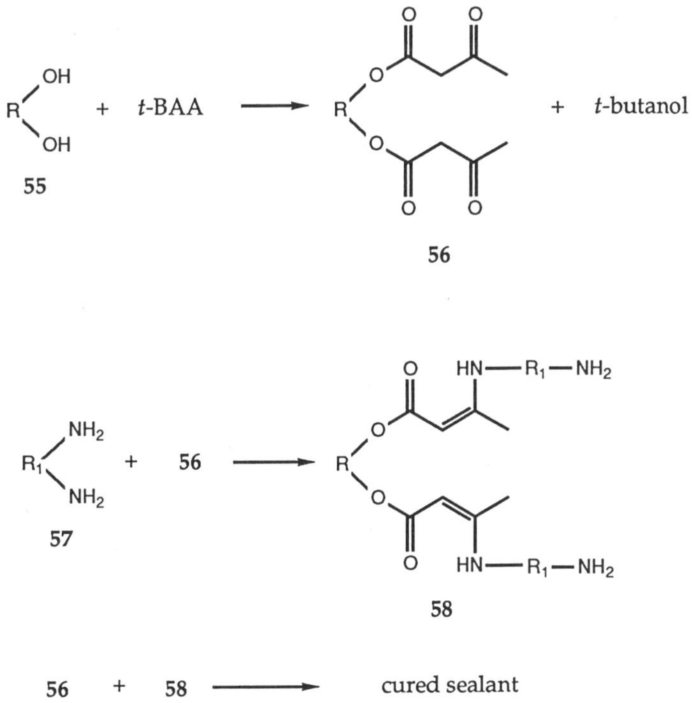

Compound 53: $CH_3(CH_2)_{12}$-CH_2-CH_2-CH_2-N-pyridinium Cl^-
rat oral LD_{50} = 108 mg/kg

Compound 54: $CH_3(CH_2)_{12}$-C(=O)-O-CH_2-N-pyridinium Cl^-
rat oral LD_{50} > 4000 mg/kg

Hydrolysis products: $CH_3(CH_2)_{12}$-C(=O)-OH, pyridine, $CH_2=O$

Figure 13. Acetoacetate-Based Sealants and Adhesives.
t-BAA = t-acetoacetate.

55 (diol) + t-BAA → 56 (bis-acetoacetate ester) + t-butanol

57 (diamine $R_1(NH_2)_2$) + 56 → 58

56 + 58 ⟶ cured sealant

59

Sulfonated Diaminobenzanilides as Substitutes for Benzidines in Dyes.

Benzidine and many of its congeners were at one time widely used in the synthesis of dyestuffs. Their unique color and fastness properties made them particularly useful for this purpose. When it became apparent that benzidine and a number of its congeners are highly carcinogenic *(33)* their use as synthetic intermediates in dyestuffs dropped drastically. Many researchers have attempted to find non-carcinogenic alternatives to benzidines that have the same desired properties as benzidines. Sulfonated diaminobenzanilides where recently reported to be useful substitutes for benzidine in the synthesis of direct dyes *(94)*. One of these substances (**60**) is shown below. Although it is not yet known whether these substances are carcinogenic, the sulfonic acid moiety may make them non-carcinogenic because the carcinogenicity of other aromatic amines is often eliminated by the inclusion of this moiety. (The carcinogenicity of aromatic amines is related to their metabolism, and it is believed that the sulfonic acid group reduces the need for metabolism because it greatly increases water solubility and promotes rapid excretion from the kidney. See ref. *95,* and chapter by Lai, et al. for further details).

60

Designing Safer Chemicals: Elimination of the Need for Associated Toxic Substances

This approach to designing safer chemicals also differs from the approaches discussed previously in that its focus is not on the chemical itself but rather on any associated substances that are toxic. Although a chemical substance may not be toxic, its storage, transportation or use may require an associated substance that is toxic (e.g., a solvent such as carbon tetrachloride). In such instances it is the associated substance that represents the toxic component. In this approach one needs to somehow eliminate the need for the associated toxic substance. In some cases this could be accomplished by simply identifying an alternative, less toxic associated substance that will serve the same purpose as the toxic substance (e.g., switching from a toxic solvent to a less toxic, equally useful solvent). In other cases, more elaborate formulation changes may be necessary. In cases where switching to a less toxic associated substance or reformulation is not possible, the original substance may have to be structurally modified to a new substance for which a less toxic associated substance can be used or reformulation is possible. These structural modifications should not, of course, impart toxicity.

Elimination of the need for associated toxic substances has been recognized by chemists for years, and has had and continues to have particular application in coatings. Perhaps the most classic example is that of water-based paints as replacements for oil-based paints. Oil-based paints have the disadvantage of requiring toxic organic solvents. Over the years many oil-based paint products have been reformulated to water-based (environmentally friendlier) paint products. For many of these products, structural modifications to the paint pigments were not necessary; only the formulations were changed. A more recent example is the use of supercritical carbon dioxide to replace volatile organic compounds (VOC) in spray paints *(96)*. Using this technology, it has been reported that VOC emissions can be reduced up to 80%. In addition, this technology improves the finish quality of the paint. Other approaches to eliminating the need for VOCs in coatings include the use of aldimines (see chapter by Wicks and Yeske) and the use of "dibasic esters". Dibasic esters are mixtures of simple alkyl (e.g., methyl) esters of dicarboxylic acids such as adipic, succinic, and glutaric acids. The DuPont Company promotes the use of dibasic esters in coatings as less toxic replacements for solvents such as glycol ethers, cyclohexanone, isophorone, cresylic acid, methylene chloride, and others. Dibasic esters appear to have equal or better performance capacity compared to these solvents.

Obtaining Toxicity-Related Information

The earlier sections of this chapter discuss how information pertaining to toxic mechanisms, structure-activity relationships, isosterism, and retrometabolism can be used for designing safer chemicals. Armed with such information, the chemist may use one or more of the approaches discussed in this chapter to design safer chemicals. In many cases the design of a safer chemical will involve deducing structural modifications that lessen the likelihood of chemical toxicity. Implicit in these sections, however, is that the chemist is already aware of the known or potential toxicity of a substance, and has obtained the necessary toxicity information (on the substance itself or its analogues) that can be used to design a safer chemical. It is important to stress that chemists need to have a "feel" for any toxic properties associated with the classes of substances that they develop. To do this, and to design safer chemicals, chemists need to familiarize themselves with as much toxicity information as possible that pertains to a class of substances *before* they design and market substances within the class. Chemists who develop glycol ethers or carboxylic acids, for example, need to know that many of these substances can cause developmental toxicity and, when designing new substances of these classes, should include structural modifications that lessen the likelihood of developmental toxicity, as discussed earlier.

But how does one go about acquiring toxicity-related information? There are a number of excellent toxicology reference sources that the chemist will find useful for obtaining relevant information. Table V provides a list of textbooks that contain information related to toxic mechanisms, structure-activity relationships, metabolism, and toxicophoric/toxicogenic substituents, as well as other health and toxicity-related data on many specific substances and classes of substances.

Table VI provides a list of databases (most of which are now available online) that also contain information related to chemical toxicity. These databases contain enormous amounts of information that can be easily retrieved. For example if one wanted to find toxicity information on a specific substance, one could search these databases using the chemical name or the substance's Chemical Abstracts Service (CAS) registry number to obtain literature citations and other information. If one wanted to identify structural analogs of a substance, one could perform a substructure search of the Chemical Abstracts Registry File (which contains the structures of over

Table V. Examples of Textbooks Containing Toxicity-Related Information

---

Casarett and Doull's Toxicology. The Basic Science of Poisons, Fifth Edition (McGraw Hill: New York, NY, 1996)

Goodman and Gilman's The Pharmacological Basis of Therapeutics, Ninth Edition (McGraw Hill: New York, NY, 1996)

Drug Toxicokinetics (Marcel Dekker, Inc.: New York, NY, 1993)

Burger's Medicinal Chemistry and Drug Discovery, Fifth Edition (a multivolume set currently in press, John Wiley & Sons, Inc.: New York, NY)

Comprehensive Medicinal Chemistry (a six volume set, Pergamon Press: New York, NY, 1989)

Tissue-Specific Toxicity (Academic Press: New York, NY, 1992)

Conjugation-Dependent Carcinogenicity and Toxicity of Foreign Compounds (Academic Press: New York, NY, 1994)

Bioactivation of Foreign Compounds (Biochemical Pharmacology and Toxicology Series; Academic Press: New York, NY, 1985)

Chemical Carcinogens: Activation Mechanisms, Structural and Electronic Factors, and Reactivity (Elsevier: New York, NY, 1988)

Chemical Induction of Cancer, volumes IIa (1974), IIb (1974), IIIa (1982), IIIb (1985), and IIIc (1988) (Academic Press: New York, NY, 1988)

Selective Toxicity, The Physico-Chemical Basis of Therapy, Seventh Edition (Chapman and Hall: New York, NY, 1985)

Biological Alkylating Agents, Fundamental Chemistry and the Design of Compounds for Selective Toxicity (London Butterworths: London, 1962)

Toxicology (CRC Press: Boca Raton, FL, 1996)

CRC Hanbook of Toxicology (CRC Press: Boca Raton, FL, 1995)

Table VI. Examples of Databases Containing Toxicity-Related Information[a]

Registry of Toxic Effects of Chemical Substances (RTECS)

MEDLINE, TOXLINE, TOXLIT, CHEMLIST

Hazardous Substances Data Bank (HSDB)[b]

Toxic Substances Control Act Test Submissions (TSCATS)[b]

Chemical Abstracts, Chemical Abstracts Registry File[c]

[a] For further information regarding these databases, contact STN International, c/o Chemical Abstracts Service, Columbus, Ohio (phone: 614-447-3600). [b] TSCATS contains test data submitted to the U.S. Environmental Protection Agency under sections 4 and 8 of the Toxic Substances Control Act (TSCA). The online version is the most up-to-date version. For additional information, contact the Syracuse Research Corporation, Merrill Lane, Syracuse, New York (phone: 315-426-3429).
[c] Chemical Abstracts Registry File is not a toxicity database, it is a substance database. It is included here because it is quite useful for identifying structural analogs and other information. See text for details.

12 million different substances) to obtain the structures, names, and CAS registry numbers, of analogous substances. The CAS registry numbers of these substances could then be used to search other databases. (The Registry File also has a field that lists other databases in which a substance is mentioned). The Chemical Abstracts database, which began in 1907, is exceptionally useful because it abstracts studies published in thousands of journals and technical reports and contains a great deal of toxicity information on many substances. (This fact is unknown to many scientists, including toxicologists!)

A word of caution in using online versions of the databases in Table VI (and many others) is that they are not necessarily as comprehensive as the printed versions. With the exception of the Chemical Abstracts Registry File, the online versions (and in some cases the printed versions) of the databases listed in Table VI contain information published only within the past 2 to 3 decades. Studies published prior to the 1960s will not be found in any of the online versions of the databases in Table VI. One must be careful, therefore, not to overlook useful information that is available but not cited in on-line databases. For example if one wanted to find toxicity information on sulfonium compounds, very little will be found by searching online databases. Yet many toxicity studies pertaining to sulfonium compounds appear in the "older" literature and can only be found by manually searching the printed versions of databases. To conduct a comprehensive literature search on a substance or class of substances, it is best to start with the first decenial index (1907-1916) of Chemical Abstracts and search (manually) the substance by name or molecular formula, or the class of substances under subject heading (e.g., "sulfonium") and continue with each subsequent decenial index up to about 1970. The search at this point can be continued using the online version of Chemical Abstracts as well as other online databases.

Conclusion

The evolution of pollution prevention and therewith the concept of green chemistry as the preferred strategy for reducing the risks posed by commercial chemical substances has changed the way chemists think about how chemicals are designed and synthesized. Chemists are now trying to design chemical substances such that they are not only useful from a commercial standpoint, but are safe to humans and the environment as well. In addition, chemists are now trying to identify less polluting syntheses for the chemicals they design. The ultimate environmental goal for chemists is to design safe chemicals and safe chemical syntheses.

Few chemists, however, receive very little training in green chemistry during their graduate education, and generally are not well prepared to design safer chemicals and syntheses. This chapter discusses briefly several approaches that can be used by chemists to design safer commercial chemical substances, and is intended to provide a general framework to do so. (The synthesis aspect of green chemistry is discussed in another book, 97.) Many of the approaches discussed in this chapter have been used for many years by medicinal chemists for the design of safer drug substances. It is the author's hope that chemists who strive to design safer chemicals will find this chapter useful. The remaining chapters of this book provide detailed discussions of these approaches and additional examples of their succesful application. Other important aspects for the design of safer chemicals not mentioned in this chapter due to space limitations include: biodegradation; aquatic toxicity; and the use of computers to predict chemical toxicity. These aspects are discussed in detail in the chapters by Boethling; Newsome, et al.; and Milne, et al., respectively.

To implement the approaches discussed in this chapter, the chemist should try to answer the following questions during the design stages of a substance:

- *Are there any toxicities associated with the chemical class to which the substance belongs?*

    - Does the substance contain any toxicophoric substituents?

    - Are any structure-activity (toxicity) data available?

    - Is the substance related to any pharmacologically active (e.g, drug) substances ?

- *Have any toxicity-related (e.g., mechanistic) studies been conducted on the substance or analogous substances ?*

- *Are metabolism data available?*

    - What are the known or likely products of metabolism? Are they toxic?

    - are toxicogenic substituents present ?

- *Can the substance be absorbed from the skin, lung, or gastrointestinal tract?*

    - If so, which is the most likely route?

- *From answering the above questions, is the substance that I plan to make likely to be toxic?*

    - If so, what can be done to reduce the toxicity?

- *Will the substance require the use of any associated toxic substances?*

If it appears after answering the above questions that the substance is not likely to be toxic, no structural modification is necessary provided that the substance does not require associated toxic substances (e.g., toxic solvents). If associated toxic substances are required, the chemist should consider structural modifications that obviate their need. If, on the other hand, it seems likely that the chemical substance will be toxic (or is known to be toxic), the chemist should try to design (or redesign) the substance such that toxicity is minimized but commercial usefulness is maintained. The information obtained in answering the questions above should be helpful for inferring structural modifications to reduce toxicity. If it seems likely that the substance will be toxic and the toxicity cannot be reduced through structural modification, the chemist should consider identifying a less toxic, structurally-unrelated substance that has comparable commercial usefulness.

Acknowledgment

The author would like to thank Dr. Carol A. Farris, his close friend and colleague, for her careful editorial review of this paper.

Disclaimer

This chapter was prepared by Dr. Stephen C. DeVito in his private capacity. The contents of this chapter do not necessarily reflect the views, rules or policies of the U.S. Environmental Protection Agency, nor does mention of any chemical substance necessarily constitute Agency endorsement or recommendation for use.

Literature Cited

1. Ariens, E.J. *Design of Safer Chemicals.* In *Drug Design;* Ariens, E.J., Ed.; Medicinal Chemistry, A Series of Monographs Volume IX; Academic Press: New York, NY, 1980; pp 1-46.
2. Ariens, E.J.; Simonis, A.M. *General Principles of Nutritional Toxicology.* In *Nutritional Toxicology;* Academic Press: New York, NY, 1982; Vol. 1, pp 17-80.
3. Ariens, E.J. *Drug Metab. Rev.* **1984** *15,* pp 425-504.
4. Finch, N. *Med. Res. Rev.* **1981,** *1,* pp 337-372.
5. Bodor, N. *Med. Res. Rev.* **1984,** *4,* pp 449-469.
6. Baumel, I.P. *Drug Metab. Rev.* **1984,** *15,* pp 415-424.
7. *Safer Chemicals Through Molecular Design,* A Symposium, DiCarlo, F.J., Chairman; *Drug Metab. Rev.,* **1984,** *15(3-7).*
8. Klaassen, C.D.; Rozman, K. In *Casarett and Doull's Toxicology. The Basic Science of Poisons, Fourth Edition;* Amdur, M.O.; Doull, J.; Klaassen, C.D. Eds.; Pergamon Press: New York, NY, 1991, pp 50-87.
9. Sipes, G.; Gandolfi, J. In *ibid,* pp 88-126.
10. Benet, L.Z.; Mitchell, J.R.; Sheiner, L.B. In *Goodman and Gilman's The Pharmacological Basis of Therapeutics, Eighth Edition;* Goodman Gilman, A.; Rall, T.W.; Nies, A.S.; Taylor, P. Eds.; Pergamon Press: New York, NY, 1990, pp 3-32.
11. Dethloff, L.A.; In *Drug Toxicokinetics;* Welling, P.G.; De La Iglesia, F.A., Eds.; Marcel Dekker, Inc.: New York, NY, 1993, pp 195-219.
12. Valberg, P.A.; In *Principles of Route-to-Route Extrapolation for Risk Assessment;* Gerrity, T.R.; Henry, C.J. Eds.; Elsevier: New York, NY, 1990, pp 61-70.
13. Overton, J.H.; In *ibid,* pp 71-91.
14. Flynn, G.L.; In *ibid,* pp 93-127.
15. Bronaugh, R.L.; In *ibid,* pp 185-191.
16. D'Souza, R.W.; In *ibid,* pp 173-183.
17. Williams, D.A.; In *Principles of Medicinal Chemistry, Third Edition;* Foye, W.O. Ed.; Lea & Febiger: Philadelphia, PA, 1989, pp 79-117.
18. Koymans, L.; Donne-Op den Kelder, G.M.; Koppele Te, J.M.; Vermeulen, N.P.E. *Drug Metab. Rev.,* **1995,** *25,* pp 325-387.
19. Goeptar, A.R.; Scheerens, H.; Vermeulen, N.P.E. *Crit. Rev. Toxicol.,* **1995,** *25,* pp 25-65.
20. *Bioactivation of Foreign Compounds;* Anders, M.W., Ed.; Biochemical Pharmacology and Toxicology Series; Academic Press: New York, NY, 1985.
21. Castagnoli, N.; Castagnoli, K.P. *Biotransformation of Xenobiotics to Chemically Reactive Metabolites.* In *Drug Toxicokinetics;* Welling, P.G., De La Iglesia, F.A. Eds.; Drug and Chemical Toxicology Series, 9; Marcel Dekker: New York, NY, 1993, pp 43-68.
22. Guengerich, P.F.; Shimada, T. *Chem. Res. Toxicol.,* **1991,** *4,* pp 391-407.
23. Vermeulen, N.P.E.; Donne-Op den Kelder; Commandeur, J.N.M. *Formation of and Protection Against Toxic Reactive Intermediates.* In *Perspectives in Medicinal Chemistry;* Testa, B., Kyburz, E., Fuhrer, W., Giger, R. Eds.; VCH: New York, NY, 1993, pp 573-593.

24. Commandeur, J.N.M.; Vermeulen, N.P.E. *Chem. Res. Toxicol.*, **1990**, *3*, pp 171-194.
25. Vermeulen, N.P.E.; te Koppele, J.M. *Stereoselective Biotransformation. Toxicological Consequences and Implications.* In *Clinical Pharmacol.*, **1993**, *18 (Drug Stereochemistry, Second Edition)*, Wainer, I.W. Ed.; pp 245-280.
26. DeCaprio, A.P. In *Selectivity and Molecular Mechanisms of Toxicity.* DeMatteis, F.; Lock, E.A. Eds.; MacMillan Press: New York, NY, 1987, pp 249-263.
27. Chung, F-L.; Tanaka, T.; Hecht, S.S. *Cancer Res.*, **1986**, *46*, pp 1285-1289.
28. Karol, M.H.; Jin, R. *Chem. Res. Toxicol.* **1991**, *4*, pp 503-509.
29. Hemminki, K.; Falck, K.; Vaino, H. *Arch. Toxicol.*, **1980**, *46*, pp 277-285.
30. Sipes, G.I.; Gandolfi, A.J. In *Casarett and Doull's Toxicology. The Basic Science of Poisons, Third Edition,* Klaassen, C.D.; Amdur, M.O.; Doull, J. Eds.; MacMillan Press: New York, NY, 1986, pp 94-95.
31. U.S. Environmental Protection Agency. *New Chemicals Program Categories of Concern* (revised February, 1992). U.S. EPA, Washington, D.C. Available through the TSCA Assistance Information Service, U.S. EPA (TS-799), 401 M Street, S.W., Washington, DC 20460 (phone 202-554-1404).
32. Lipnick, R.L. *Sci. Total Environ.*, **1991**, *109/110*, pp 131-153.
33. *Seventh Annual Report on Carcinogens.* **1994**. U.S. Dept. of Health and Human Services, National Toxicology Program. Research Triangle Park, N.C.
34. Osman, R.; Namboodiri, K.; Weinstein, H.; Rabinowitz, J.R. *J. Am. Chem. Soc.*, **1988**, *110*, pp 1701-1707.
35. National Toxicology Program Report TR-259: Carcinogenesis Studies of Ethyl Acrylate (CAS No. 140-88-5) in F344/N Rats and B6C3F$_1$ Mice (Gavage Studies). **1986**. U.S. Dept. of Health and Human Services, National Toxicology Program. Research Triangle Park, N.C.
36. National Toxicology Program Report TR-314: Carcinogenesis Studies of Methyl methacrylate (CAS No. 80-62-6) in F344/N Rats and B6C3F$_1$ Mice (Inhalation Studies). **1986**. U.S. Dept. of Health and Human Services, National Toxicology Program. Research Triangle Park, N.C.
37. Coles, B. *Drug Metab. Rev.* **1984-85**, *15*, pp 1307-1334.
38. Thompson, D.C.; Thompson, J.A.; Sugumaran, M.; Moldeus, P. *Chem-Biol. Interactions* **1992**, *86*, pp 129-162.
39. Bolton, J.L.; Valerio, L.G.; Thompson, J.A. *Chem. Res. Toxicol.* **1992**, *5*, pp 816-822.
40. Mizutani, T.; Nomura, H.; Nakanishi, K.; Fujita, S. *Toxicol. Applied Pharmacol.* **1987**, *87*, pp 166-176.
41. Mizutani, T.; Ishida, I.; Yamamoto, K.; Tajima, K. *Toxicol. Applied Pharmacol.* **1982**, *62*, pp 273-281.
42. Bolton, J.L.; Sevestre, H.; Ibe, B.O.; Thompson, J.A. *Chem. Res. Toxicol.* **1990**, *3*, pp 65-70.
43. Thompson, D.C.; Perera, K.; Krol, E.S.; Bolton, J.L. *Chem. Res. Toxicol.* **1995**, *8*, pp 323-327.
44. Thompson, D.C.; Perera, K.; Fisher, R.; Brendel, K. *Toxicol. Applied Pharmacol.* **1994**, *125*, pp 51-58.
45. Thompson, D.C.; Perera, K.; London, R. *Chem. Res. Toxicol.* **1995**, *8*, pp 55-60.
46. Lipnick, R.L.; Johnson, D.E.; Gilford, J.H.; Bickings, C.H.; Newsome, L.D. *Environ. Toxicol. Chem.* **1985**, *4*, pp 281-296.
47. Lipnick, R.L.; Watson, K.R.; Strausz, A.K. *Xenobiotica* **1987**, *17*, pp 1011-1025.
48. DeMaster, E.G.; Dahlseid, T.; Redfern, B. *Chem. Res. Toxicol.* **1994**, *7*, pp 414-419.

49. Tsai, R-S.; Carrupt, P-A.; Testa, B.; Caldwell, J. *Chem. Res. Toxicol.* **1994**, *7*, pp 73-76.
50. Wilkinson, C.F.; Murray, M. *Drug Metab. Rev.* **1984**, *15*, pp 897-917.
51. Ortiz de Montellano, P.R. *Alkenes and Alkynes.* In *Bioactivation of Foreign Compounds;* Anders, M.W., Ed.; Biochemical Pharmacology and Toxicology Series; Academic Press: New York, NY, 1985, pp 121-155.
52. Henschler, D. *Halogenated Alkenes and Alkynes.* In ibid, pp 317-347.
53. Fischer, I.U.; von Unruh, G.E.; Dengler, H.J. *Xenobiotica,* **1990**, *20*, pp 209-222.
54. Trush, M.A.; Mimnaugh, E.G.; Gram, T.E. *Biochem. Pharmacol.,* **1982**, *21*, pp 3335-3346.
55. Clark, I.A.; Cowden, W.B.; Hunt, N.H. *Med. Res. Rev.* **1985**, *5*, pp 297-332.
56. Aust, S.D.; Chignell, C.F.; Bray, T.M.; Kalyanaraman, B.; Mason, R.P. *Toxicol. Appl. Pharmacol.* **1993**, *120*, pp 168-178.
57. Goldstein, B.D. Hematotoxicity in Humans. In *Benzene Toxicity: A Critical Evaluation;* Laskin, S., Goldstein, B.D., Eds.; McGraw-Hill: New York, NY, 1977, pp 69-105.
58. Rinski, R.A.; Smith, A.B.; Hornung, R.; Filloon, T.G.; Young, R.J.; Okun, A.H.; Landrigan, P.J. *N. Engl. J. Med.* **1987**, *316*, pp 1044-1050.
59. Sammett, D.; Lee, E.W.; Kocsis, J.J.; Snyder, R. *J. Toxicol. Environ. Health,* **1979**, *5*, pp 785-792.
60. Andrews, L.S.; Snyder, R. In *Casarett and Doull's Toxicology. The Basic Science of Poisons;* Amdur, M.O., Doull, J., Klaassen, C.D., Eds.; Pergamon Press: New York, NY, 1991, 4th Edition, pp 685-690.
61. Chapman, D.E.; Moore, T.J.; Michener, S.R.; Powis, G. *Drug Metab. Dispos.* **1990**, *18*, pp 929-936.
62. Reference 60, pp 705-708.
63. DeCaprio, A. *Neurotoxicology,* **1987**, *8*, pp 199-210.
64. Serve, M.P.; Bombick, D.D.; Roberts, J.; McDonald, G.A.; Matie, D.R.; Yu, K.O. *Chemosphere,* **1991**, *22*, pp 77-84.
65. Burger, A. *A Guide to the Chemical Basis of Drug Design.* John Wiley and Sons, Inc.:New York, NY 1983.
66. Auer, C.M.; Nabholz, J.V.; Baetcke, K.P. *Environ. Health Perspect.* **1990**, *87*, pp 183-197.
67. Unpublished (in-house) data. United States Environmental Protection Agency. Office of Pollution Prevention and Toxics, Washington, DC, 20460.
68. Doerge, D.R.; Takazawa, R.S. *Chem. Res. Toxicol.* **1990**, *3*, pp 98-101.
69. Christian, J.D. *Imidazolidinethiones.* U.S. Patent 2,842,553. July 8, 1958. Monsanto Chemical Company. (See Chemical Abstract 52:20202c).
70. Haynes, R.C. *Thyroid and Antithyroid Drugs.* In *Goodman and Gilman's The Pharmacological Basis of Therapeutics, Eighth Edition;* Goodman Gilman, A.; Rall, T.W.; Nies, A.S.; Taylor, P. Eds.; Pergamon Press: New York, NY, 1990, pp 1373-1377.
71. Kumamoto, T.; Toyooka, K.; Nishida, M.; Kuwahara, H.; Yoshimura, Y.; Kawada, J.; Kubota, S. *Chem. Pharm. Bull.* **1990**, *38*, pp 2595-2596.
72. Kassahun, K.; Farrell, K.; Abbott, F. *Drug Metab. Dispo.* **1991**, *19*, pp 525-534.
73. Camper, D.L.; Loew, G.H.; Collins, J.R. *Int. J. Quantum Chem., Quantum Biol. Symp.* **1990**, *17*, pp 173-187.
74. Narotsky, M.G.; Francis, E.Z.; Kavlock, R.J. *Fundam. Appl. Toxicol.* **1994**, *22*, pp 251-265.
75. DiCarlo, F.J.; Bickart, P.; Auer, C.M. *Drug Metab. Rev.* **1986**, *17*, pp 187-220.
76. DiCarlo, F.J. *Drug Metab. Rev.* **1990**, *22*, pp 411-449.

77. *ECOSAR. A Computer Program for Estimating the Ecotoxicity of Industrial Chemicals Based on Structure-Activity Relationships.* To obtain a copy of ECOSAR, contact; ECOSAR Program, Environmental Effects Branch, Health and Environmental Review Division (7403), U.S. Environmental Protection Agency, 401 M St., SW, Washington DC, 20460.
78. Martin, Y.C. *Quantitative Drug Design;* Medicinal Research Series No. 8; Marcel Dekker, Inc.: New York, NY, 1978.
79. *Quantitative Drug Design;* Ramsden, C.A. Ed. In *Comprehensive Medicinal Chemistry,* Pergamon Press: New York, NY 1990; Vol. 4.
80. *Burger's Medicinal Chemistry and Drug Discovery;* Volume 1, *Principles and Practice;* Wolff, M.E. Ed. Wiley: New York, NY 1994.
81. *Exploring QSAR;* Volume 1, *Fundamentals and Applications in Chemistry and Biology;* Hansch, C.; Leo, A. Eds. American Chemical Society, Washington DC, 1995.
82. *Exploring QSAR;* Volume 2, *Hydrophobic, Electronic and Steric Constants;* Hansch, C.; Leo, A.; Hoekman, D. Eds. American Chemical Society, Washington DC, 1995.
83. *Practical Applications of Quantitative Structure-Activity Relationships (QSAR) in Environmental Chemistry and Toxicology;* Karcher, W.; Devillers, J. Ed. Kluwar Academic Publishers: Boston, MA 1990.
84. Langmuir, I. *J. Am. Chem. Soc.*, **1919**, *41(868),* pp 1543-
85. Burger, A. *Progress Drug Res.*, **1991**, *37,* pp 287-371.
86. Burger, A. *Med. Chem. Res.*, **1994**, *4,* pp 89-92.
87. *Biochemistry for the Medical Sciences.* Newsholme, E.A.; Leech, A.R., Eds. Wiley: New York, NY 1983. pp 100-103.
88. Freeman, H.S.; Sokolowska-Gajda, J.; Reife, A. *Int. Conf. Exhib.*, *American Association of Textile Chemists and Colorists*, **1993**, pp 254-259. (See Chemical Abstracts, 121:159353).
89. Sokolowska-Gajda, J.; Freeman, H.S.; Reife, A. *Text. Res. J.*, **1994**, *64(7),* pp 388-396.
90. Sieburth, S. McN.; Manly, C.J.; Gammon, D.W. *Pestic. Sci.*, **1990**, *28,* pp 289-307.
91. Sieburth, S. McN.; Langevine, C.N.; Dardaris, D.M. *Pestic. Sci.*, **1990**, *28,* pp 309-319.
92. Bodor, N. *Chemtech*, October **1995**, pp 22-32.
93. DePompei, M.F. Non Isocyanate Sealants. In *Preprints of Papers Presented at the 208th American Chemical National Meeting,* **1994**, pp 393-394 (paper no. 276).
94. Czajkowski, W. *Dyes Pigm.*, **1991**, *17,* pp 297-301.
95. Clarke, E.A. *Drug Metab. Rev.* **1984**, *15,* pp 997-1009.
96. Donohue, M.D.; Geiger, J.L. Reduction of VOC Emission During Spray Painting Operations: A New Process Using Supercritical Carbon Dioxide. In *Preprints of Papers Presented at the 208th American Chemical National Meeting,* **1994**, pp 218-219 (paper no. 54).
97. *Benign by Design, Alternative Synthetic Design for Pollution Prevention;* Anastas, P.T.; Farris, C.A., Eds.; American Chemical Society Symposium Series 577; American Chemical Society: Washington, DC, 1994.

METHODS

Chapter 3

Cancer Risk Reduction Through Mechanism-Based Molecular Design of Chemicals

David Y. Lai[1], Yin-tak Woo[1], Mary F. Argus[1,2], and Joseph C. Arcos[1,2]

[1]Office of Pollution Prevention and Toxics, U.S. Environmental Protection Agency, Mail Code 7406, 401 M Street, SW, Washington, DC 20460
[2]School of Medicine, Tulane University, New Orleans, LA 70112

> Increased understanding of the mechanistic basis of chemical carcinogenesis and the relationship between molecular structure and carcinogenic activity provides opportunities not only for identifying suspect carcinogens but also for designing chemicals with lower carcinogenic potential. One of the chemical classes in which the structural and molecular basis of carcinogenicity is the most clearly understood is the aromatic amines. This paper summarizes the bioactivation mechanisms and structural criteria for aromatic amine carcinogenesis; a strategic approach of risk reduction through mechanism-based molecular design of aromatic amine dyes with lower carcinogenic potential is discussed. With our increasing knowledge of the mechanisms and structure-activity relationships, it should be possible to develop safer products for other chemical classes using similar molecular design approaches.

Mechanisms of Chemical Carcinogenesis and Structure-Activity Relationships

Considerable knowledge of the mechanisms of chemical carcinogenesis has accrued since the pioneer work by James A. and Elizabeth C. Miller at the University of Wisconsin beginning in the 1940's, particularly concepts on the metabolic activation of chemicals to reactive electrophilic intermediates that interact with cellular nucleophiles to initiate carcinogenesis (1,2). For many carcinogen classes, the molecular basis of carcinogenic activity is now known in considerable detail and the concept of electrophiles provides the most probable rationale for their carcinogenic action. Some of the electrophilic, reactive intermediates believed to be responsible for the carcinogenicity of genotoxic chemicals include: carbonium, aziridium, episulfonium, oxonium, nitrenium or arylamidonium ions, free radicals, epoxides, lactones, aldehydes, semiquinones/quinoneimines, and acylating moieties. The metabolic activation of various genotoxic chemicals to their ultimate carcinogenic metabolites has been reviewed (3).

Although a chemical may have the potential to produce electrophilic reactants, it is also known that its carcinogenic activity depends on a host of other factors. Several significant molecular parameters that may affect the carcinogenic potential of electrophiles include: (i) molecular size and shape, (ii) substituent effects (electronic or steric), (iii) molecular flexibility, and (iv) polyfunctionality and spacing/distance between reactive groups. For instance, the molecular size (as measured by incumbrance area of planar molecule) of most potent carcinogenic polycyclic aromatic hydrocarbons (PAHs) lies within a certain range. Virtually all PAHs with highly elongated shape tend to be inactive. Ring substitution with methyl group(s) at detoxification sites (e.g. L-region) tends to increase carcinogenic activity whereas ring substitution with bulky or hydrophilic substituents invariably decreases activitiy particularly near or at proelectrophilic regions such as the bay region (4-6). The importance of molecular flexibility can be illustrated by the findings that epoxides on rigid cycloaliphatic rings tend to be considerably less active than epoxides on more flexible noncyclic aliphatic chains. Diepoxides are invariably more carcinogenic than monoepoxides particularly if the two epoxy groups are favorably apart to impart crosslinking activity (7,8).

The knowledge of molecular mechanisms and of other factors that affect carcinogenesis by various types of chemical carcinogens has established the basis for identifying suspect carcinogens by Structure-Activity Relationships (SAR) analysis, which has routinely and effectively been used by the U.S. Environmental Protection Agency (EPA) to evaluate Premanufacture Notification (PMN) chemicals regulated under Section 5 of the Toxic Substances Control Act (TSCA) (9). The state-of-the-art of predicting suspect carcinogens based on SAR analysis has previously been presented (7,8). Increased understanding of the mechanistic basis of chemical carcinogenesis and the relationship between molecular structure and carcinogenic activity provides opportunities not only for identifying suspect carcinogens but also for designing chemicals with lower carcinogenic potential. One of the chemical classes in which the structural and molecular basis of carcinogenicity is the most clearly understood is the aromatic amines. This paper summarizes the bioactivation mechanisms and structural criteria for aromatic amine carcinogenesis. A strategic approach of risk reduction through mechanism-based molecular design of aromatic amine dyes with lower carcinogenic potential is discussed.

Aromatic Amine Dyes and Structural Criteria for Carcinogenicity

Aromatic amine derived dyes are synthetic organic colorants of considerable industrial and commercial importance. They are widely used in textile, paper, leather, plastics, cosmetics, drug and food industries. Several aromatic amines, such as benzidine, 2-naphthylamine, 4-aminobiphenyl are known human carcinogens (10). The recognition that these and other aromatic amines are
carcinogenic led to the abandonment of many of these compounds by the dye industry and prompted a search for safer replacement products.

Since the 1950's, a considerable number of aromatic amines have been tested for carcinogenicity. Carcinogenicity data are available for many aromatic amines with diverse ring systems and different substituents, thus allowing SAR analysis and the determination of structural criteria for carcinogenic activity (11, 12).

The basic requirements for an aromatic amine to be carcinogenic are: the presence of an aromatic ring system (a single ring or more than one ring forming a conjugated system, fused or non-fused) and of the amine/amine-generating group(s). **Figure 1** shows the aromatic ring systems in some well-known carcinogenic aromatic amines. The unattached bonds on these ring systems indicate the positions where attachment of amine or amine-generating group(s) gives rise to carcinogenic compounds. Owing to the possible metabolic interconversion of the amino group with hydroxylamino and nitroso groups and the metabolic reduction of the nitro to nitroso, the latter three groups (*i.e.* hydroxylamino, nitro and nitroso) are often termed "amine-generating groups". In some cases, replacement of an amino group with a dimethylamino group does not result in a significant loss of the carcinogenic activity of aromatic amine compounds since metabolic N-demethylation readily occurs *in vivo*. Therefore, some monoalkylamino/dialkylamino groups are also amine-generating groups.

Beyond the presence of an aromatic ring system and of amine/amine-generating group(s), SAR analysis has revealed several structural features important for predicting the carcinogenicity of aromatic amines. These include: (i) number and nature of aromatic rings; (ii) nature of amine/amine-generating group(s); (iii) position of amine/amine-generating group(s); (iv) nature, number and position of other ring substituent(s); and (v) size, shape and planarity of the molecules (*12*). Interestingly, many of the structural features that are important for the carcinogenicity of aromatic amines also have important influences on their bioactivation mechanisms.

Bioactivation Mechanisms in Relation to Structural Criteria for Carcinogenicity of Aromatic Amine Dyes

The aromatic amines require a two-stage metabolic activation for carcinogenicity, and the mechanism of activation is now known in considerable detail (*rev. 3, 13, 14*). The first stage of metabolic activation involves N-hydroxylation and/or N-acetylation to yield the respective N-hydroxylated and/or N-acetylated derivatives. The second stage involves O-acylation (where the acyl group can be sulfonyl, phosphoryl, or acetyl) and N-, or O-glucuronidation in the formation of acyloxyamines and glucuronides. Many of the enzymes (*e.g.* acetyltransferase, sulfotransferase) involved in these reactions have been identified and recently been characterized at the molecular level (*15,16*). The multiple metabolic activation pathways of aromatic amines are summarized in **Figure 2**.

The acyloxyamines are highly reactive and, upon departure of the acyloxy anion, the arylamidonium/arylnitrenium ion formed may readily bind covalently to cellular nucleophiles such as DNA to initiate carcinogenesis (*13,17*). The glucuronides, on the other hand, are essentially unreactive and represent stable excretion products from the liver (the site of their formation). In rodents, glucuronide formation is actually a detoxification pathway. However, at the acidic pH found in the urinary bladder of dogs and humans, and in the presence of urinary ß-glucuronidase, they are cleaved to give the electrophilic arylnitrenium ion which interacts with cellular macromolecules such as DNA to induce bladder tumors (*18*).

There is also evidence that some aromatic amines may induce bladder tumors via the formation of a reactive quinoneimine metabolite after bioactivation by

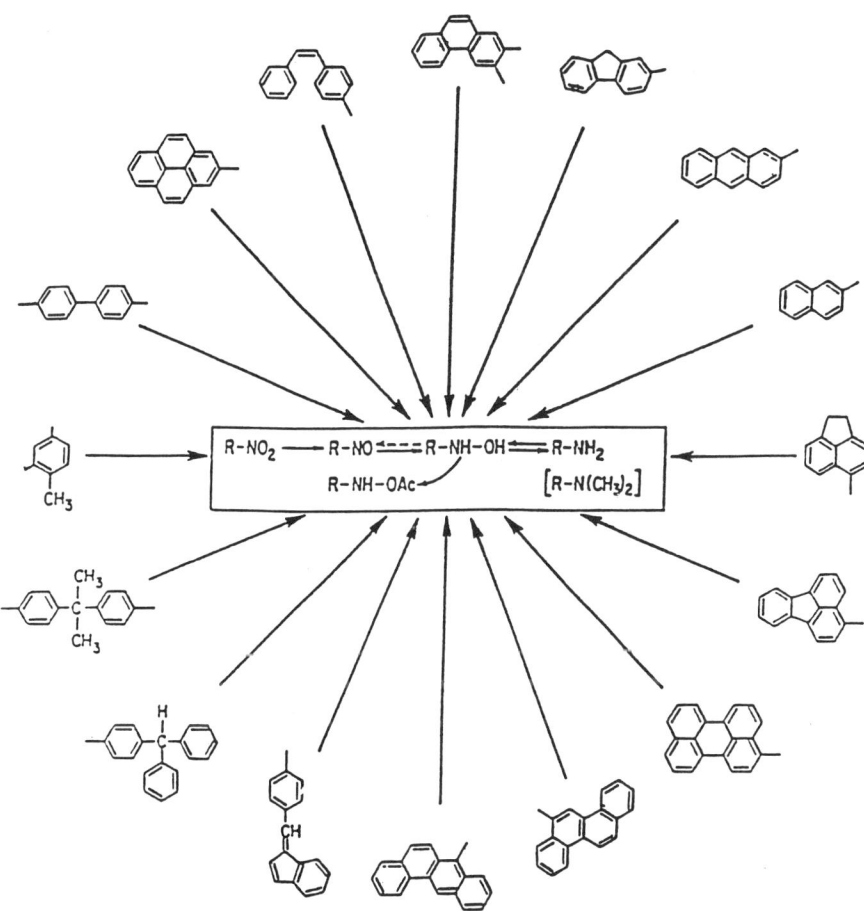

Figure 1. Aromatic Ring Systems in Some Carcinogenic Aromatic Amines. The unattached bonds on these ring systems indicate the positions where attachment of amine or amine generating group(s) gives rise to carcinogenic compounds. (Reproduced with permission from reference 8. Copyright 1978, Begell House, Inc.)

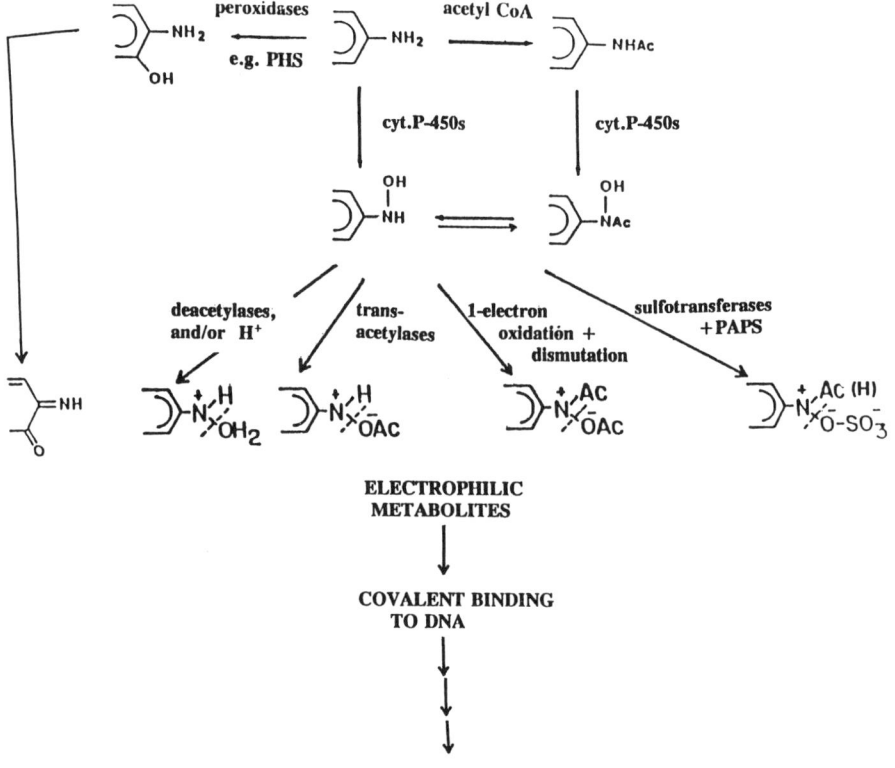

Figure 2. Multiple Metabolic Activation Pathways of Aromatic Amines.

peroxidases, such as prostaglandin H synthase (PHS) which has been detected at high levels in dog bladder (*19,20*).

Number and Nature of Aromatic Ring(s). The bioactivation mechanisms of aromatic amines have previously been discussed (*4,9*). The carcinogenic potency of aromatic amines is a function of the leaving potential of the acyloxy anion which is governed by the force of conjugation of the respective aryl moieties. The roles of the aromatic moiety are: (i) to provide π-electron shift of sufficient strength to facilitate the departure of the acyloxy anion; and (ii) to stabilize the electrophilic nitrenium ion. Therefore, an aromatic ring system is essential to the carcinogenicity of aromatic amines; many heterocyclic aromatic amines (*e.g.* amino-imidazoles, amino-carbolines formed during the cooking of food; *21*) but not non-aromatic cyclic amines are also carcinogenic. The force of conjugation, facilitating the departure of the acyloxy anion, increases from phenyl toward higher aryl groups. This is consistent with the findings that aniline (single phenyl ring) is a weaker carcinogen than benzidine or ß-naphthylamine (two phenyl rings).

Nature of Amine/Amine-generating Group(s). An amine/amine-generating group is required since the first stage of metabolic activation involves N-hydroxylation and/or N-acetylation of the amine/amine-generating group(s) to yield the respective N-hydroxylated and/or N-acetylated derivatives, and subsequently the reactive eletrophilic nitrenium ion. The nature of the amine-generating groups is important, since for dialky amino groups with bulky or long alkyl substitution N-dealkylation does not readily occur to allow further bioactivation. Replacement of the dimethylamino ($-NMe_2$) group of 4-dimethylaminoazobenzene by a diethylamino ($-NEt_2$) or a higher dialkylamino group has been shown to lead to marked attenuation of its carcinogenicity (*11*) and mutagenicity (*22*).

Position of Amine/Amine-generating Group(s). The position of the amine/ amine-generating group(s) is important because it may affect the molecular shape/planarity of the molecules and the susceptibility of the group(s) being substrates for the oxidizing or acetylating enzymes (*see* discussions on "planarity of molecules" below).

In addition, the force of conjugation which facilitates the departure of the acyloxy anion to form the ultimate carcinogenic metabolites, increases with the length of the conjugated double-bonded system involved. Within a given molecule, the force of conjugation is greatest when the amine/amine-generating group is linked to the aromatic frame in position(s) corresponding to the terminal end(s) of the longest conjugated system in the molecule (*8*). As shown in **Figure 1**, consistent with this bioactivation mechanism most potent carcinogenic aromatic amines have the amine/amine-generating group(s) linked to the aromatic frame in position(s) corresponding to the terminal end(s) of the longest conjugated system of the molecule (*e.g.* 2-position of ß-naphthylamine, 2-acetylaminofluorene and 2-aminoanthracene; 4-position of 4-aminobiphenyl and 4-aminostilbene; as well as the amino groups in benzidine and 2,7-bisaminofluorene).

Nature, Number and Position of Other Ring Substituent(s) -- Substituents Effect. Substitution on the rings may exert electronic or steric effects on the reactivity and stability of the ultimate carcinogenic metabolite(s). There is evidence from SAR analysis on analogues of aniline, phenylenediamine and methylene-bis-aniline that substitution of a chloro group or a methyl/methoxy group *ortho* to the amino group often enhance reactivity (carcinogenicity and mutagenicity) (*23-25*). However, the effect of *ortho* substitution is related to the *size* of the alkyl groups; the larger the substituents, the less potent the mutagenic/carcinogenic activity. If both *ortho* positions of an amino group are substituted by large alkyl groups, an additional decrease in activity occurs; this is because large alkyl substituents at the *ortho* position provide steric hindrance around the amino group thus inhibiting hydroxylation/acetylation -- the necessary steps required for metabolic activation to the ultimate mutagenic/ carcinogenic species.

Arylamines are oxidized to form a number of ring hydroxylated products in addition to a N-hydroxylated metabolite. The ring hydroxylated products are generally considered products of detoxification. However, ring hydroxyl substituents *ortho/para* to amino groups can form quinoneimines which are reactive and can bind to cellular nucleophiles to initiate carcinogenesis. The formation of a reactive quinoneimine metabolite after bioactivation of benzidine and other aromatic amines by peroxidases such as prostaglandin H synthase (PHS) has been demonstrated (*20,26*).

Planarity of Molecules. SAR analysis has shown that the planarity of the molecule is an important determinant in the metabolic activation of aromatic amines. For instance, studies by Ioannides and coworkers (*27*) on the three isomeric forms of aminobiphenyl show good correlations among mutagenic/ carcinogenic activity, the capability of N-hydroxylation, and the planarity of the molecules. Both the 3- and 4- isomers are planar and can be N-hydroxylated; 3-aminobiphenyl is a weak carcinogen, whereas 4-amino-biphenyl is a potent carcinogen. The presence of an amino group/substituent at the 2-position of aminobiphenyl results in marked loss of planarity; this is because 2-aminobiphenyl, being nonplanar, does not form any N-hydroxylation products *in vitro*, and is noncarcinogenic and nonmutagenic. Isozymes of the cytochrome P-450 families such as cytochrome P-450 1A2, which catalyse the N-hydroxylation of many aromatic amines including 4-aminobiphenyl cannot N-hydroxylate 2-aminobiphenyl since these enzymes are unable to carryout oxygenation at conformationally hindered positions. The non-planarity of the compound, therefore, precludes it from being a substrate for the cytochrome P-450 oxidizing enzymes of N-hydroxylation.

Similarly, the presence of substituent(s) *ortho* to the intercyclic linkage (at the 2- or 2'-position) of benzidine will result in marked loss of planarity and will render it a poor substrate of the cytochrome P-450 oxidizing enzymes for N-hydroxylation. Hence, the amino groups of *m*-tolidine and of other 2-substituted or 2-,2'-disubstituted benzidines may not be readily activated metabolically. A recent National Toxicology Program carcinogenesis bioassay has shown that 4,4'-diamino-2,2'-stilbenedisulfonic acid was inactive in male and female rats and mice (*28*). This may be partly because of the sulfonation effect (*see* discussions below) and partly because of the substituent effect of the sulfonic acid group on the planarity of the molecule.

In addition to affecting metabolic activation, planarity is also expected to affect the potential of aromatic amines to serve as intercalating agents.

Molecular Design of Aromatic Amine Dyes of Low Carcinogenic Potential

With the understanding of the bioactivation mechanisms and the structural requirements for the carcinogenicity of aromatic amine dyes, safer dyes with low carcinogenic potential can be designed through modifications of their chemical structures by a number of approaches. These approaches along with their rationale are discussed below and are summarized in **Table I**.

Introduce Bulky Substituent(s) *ortho* to the Amine/Amine-generating Group(s). Since the first stage of metabolic activation of aromatic amine dyes involves N-hydroxylation and/or N-acetylation to yield the respective N-hydroxylated and/or N-acetylated derivatives, introducing bulky substituent(s) *ortho* to the amine/amine-generating group(s) will provide steric hindrance to inhibit bioactivation by the oxidizing/acetylating enzymes.

Introduce Long-chain Alkyl or Bulky N-substituent(s) to the Amino Group(s). Introducing long-chain alkyl or bulky N-substituent(s) to the amino group(s) will make the chemicals poor substrates for the oxidizing/acetylating enzymes. This is because prior to bioactivation, N-dealkylation is required and for dialkyamino groups with bulky or long alkyl substitution, N-dealkylation may not readily occur to allow further bioactivation.

Introduce Bulky Groups *ortho* to the Intercyclic Linkage(s). For aromatic amine dyes with more than one, non-fused rings, introduction of bulky groups *ortho* to the intercyclic linkages (2,2'-positions) will distort the planarity of the molecule and render it less favorable for DNA intercalation and a poorer substrate for the bioactivation enzymes.

Alter the Position of the Amine/Amine-generating Group(s). The carcinogenic potential of aromatic amine dyes can be reduced by altering the position of the amine/amine-generating group(s) on the aromatic ring(s) of carcinogenic aromatic amines (shown in **Figure 1**) because it will: (i) reduce the length of the conjugation path and thus the force of conjugation which facilitates the departure of the acyloxy anion; (ii) distort the planarity of the molecule making it less favorable for DNA intercalation and a poorer substrate for the bioactivation enzymes; and (iii) render the conjugation path non-linear and thus less resonance stabilization of the electrophilic nitrenium ion will occur. Resonance stabilization of reactive intermediates is important for carcinogenicity of chemicals since stabilized reactive intermediates have a better chance of remaining reactive during transport from the site of activation to target macromolecules; compounds which yield reactive intermediates that can be stabilized by resonance often have higher carcinogenic potential (9).

Table I. Molecular Design of Aromatic Amine Dyes with Lower Carcinogenic Potential

Approaches	Rationale
1. Introduce bulky substituent(s) *ortho* to the amine/amine-generating group(s).	Provide steric hindrance to inhibit bioactivation.
2. Introduce bulky N-substituent(s) to the amine/amine-generating group(s).	Make the dye a poor substrate for the bioactivation enzymes.
3. Introduce bulky groups *ortho* to the intercyclic linkages.	Distort the planarity of the molecule making it less accessible and a poorer substrate for the bioactivation enzymes.
4. Alter the position of the amine/aromatic ring(s).	A. Reduce the length of the conjugation path and thus the force of conjugation which facilitates departure of the acyloxy anion. B. Non-linear conjugation path; less resonance stabilization of the electrophilic nitrenium ion. C. In some cases, distort the planarity of the amine-generating group(s) in the molecule making it less favorable for DNA intercalation and a poorer substrate for the bioactivation enzymes.
5. Replace electron-conducting intercyclic linkages by electron-insulating intercyclic linkages.	A. Disrupt the conjugation path and thus reduce the force of conjugation which facilitates the departure of the acyloxy anion. B. Less resonance stabilization of the electrophilic nitrenium ion.
6. Ring substitution with hydrophilic groups (*e.g.* sulfonic acid); especially at the ring(s) bearing the amine/amine-generating group(s).	Render the molecule more water soluble thus reducing absorption and accelerating excretion.

Replace Electron-conducting Intercyclic Linkages by Electron-insulating Intercyclic Linkages. For aromatic amine dyes with more than one, non-fused rings, replacing electron-conducting intercyclic linkages (*e.g.* single bond, $-CH_2-$, $-CH=CH-$) by electron-insulating intercyclic linkages [*e.g.* $-CO-CH_2-$, $-(CH_2)_n-$ for n>1] will disrupt the conjugation path and thus the force of conjugation. Further, it will provide less resonance stabilization for the electrophilic nitrenium ion.

Introduce Hydrophilic Groups into the Ring(s). Ring substitution with hydrophilic groups (*e.g.* sulfonic acid, carboxylic acid), especially at the ring(s) bearing the amino/amine-generating group(s) will reduce their carcinogenic potential since it renders the molecule more water soluble and thus reduces their absorption and accelerates their excretion. A review of the genotoxicity and carcinogenicity data by Jung and coworkers (*29*) on a large number of aromatic aminosulfonic acids has confirmed that aromatic aminosulfonic acids, in contrast with some of their unsulfonated analogues, generally have no or very low genotoxic and tumorigenic potential.

Conclusion

While there are a number of approaches to reduce cancer risk of chemicals during their manufacture, processing, or use, including exposure reduction (*30*), the application of the growing knowledge on the mechanisms and structure-activity relationships to the development of products with lower carcinogenic potential appears to be a rational one. For some chemical classes such as the aromatic amine dyes, the structural and molecular basis of carcinogenicity is now known in considerable detail, thus providing opportunities for designing replacement dyes of lower carcinogenic potential. It is expected, however, that some of the dyes with the proposed structural modifications may possess different application properties. Therefore, the application properties of various replacement dyes may have to be examined and compared to determine which approach is the most appropriate for specific applications of the dyes.

For the approximately 30 carcinogenic chemical classes that have been systematically reviewed, there are various metabolic activation pathways and somewhat unique structural requirements for carcinogenicity (*rev. 4,11,31-33*). With our increasing knowledge of the mechanisms and structure-activity relationships, it should be possible to develop safer products for other chemical classes using similar molecular design approaches.

Disclaimer: The contents of this chapter have been presented in part at the Society of Toxicology 34th Annual Meeting, Baltimore, MD, March 1995. The views in this paper are solely those of the authors and do not necessarily reflect the views or policies of the U.S. Environmental Protection Agency. Mention of trade-names, commercial products or organizations does not imply endorsement by the U.S. government.

Literature Cited

1. Miller. J.A. *Drug Metabolism Rev.* **1994**, *26(1&2)*, 1-36.
2. Miller. E.C. *Cancer Res.* **1978**, *38*, 1479-1496.
3. Woo, Y.-t.; Arcos, J.C.; Lai, D.Y., in *Chemical carcinogens: Activation Mechanisms, Structural and Electronic Factors, and Reactivity;* Politzer, P., Martin, F.J. Jr. Eds., Elsevier, New York, 1988, Chapter 1, pp.1-31.
4. Arcos, J.C.; Argus, M.F. *Chemical Induction of Cancer;* Academic Press, New York, 1974, Vol. IIA.
5. *Polycylic Aromatic Hydrocarbon Carcinogenesis: Structure-Activity Relationship;* Yang, S.K.; Silverman, B.D. Eds.; CRC Press, Boca Raton, Florida, 1988; Vol. I, 213 pp., and Vol. II, 210 pp.
6. Richard, A.M.; Woo, Y.-t. *Mutat. Res.* **1990**, *242*, 285-303.
7. Woo, Y.-t.; Arcos, J.C.; Lai, D.Y., in *Handbook of Carcinogen Testing;* Milman, H.A.; Weisburger, E.K. Eds.; Noyes Publication, Park Ridge, NJ., 1985, Chapter 1, pp.1-25.
8. Arcos, J.C. *J. Environ. Pathol. Toxicol.* **1978**, *1*, pp. 433-458.
9. Arcos, J.C. *J. Amer. Coll. Toxicol.* **1983**, 2, pp. 131-145.
10. International Agency for Research on Cancer. *IARC Monographs on the Evaluation of Carcinogenic Risks of Chemicals to Humans;* Lyon, France, 1987, *Suppl. 7.*
11. Arcos, J.C.; Argus, M.F. *Chemical Induction of Cancer;* Academic Press, New York, 1974, Vol. IIB.
12. Lai, D.Y.; Woo, Y.-t.; M.F. Argus; Arcos, J.C. *Toxicologist* **1994** *14(1)*, p 134.
13. Beland, F.A.; Kadlubar, F.F. *Environ. Health Persp.* **1985**, *62*, pp. 19-30.
14. Boeniger, M. *The carcinogenicity and metabolism of azo dyes, especially those derived from benzidine.* Technical Report, 1980, DHHS (NIOSH) Publication No. pp. 80-119.
15. Land, S. King, C.M. *Environ. Health Persp.* **1994** *102(Suppl. 6)*, pp. 91-93.
16. Yamazoe, Y.; Azawa, S.; Nagata, K.; Gong, D.-W.; Kato, R. *Environ. Health Persp.* **1994** *102(Suppl. 6)*, pp. 99-103.
17. Meerman, J.H.N; van de Poll, M.L.M. *Environ. Health Persp.* **1994** *102(Suppl. 6)*, pp. 153-159.
18. Kadlubar, F.F.; Miller, J.A.; Miller, E.C. *Cancer Res.* **1977**, *37*, pp. 805-814.
19. Boyd, J.A.; Eling, T.E. *Environ. Health Persp.* **1985**, *64*, pp. 5-51.
20. Zenser, T.V.; Cohen, S.M.; Mattammal, M.B.; Wise, R.W.; Rapp, N.S.; Davis, B.B. *Environ. Health Persp.* **1983**, *49*, pp. 33-41.
21. Sugimura, T. *Mutat. Res.* **1985**, *150*, pp. 33-41.
22. Ashby, J.; Styles, J.A.; Patton, D. *Br. J. Cancer* **1978**, *38*, pp. 34-50.
23. Milman, H.A.; Peterson, C. *Environ. Health Persp.* **1984**, *56*, pp. 261-273.
24. Pullin, T.G.; Mulholland, A.; Ter Haar, G.L., presented in *Sixth Annual Meeting of the American College of Toxicology*, Washington DC, Nov. 5-7, **1985**. Abstract No. 109.

25. Esancy, J.F.; Freeman, H.S,; Claxton, L.D. *Mutat. Res.* **1990**, *238*, pp. 23-38.
26. Boyd, J.A.; Eling, T.E. *Environ. Health Persp.* **1985**, *64*, pp. 45-51.
27. Ioannides, C.; Lewis, D.F.V.; Trinick, J.; Neville,S.; Sertkaya, N.N.; Kajbat, M.; Gorrod, J.W. *Carcinogenesis* **1989**, *10(8)*, pp. 1403-1407.
28. National Toxicology Program, *Toxicology and carcinogenesis Studies of 4,4'-Diamino-2,2'-Stilbenedisulfonic Acid, Disodium Salt in F344/N Rats and B6C3F1 Mice.* Technical Report Series No. 412, Research Triangle Park, NC, 1992.
29. Jung, R.; Steinle, D.; Anliker, R. *Fd. Chem. Toxicol.* **1992**, *30(7)*, pp. 635-660.
30. Clarke, E.A. *Drug Metabolism Rev.* **1984**, *15(5&6)*, pp. 997-1009.
31. Arcos, J.C., Y.-t. Woo; Argus, M.F. with the collaboration of Lai, D.Y. *Chemical Induction of Cancer.* Academic Press, New York, 1982, Vol. IIIA.
32. Woo, Y.-t.; Lai, D.Y.; Arcos, J.C.; Argus, M.F. *Chemical Induction of Cancer.* Academic Press, New York, 1985, Vol. IIIB.
33. Woo, Y.-t.; Lai, D.Y.; Arcos, J.C.; Argus, M.F. *Chemical Induction of Cancer.* Academic Press, New York, 1988, Vol. IIIC.

Chapter 4

Isosteric Replacement of Carbon with Silicon in the Design of Safer Chemicals

Scott McN. Sieburth

Department of Chemistry, State University of New York, Stony Brook, NY 11794-3400

Organosilanes are inexpensively made, lack intrinsic toxicity, and have many similarities to comparable carbon compounds. Isosteric replacement of carbon with silicon in compounds with desirable properties can lead to products with enhanced environmental degradation, or other safety features. Many examples of silicon substitution have been studied in agrochemicals and pharmaceuticals. Application of this strategy to agrochemicals, where environmental concerns are paramount, has been particularly successful. An unsuccessful application of this isosteric replacement strategy is also instructive. Three case studies, and the current understanding of the environmental fate of organosilanes, are discussed.

Isosterism and the Design of Safer Chemicals

The process of designing safer synthetic chemicals implies an application of chemical principles to eliminate one or more undesirable feature found in a material, while retaining the desirable properties. For agrochemicals and pharmaceuticals, that subset of chemical research concerned with biologically active molecules, the science of fine-tuning molecular properties has been under development for several decades (1,2). Many broadly useful concepts have emerged from this research.

One of the early concepts for rationally modifying chemical properties was isosterism (Figure 1), an idea originally advanced by Langmuir to show broad trends among disparate molecules with an identical arrangements of bonds (3). Isosteric molecular fragments have a similarity of shape, the shape yields certain properties, and therefore isosteric fragments are potentially interchangeable. In the application of isosterism to biologically active molecules, the term 'bioisosterism' has developed for

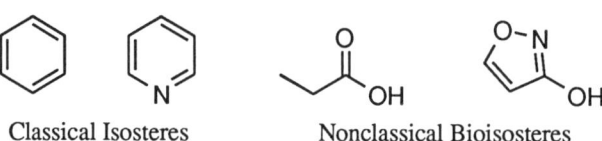

Classical Isosteres   Nonclassical Bioisosteres

Figure 1. Isosteric and bioisosteric pairs (4).

interchangeable groupings of atoms that elicit a similar biological response, even if they are sterically dissimilar (4). Because bioisosterism includes groups that are not isosteric, the terms 'molecular metaphor' and 'bioanalogy' have recently been nominated to take its place (5,6).

Silicon as an Isostere of Carbon

One of the now classic isosteric replacements is the substitution of silicon for carbon. From a chemistry perspective this is a rather natural replacement, as silicon and carbon are grouped in column 4A of the Periodic Table, and therefore have many chemical similarities. Like all Group 4A elements, silicon and carbon are tetravalent, tetrahedral, and form stable bonds with carbon. Notwithstanding the current issues involving silicone implants (7), organic derivatives of silicon and carbon generally have no intrinsic toxicity (8, 9), in contrast with the other Group 4A elements germanium (10), tin (11), and lead (12). From a toxicity perspective, silicon is the *only* Group 4A element that is a suitable replacement for carbon. In addition, silicon is an abundant, inexpensive element and one that is available in a variety of forms.

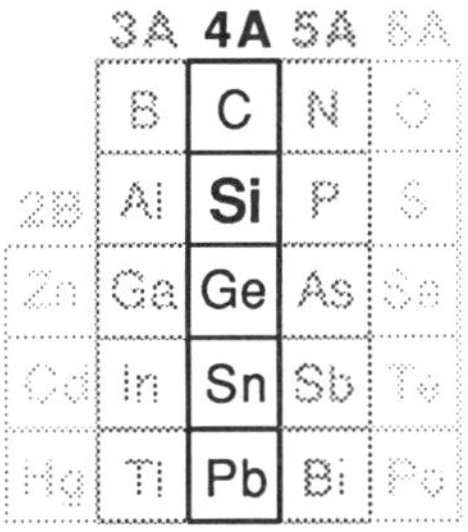

An intriguing aspect of organosilanes is the environmental novelty of the silicon-carbon bond. Naturally occurring silicon-carbon bonds are unknown (13). Undoubtedly this is due to the high relative strength of the silicon-oxygen bond and it suggests that organosilanes would be expected to have a finite environmental lifetime.

Some early examples of silicon substitution for carbon in biologically active molecules are shown in Figure 2. Urethanes 2 and 3, neutral analogs of the neurotransmitter acetylcholine (1), were found to be antagonists of 1 with identical dose-response curves (14). Interestingly, silane 3 was much less toxic to mice than 2, and exhibited muscle relaxant properties (15). Carbamate insecticide 4 and its silicon analog 5 were found to have similar toxicity to the house fly (16). In both of these examples, the *tert*-butyl and trimethylsilyl groups function simply as isosteres of trimethylammonium.

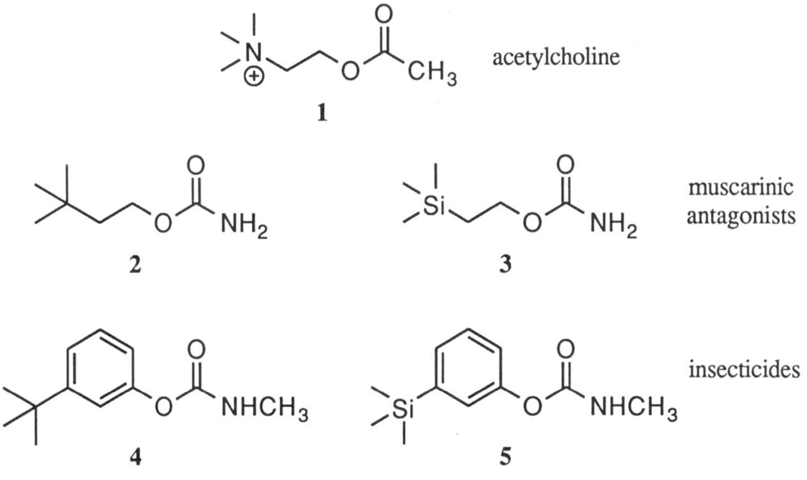

Figure 2. Silicon substitution in acetylcholine analogs.

Similarities and Differences of Silicon- and Carbon-based Chemicals

Silicon is the element most similar to carbon, however it is not a generic replacement for all carbon atoms and certain strict limitations apply (17). Double bonds to silicon, and three-membered rings containing silicon, are unstable to air and moisture (18, 19). Single bonds from silicon to heteroatoms such as nitrogen and oxygen are strong but can hydrolyze readily. The silicon-hydrogen bond is more polarized than the carbon-hydrogen bond. In contrast to the carbon case, increasing the number of hydrogen on a silicon increases the ease of oxidation, and the parent silane (SiH_4) is pyrophoric. Nevertheless, an analog of polyethylene that is rich in SiH_2 groups (Figure 3) has recently been reported to be air stable (20).

$H_2Si=CH_2$
1-silaethylene
(unstable)

poly(1-silaethylene)
(stable)

$H_2C=CH_2$
ethylene
(stable)

polyethylene
(stable)

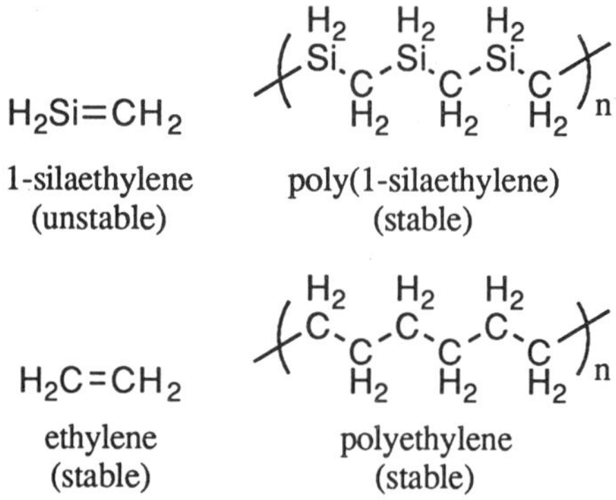

Figure 3. A silane analog of polyethylene is reported to be air stable (20).

A subtle yet influential difference in atomic size exists (Figure 4) for silicon and carbon (21). Important differences in chemical reactivity also exist. When silicon is proximal to unsaturation, as in a vinyl or allyl silane, the compounds are stable, but unlike their carbon analogs they are subject to acid catalyzed silicon-carbon bond cleavage (22). Herein lies a potential avenue for the design of environmentally degradable products (Figure 5).

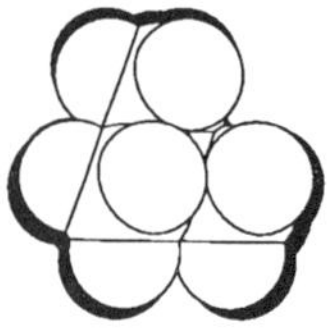

Space filling models
White: $C(CH_3)_4$
Black: $Si(CH_3)_4$

Figure 4. Size comparison of tetramethyl silicon and carbon (21).

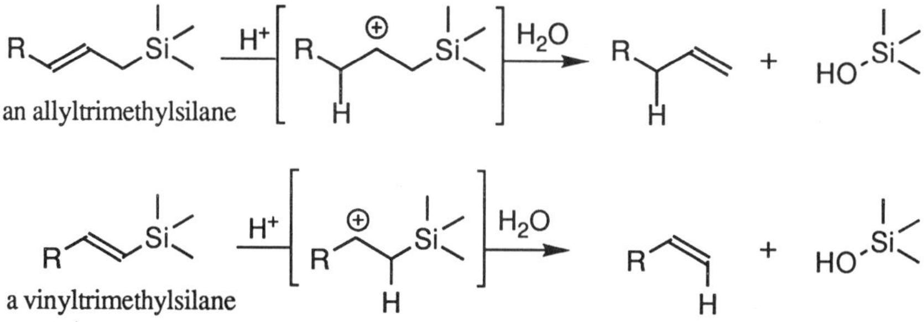

an allyltrimethylsilane

a vinyltrimethylsilane

Figure 5. Allyl and vinyl silanes break down under acid catalysis.

Degradation and Oxidative Metabolism of Organosilanes

An important component of designing safer chemicals is predicting their environmental fate, and both abiotic degradation and biological oxidation can play a role. Designing chemicals that will biodegrade to innocuous products is highly desirable, and isosteric substitution of carbon with silicon in many cases may enhance abiotic degradation and biological oxidation. The environmental fate of organosilanes has been discussed (23). Metabolism of organosilanes by mammals has been comprehensively reviewed (24).

Abiotic Degradation. Currently the major environmental source of organosilanes is silicone polymer (siloxanes, see Figure 6), primarily polymers of 1,1-dimethylsilanediol. Siloxanes were once thought to be environmentally stable, but are now known to depolymerize in the presence of water and soil (25-28). Furthermore, labeling studies have shown that the methyl groups can be photochemically cleaved from the silicon, with the final product being the naturally occurring silicates (29).

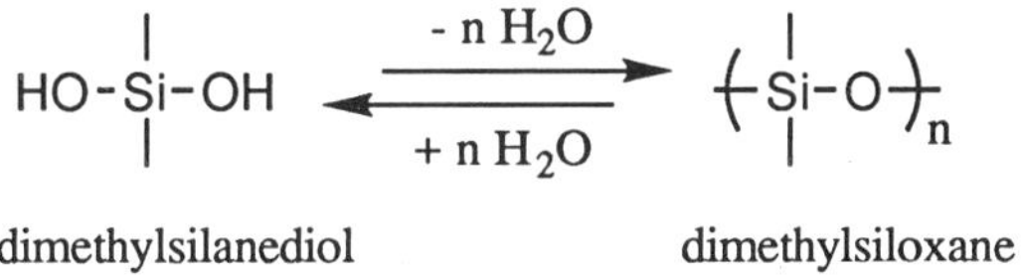

Figure 6. In the presence of water and/or soil, siloxanes hydrolyze to smaller oligomers and monomeric 1,1-dimethylsilanediol (26-28).

Biological Oxidation. Early studies of microbial growth in the presence of permethyl siloxanes suggested that biological cleavage of silicon-carbon bonds could occur (30). More recent work has shown that microorganisms can utilize dimethylsiloxanes as a source of carbon (31), and soil incubation of ^{14}C labeled siloxanes have been found to release $^{14}CO_2$ (32).

Pioneering work by Fessenden on the metabolism of organosilanes by mammals found that phenyl and alkyl silanes were oxidized very much like their carbon analogs (33). A notable difference, however, was found for dimethylphenylsilane (**8**) in which the silicon-hydrogen bond was rapidly oxidized *in vivo* (Figure 7).

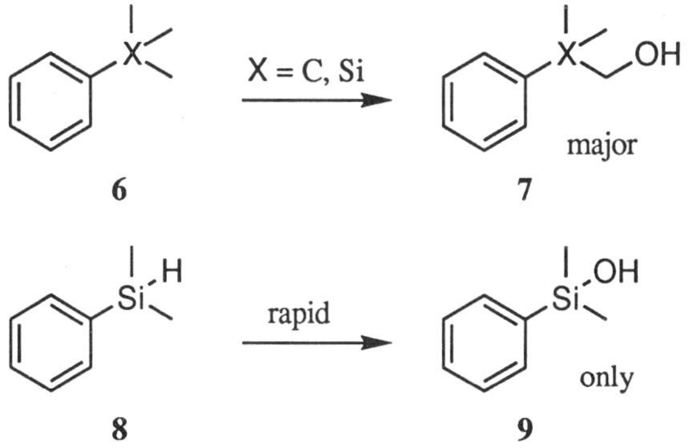

Figure 7. Mammals rapidly oxidize the silicon-hydrogen bond, whereas the carbons attached to silicon are metabolized much like simple hydrocarbons (34).

Environmentally Safer Chemicals by Silicon Substitution

The silicon-for-carbon isosteric replacement has been extensively studied with pharmaceuticals, and this area has been the subject of several excellent reviews (*24, 33, 35-37*). A contemporaneous effort in agrochemical research has also been reviewed (*38*). The environmental fate of pesticides has received great scrutiny, and so pesticide examples of the use of silicon-for-carbon substitution are particularly relevant to the design of environmentally safer chemicals. Three examples are discussed below.

Silane Analogs of DDT. Despite the relative safety of DDT (**10**) to mammals, its toxicity to other species and environmental persistence led to the ban of this important pesticide (*39*). In an early effort to design a more benign version of DDT, a number of silane analogs such as the DDD analog **12** (Figure 8) were prepared with the anticipation that these would be less environmentally persistent (*40*). While this would undoubtedly be true, all of the silanes proved to be nontoxic to insects. The presence of the readily oxidized silicon-hydrogen bond would have been one source of instability, both environmentally and *in vivo* (*33*). More significant for this research however, was an SAR study that found the overall size of the DDT molecule strongly correlated with bioactivity, implicating the atomic size of the central silicon for the lack of insect toxicity for **12** and congeners (see Figure 4).

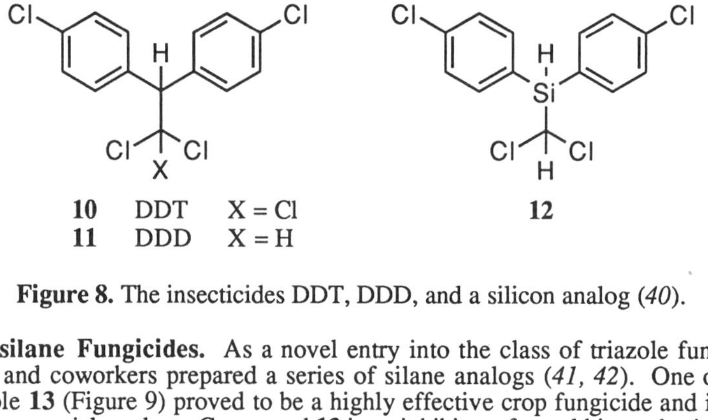

```
10   DDT   X = Cl
11   DDD   X = H
                      12
```

Figure 8. The insecticides DDT, DDD, and a silicon analog (*40*).

Organosilane Fungicides. As a novel entry into the class of triazole fungicides, Moberg and coworkers prepared a series of silane analogs (*41, 42*). One of these, flusilazole **13** (Figure 9) proved to be a highly effective crop fungicide and is now a major commercial product. Compound **13** is an inhibitor of sterol biosynthesis, similar to other triazole fungicides. The properties of this compound responsible for its field performance, including volatility, solubility, and movement within plants, have been described (*43*). The primary metabolite of **13** is the silanol **14**, resulting from cleavage of the triazole-substituted methyl group from silicon (*44*). Presumably **14** has little or now biological activity, and the higher oxidation level **14** will enhance the rate of its further degradation (relative to **13**). Flusilazole is an excellent example of the commercial potential for biologically active organosilanes.

Organosilane Pyrethroids. A major new class of insecticides to emerge during the 1970s were the synthetic pyrethroids, analogs of the naturally occurring pyrethrins (*45*). The pyrethrin analogs ethofenprox **15** and MTI-800 **16** (Figure 10) (*46*), described in the 1980s, were the first pyrethroids with a central, quaternary carbon that could be reasonably altered by an isosteric substitution with silicon (*47*). This *gem*-dimethyl group is critical for biological activity and is very sensitive to modifications (*47*). Silafluofen (**17**) slightly increases the size of the *gem*-dimethyl group. In laboratory tests, silafluofen **17** proved to be somewhat less active as an insecticide than

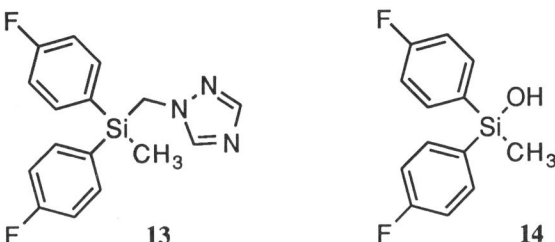

Figure 9. Fungicide flusilazole (**13**) and its major metabolite **14** (*43*).

MTI-800, and more active than **15**. As **15** was to become a commercial product, **17** was the first silane with a commercial level of insecticidal activity.

One feature of **15** and **16** that differentiated them from the earlier pyrethroids, was the relatively low fish toxicity of these structures. Previously prepared pyrethroids, like the natural product itself, were very low in toxicity to birds and mammals but were extremely toxic to fish. This made them unsuitable for crops such as rice. Silafluofen **17** had appreciable insecticidal activity and proved to be *nontoxic* to fish. Steric and aqueous solubility changes have been proposed to account for this fish safety (*47, 48*).

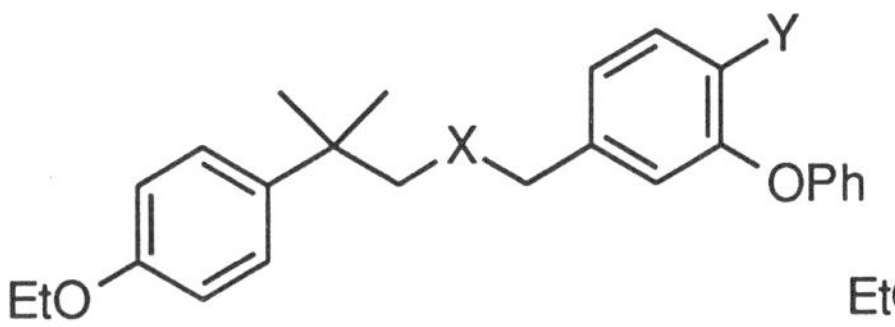

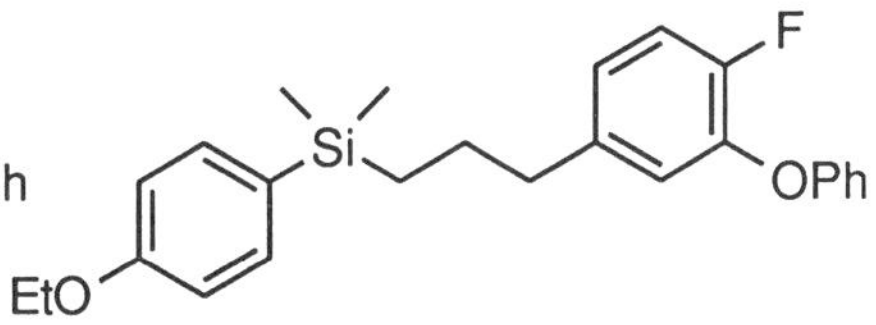

15 ethofenprox   X = O   Y = H
16 MTI-800   X = CH$_2$ Y = F
17 silafluofen

Figure 10. Two carbon-based pyrethroids and a silane analog.

As with DDT, the degradation rate of pesticides is one of the environmentally critical parameters: the substance must last as long as required, but should last no longer. For aryl silane **17**, as well as the non-silanes **15** and **16**, a potential route of decomposition is protodearylation (Figure 11, related to the chemistry shown in Figure 5). Ipso protonation of **18** by an acid, such as humic acids found in soil (*49-51*), would lead to cationic intermediate **19**. Capture of this intermediate by a nucleophile like water would lead to cleavage of the carbon-X bond, releasing the aromatic ring and an alcohol (or silanol) **21**. Acid catalyzed dealkylation and desilylation of *tert*-butylarenes and arylsilanes, respectively, are well known synthetic transformations. The rates of desilylation (X = Si) were estimated by Eaborn (*52*) to be faster than dealkylation (X = C) by four orders of magnitude! Thus, silafluofen (**17**) might be expected to have a shorter half-life than its carbon analog **16**. Although the details of the environmental persistence of these two proprietary materials are not yet public knowledge, notably the silane **17** (*53-56*) is now a commercial insecticide in Japan, whereas **16**, more potent and invented earlier (*57*), is not.

Field Activity of Organosilane Insecticide Silafluofen. Silane **17** proved to be very effective under field conditions, with an excellent level of activity for several weeks following a single application. Figure 12 shows the use of **17** to control the gypsy moth (*Lymantria dispar*) on oak trees (*48*). The silane was tested at two

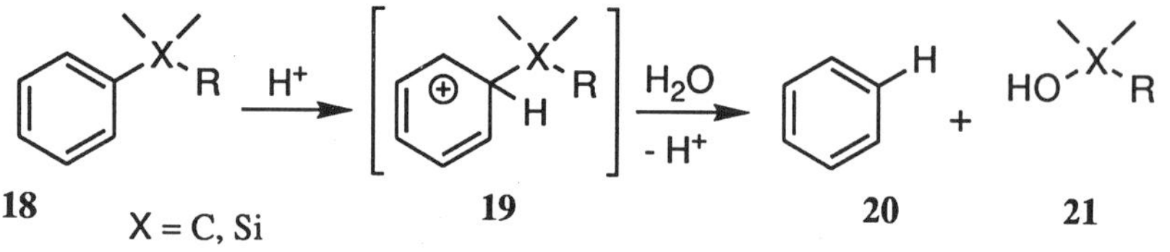

Figure 11. Protodearylation of carbon and silicon. The rate for silicon is faster by a factor of about 10,000 (52).

different rates, and compared to the standard chemical treatment for gypsy moth infestations, carbaryl (**22**, commercially known as SEVIN), used at a substantially higher rate.

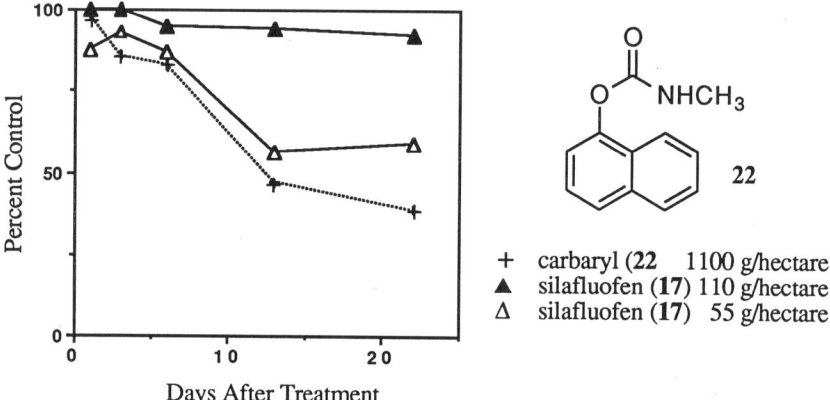

+ carbaryl (**22**) 1100 g/hectare
▲ silafluofen (**17**) 110 g/hectare
△ silafluofen (**17**) 55 g/hectare

Figure 12. Results of a field test of silane **17**, in comparison with **22**, for the control of the gypsy moth (Lymantri dispar) on oak.

Conclusions

Isosteric replacement of carbon with silicon is an appropriate strategy for modifying quaternary carbons, as well as methines (CH) and perhaps methylene groups (CH$_2$). Organosilanes lack intrinsic toxicity and can be expected to oxidatively and hydrolytically decompose in the environment with the silicon ultimately becoming silicate, one of the most broadly distributed, naturally occurring chemical entities (58, 59).

For the fields of fungicides and insecticides, isosteric replacement of silicon for carbon has produced a new commercial product in a well established area of research (triazole fungicides, pyrethroid insecticides). The fish safety of silafluofen (**16**) gives this product a significant environmental advantage.

The use of isosteric substitution other than silicon for carbon in the design of safer drugs has been comprehensively reviewed (60). The application of these types of isosteric substitutions for the design of safer industrial chemicals presents an intriguing and promising area of research.

Acknowledgment. The author is grateful to Dr. Jim Cella of General Electric Corporation, Dr. Gert Volpp of FMC Corporation, and Dr. William Moberg of DuPont for helpful discussions.

Literature Cited

1. Silverman, R. B. *The Organic Chemistry of Drug Design and Drug Action*; Academic: New York, 1992.
2. Cannon, J. G. In *Burger's Medicinal Chemistry and Drug Discovery*; 5th ed.; Wolff, M. E., Ed.; John Wiley and Sons: New York, 1995; Vol. 1; pp 783-802.
3. Langmuir, I. *J. Am. Chem. Soc.* **1919**, *41*, 1543-1559.
4. Reference (*1*) pp. 19-23.
5. Pirrung, M. C.; Han, H.; Ludwig, R. T. *J. Org. Chem.* **1994**, *59*, 2430-2436.
6. Burger, A. *Med. Chem. Res.* **1994**, *4*, 89-92.
7. van de Kamp, J.; Hunt, J. P. *Silicone Implants*; National Library of Medicine, National Institutes of Health: Bethesda, Maryland, 1994; Vol. 94-8, pp 75.
8. Friedberg, K. D.; Schiller, E.; Farnham, W. B., in *Handbook on Toxicity of Inorganic Compounds* Seiler, H. G. and Sigel, H., Eds.; Marcel Dekker: New York, 1988 pp. 595-617.
9. Cassidy, S. *Manufacturing Chemist* **1989**, *December*, 51.
10. Lukevics, E.; Germane, S.; Ignatovich, L. *Adv. Organomet. Chem.* **1992**, *6*, 543-564.
11. Davies, A. G.; Smith, P. J., in *Comprehensive Organometallic Chemistry* Wilkinson, G.; Stone, F. G. A. and Abel, E. W., Eds.; Pergamon: New York, 1983, Vol. 2, pp. 519-627.
12. Harrison, P. G., in *Comprehensive Organometallic Chemistry* Wilkinson, G.; Stone, F. G. A. and Abel, E. W., Eds.; Pergamon: New York, 1983, Vol. 2, pp. 629-680.
13. Sakurai, H., in *Encyclopedia of Inorganic Chemistry* King, R. B., Eds.; John Wiley & Sons: New York, 1994, Vol. 7, pp. 3805-3821.
14. Fessenden, R. J.; Rittenhouse, R. *J. Med. Chem.* **1968**, *11*, 1070-1071.
15. Fessenden, R. J.; Coon, M. D. *J. Med. Chem.* **1965**, *8*, 604-608.
16. Metcalf, R. L.; Fukuto, T. R. *J. Econ. Entomol.* **1965**, *58*, 1151.
17. Fleming, I., in *Comprehensive Organic Chemistry* Barton, D. and Ollis, W. D., Eds.; Pergamon: New York, 1979, Vol. 3, pp. 541-686.
18. Raabe, G.; Michl, J., in *The Chemistry of Organic Silicon Compounds* Patai, S. and Rappoport, Z., Eds.; John Wiley & Sons: New York, 1989, Vol. 2, pp. 1015-1142.
19. Seyferth, D.; Annarelli, D. C.; Vick, S. C. *J. Organomet. Chem.* **1984**, *272*, 123-139.
20. Interrante, L. V.; Wu, H.-J.; Apple, T.; Shen, Q.; Ziemann, B.; Narsavage, D. M.; Smith, K. *J. Am. Chem. Soc.* **1994**, *116*, 12085-12086.
21. Hwu, J. R.; Wang, N. *Chem. Rev.* **1989**, *89*, 1599-1615.
22. Fleming, I.; Dunoguès, J.; Smithers, R. *Org. React. (N.Y.)* **1989**, *37*, 57-575.
23. Frye, C. L. *Sci. Total Environ.* **1988**, *73*, 17-22.
24. Tacke, R.; Linoh, H., in *The Chemistry of Organic Silicon Compounds* Patai, S. and Rappoport, Z., Eds.; John Wiley & Sons: New York, 1989, Vol. 2, pp. 1143-1206.
25. Frye, C. L., in *Organometallic Chemistry Reviews* Seyferth, D., Eds.; Elsevier: New York, 1980, Vol. 9, pp. 253-260.
26. Buch, R. R.; Ingebrigtson, D. N. *Environ. Sci. Technol.* **1979**, *13*, 676-679.
27. Lehmann, R. G.; Varaprath, S.; Frye, C. L. *Environ. Toxicol. Chem.* **1994**, *13*, 1061-1064.
28. Carpenter, J. C.; Cella, J. A.; Dorn, S. B. *Environ. Sci. Technol.* **1995**, *29*, 864-868.

29. Anderson, C.; Hochgeschwender, K.; Weidemann, H.; Wilmes, R. *Chemosphere* **1987**, *16*, 2567-2577.
30. Heinen, W., in *Biochemistry of Silicon and Related Problems* Bendz, G. and Lindqvist, I., Eds.; Plenum: New York, 1978 pp. 129-147.
31. Wasserbauer, R.; Zadák, Z. *Folia Microbiol.* **1990**, *35*, 384-393.
32. Lehmann, R. G.; Varaprath, S.; Frye, C. L. *Environ. Toxicol. Chem.* **1994**, *13*, 1753-1759.
33. Fessenden, R. J.; Fessenden, J. S. *Adv. Organomet. Chem.* **1980**, *18*, 275-299.
34. Fessenden, R. J.; Hartman, R. A. *J. Med. Chem.* **1970**, *13*, 52-54.
35. Fessenden, R. J.; Fessenden, J. S. *Adv. Drug Res.* **1967**, *4*, 95-132.
36. Tacke, R.; Wannagat, U. *Top. Curr. Chem.* **1979**, *84*, 1-75.
37. Tacke, R., in *Organosilicon and Bioorganosilicon Chemistry* Sakurai, H., Eds.; John Wiley & Sons: New York, 1985.
38. Sieburth, S. McN.; Lin, S. Y.; Cullen, T. G. *Pestic. Sci.* **1990**, *29*, 215-225.
39. Mellanby, K., in *Progress and Prospects for Insect Control* 1989, BCPC Monogr. 43, pp. 3-20.
40. Fahmy, M. A. H.; Fukuto, T. R.; Metcalf, R. L.; Holmstead, R. L. *J. Agric. Food Chem.* **1973**, *21*, 585-591.
41. Moberg, W. K. US Patent 4,510,136, 1985; *Chem. Abstr.* **1986**, *104*, P207438k.
42. Moberg, W. K.; Basarab, G. S.; Cuomo, J.; Liang, P. H., in *Pesticide Science and Biotechnology* Greenhalgh, R. and Roberts, T. R., Eds.; Blackwell Scientific: Boston, 1986 pp. 57-60.
43. Smith, C. M.; Klapproth, M. C.; Saunders, D. W.; Johnson, L. E. B.; Trivellas, A. E., in *Brighton Crop Prot. Conf.—Pests Dis.*1992 pp. 639-644.
44. Guinivan, R. A.; Gagnon, M. R. *J. Assoc. Off. Anal. Chem.* **1994**, *77*, 728-735.
45. Davies, J. H., in *The Pyrethroid Insecticides* Leahey, J. P., Eds.; Taylor & Francis: Philadelphia, 1985 pp. 1-41.
46. Udagawa, T.; Numata, S.; Oda, K.; Shiraishi, S.; Kodaka, K.; Nakatani, K., in *Recent Advances in the Chemistry of Insect Control* Janes, N. F., Eds.; Royal Society of Chemistry: London, 1985 pp. 193-204.
47. Sieburth, S. McN.; Manly, C. J.; Gammon, D. W. *Pestic. Sci.* **1990**, *28*, 289-307.
48. Sieburth, S. McN.; Lin, S. Y.; Engel, J. F.; Greenblatt, J. A.; Burkart, S. E.; Gammon, D. W., in *Recent Advances in the Chemistry of Insect Control II* Crombie, L., Eds.; Royal Society of Chemistry: London, 1990 pp. 142-150.
49. *Humic Substances in Soil, Sediment, and Water: Geochemistry, Isolation, and Characterization*; Aiken, G. R.; McKnight, D. M.; Wershaw, R. L.; MacCarthy, P., Ed.; Wiley: New York, 1985.
50. Perdue, E. M., in *Humic Substances in Soil, Sediment, and Water* Aiken, G. R.; McKnight, D. M.; Wershaw, R. L. and MacCarthy, P., Eds.; Wiley: New York, 1985 pp. 493-526.
51. *Aquatic Humic Substances. Influences on Fate and Treatment of Pollutants*; Suffet, I. H.; MacCarthy, P., Ed.; American Chemical Society: Washington, DC, 1989.
52. Eaborn, C.; Pande, K. C. *J. Chem. Soc.* **1960**, 1566-1571.
53. Katsuda, Y.; Hirobe, H.; Namite, Y. Japan Patent JP 61 87,687, 1986; *Chem. Abstr.* **1986**, *105*, 191385y.
54. Yamada, Y.; Yano, T. Japan Patent JP 61,229,883, 1986; *Chem. Abstr.* **1987**, *106*, 156664n.
55. Sieburth, S. McN. US Patent US 4,709,068, 1986; *Chem. Abstr.* **1988**, *108*, 94775e.
56. Franke, H.; Joppien, H. German Patent DE 3,604,781, 1986; *Chem. Abstr.* **1988**, *108*, 132041x.

57. Nakatani, K.; Numata, S.; Kodaka, K.; Oda, K.; Shiraishi, S.; Udagawa, T. UK Patent GB 2,120,664, 1983; *Chem. Abstr.* **1984**, *100*, 174420k.
58. King, R. B., in *Encyclopedia of Inorganic Chemistry* King, R. B., Ed.; John Wiley & Sons: New York, 1994, Vol. 7, pp. 3767-3770.
59. Lickiss, P. D., in *Encyclopedia of Inorganic Chemistry* King, R. B., Ed.; John Wiley & Sons: New York, 1994, Vol. 7, pp. 3770-3805.
60. Burger, A. *Progress Drug Res.* **1991**, *37*, 287-371.

Chapter 5

Design of Biologically Safer Chemicals Based on Retrometabolic Concepts

Nicholas Bodor

Center for Drug Discovery, College of Pharmacy, University of Florida, P.O. Box 100497, J. Hillis Miller Health Center, Gainesville, FL 32610-0497

> The concept of retrometabolic-based soft drug design and the approaches therewith are reviewed. Several specific examples are given to provide an understanding of the basis for using retrometabolism in the design of safer drugs. Our knowledge of metabolic pathways is now sufficient to predict or evaluate the metabolism of a variety of chemical substances from the point of view of their likelihood of undergoing toxicification or detoxification metabolism. The concept of retrometabolic-based soft drug design needs to be applied to the design of safer commercial chemicals. The rules in transformations applied for designing soft drugs could be extended to the design of soft commercial chemical substances that are less toxic than the existing commercial substances from which they were designed. In doing so, it is quite conceivable that many commercial chemicals currently implicated in environmental and human toxicity could be redesigned and replaced with equally useful but safer substitutes.

In the modern era the environment and humans are constantly bombarded by ever increasing number of various manmade chemicals, which are used for a variety of purposes. In the United States, there are over 77,000 chemical substances on the Toxic Substances Control Act list of commercial (industrial) chemical substances. As discussed in Chapter 1, the design, synthesis, and final development of most of these chemicals is strictly use-oriented: little or no consideration was (or is) given to designing these substances such that their potential for toxicity is minimized. Many of these substances are very efficacious from a use standpoint, but are also toxic. Humans and other species can be exposed to these substances (or their various degradation products) from several routes, as shown in Figure 1.

Unlike commercial chemical substances, the design and development of drug substances involves a more rational approach, that includes safety as well as usefulness as components of the design strategy. The objective is to find highly active and effective drugs that can be used to control or cure illness without being toxic to the individual. This is done by identifying a lead compound (which is a well defined structure having the desired biological action) to which structural modifications are made that enhance the biological activity without enhancing any toxic properties. Thus, substances obtained by random and/or systematic modification of the lead structure leads to a large number of analogs which are tested for the desired purpose and then the most potent, long acting, more stable, least toxic substance is selected for drug development. Most of the time, however, the undesired, toxic side effects of new biologically active compounds runs parallel with its desired activity. In other words, although more potent drugs are being obtained, the therapeutic index, the ratio of its toxicity and effectiveness, does not change much. There are numerous reasons for this. Most of the time the side effects are directly related to the intrinsic activity, specifically receptor affinity responsible for the desired activity. And thus all drug properties including pharmacodynamic and pharmacokinetic ones will change parallel with the activity changes. Even at this stage, the toxicity issues related to a potential new drug represent a black box. It is the major unknown and most expensive component of the process of drug development. This is also why most potential drug candidates fail to be developed, as at some point in time unacceptable toxic side effects develop. This is clearly not strictly related to the desired specific receptor\affinity activity. In most cases, this is due to the fact that the drug in the body will undergo metabolic conversion during which analog metabolites are formed, $(A_1, A_2 - A_n)$, which are structures with minimal structural modifications, have a similar type of activity but different pharmacokinetic properties than the original drug (D), in addition to other metabolites $(M_1 - M_n)$ and potential reactive intermediates $(I_1^*, I_2^* - I_n^*)$, which are responsible for various cell damage (Figure 2).

These toxic metabolites are in most instances formed by oxidative metabolism and include various radical epoxides, and their formation in general can only be demonstrated by some of the products derived from them. Accordingly, the drug toxicity can be described as a combination of the intrinsic toxicity-selectivity and the toxicity due to the various metabolic processes as shown in Figure 3.

In the United States, federal regulations governing Food and Drug Admininstration (FDA) approval of new drug substances require testing for potential toxicity of drug substances and their metabolites prior to submission of the substance to FDA. These data are used along with other data to decide whether drug substances are suitably safe and efficacious. If a substance or its metabolites are found to be unacceptably toxic, then

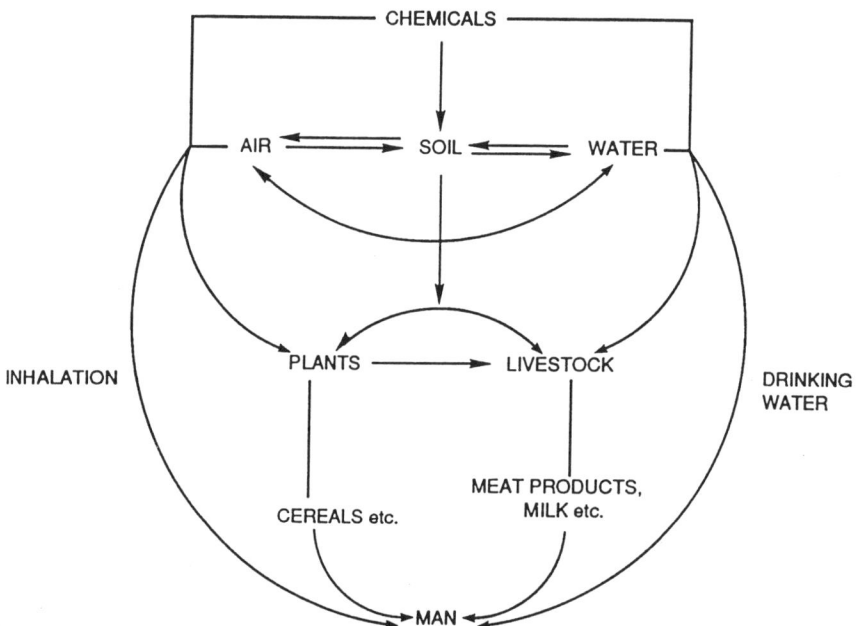

Figure 1. Schematic Representation of Routes Whereby Exposure to Chemical Substances can Occur.

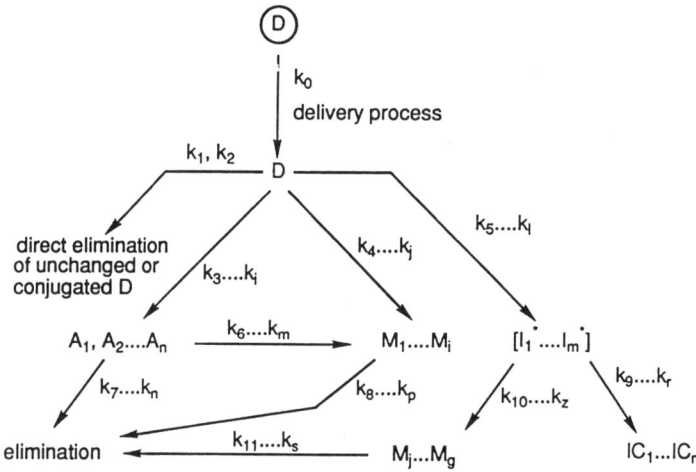

Figure 2. General Scheme for Metabolism of Drug Substances.

marketing of the substance will not be allowed. The development of a new drug substance, identification of its metabolites, and subsequent toxicity testing is a lengthy, expensive process. Many new potential drugs will undergo multiple metabolism simultaneously or sequentially, and at least the major metabolites need to be studied in detail. This process is well understood and necessary, since drugs are designed to be used specifically in humans or domestic animals for the treatment or prevention of disease. It is therefore important for drugs to be safe as well as effective. Substances intended to be used as pesticides are also required (under the Federal Insecticide, Fungicide and Rodenticide Act) to undergo extensive toxicity testing before they can be considered for approval by the Environmental Protection Agency (EPA).

The situation is much different, however, for commercial (i.e. industrial) chemical substances that are not intended to be used as pharmaceuticals or pesticides. Under the Toxic Substances Control Act (TSCA), anyone who wants to manufacture, import or use a new chemical substance for purposes other than as a pharmaceutical or an insecticide must submit a Premanufacture Notice to the EPA. TSCA **does not** require any toxicity testing prior to submission to EPA. In fact, the majority of the approximately 2,000 Premanufacture Notices reviewed annually by the EPA do not contain any toxicity data, and the toxicity of these substances are essentially unknown. Although these substances are not intended for human consumption, humans are nonetheless unintentionally exposed to them following release into the environment from their manufacture, disposal or use (Figure 1).

Accordingly, the above thoughts on drug toxicity and development can be extended to commercial chemical substances (i.e., substances not intended to be used as drugs or pesticides) as well. As shown in Figure 4, the toxicity of a commercial chemical substance in general can be described in a similar way as the toxicity of drugs, the main difference being that the intrinsic toxicity or selectivity term of the substance itself (T_c) is usually less important as many of the commercial chemical substances have no strong receptor binding capacity (with some notable exceptions). Thus the main issue becomes the intrinsic toxicity of the metabolic products of commercial substances. As the most important characteristic of a drug is its therapeutic index (TI), the safety index (SI) should be the most important property of any commercial chemical substance.

It is important to note that many commercial chemical substances have been toxicologically evaluated by groups such as the National Toxicology Program and commercial chemical manufacturers. In many instances, criteria for testing include high production volume, significant exposure to the general population and some anecdotal indication that a given substance is toxic. Chemicals meeting these criteria are generally existing chemicals that have been used commercially for many years and for which significant environmental release and subsequent toxicity may have already occurred or is occurring. Examples of such chemicals are benzene, 2-naphthylamine and certain benzidines, which for many years were released into the environment,

Drug (D) toxicity:

$$T(D) = T^D(i) + T(A_1, A_2, ... A_n) + T(M_1, M_2, ... M_n) + T(I^*_1, I^*_2, ... I^*_n)$$

$T^D(i)$; intrinsic toxicity - selectivity

$A_1, A_2, ... A_n$; analog metabolites with activities of the type of D

$M_1, M_2, ... M_n$; other metabolites

$I^*_1, I^*_2, ... I^*_n$; reactive intermediates

TI = TD/ED; therapeutic index (the ratio of the dose required to produce a toxic effect and the dose needed to elicit desired therapeutic response).

T = Toxicity

TD = Toxic Dose

ED = Effective dose (dose needed to elicit desired therapeutic response).

Figure 3. General Equation for Drug Toxicity.

Chemicals (C) in general:

$$T(C) = TC(i) + T(C_1, C_2, ... C_n) + T(M_1, M_2, ... M_n) + T(I^*_1, I^*_2, ... I^*_n)$$

$TC(i) \ll TD(i)$

$I^*_1, I^*_2, ... I^*_n$; most important

SI = TD/EmD; safety index

EmD = maximum environmental dose

Figure 4. General Equation for Chemical (nondrug) Toxicity (see Figure 3 for definitions of terms).

and are now known to be highly carcinogenic. Unlike substances intended to be used as drugs or pesticides, which are submitted in relatively small numbers to the FDA and EPA, respectively, the large number of new commercial chemical substances submitted to the EPA annually and the high cost of toxicity testing precludes routine toxicity screening of these substances. Although highly desirable, it is simply not feasible to undertake detailed toxicity studies for every new and existing commercial chemical substance. The obvious problem is, of course, that certain chemical substances (new and existing) may, in fact, be quite toxic but will only be recognized as such long after significant release to the environment and subsequent toxicity have occurred.

But, it appears that there is at least <u>one general solution to improve this situation</u>. All data accumulated on the metabolic activation-toxication and/or deactivation of drugs and other chemical substances clearly indicate the necessity of including metabolic considerations as a rational approach in the general design process of new, less toxic drugs, pesticides and commercial chemical substances. (The importance of metabolic considerations in the design of safer chemicals was discussed in great detail in Chapter 2, and is further exemplified in other chapters in this book). As applied to drugs, this means that rather than performing metabolism studies on the best drug candidate, structure-activity relationship (SAR) data of congeneric substances should be combined from the beginning and throughout the drug design process with their structure-metabolism relationships (SMR). This combination of SAR and SMR provides the basis for the <u>retrometabolic drug design (RMDD) concept</u>. The RMDD concept enables accurate prediction of metabolism of a new drug candidate. From the predicted metabolic pathways one can better assess if the new drug candidate will be metabolized (bioactivated) to toxic metabolites or metabolic intermediates, and rationalize what structural modifications need to be incorporated that redirect the metabolism of the drug candidate to nontoxic, readily excretable metabolites. The very same concept can be extended to the design of less toxic commercial chemical substances, provided that adequate structure-activity relationship and metabolism data of analogous substances are available.

The Retrometabolic Design (RMDD) Concept

Considering a drug (D) as the lead compound, the methods to improve its therapeutic index by retrometabolic drug design approaches would cover two opposite directions. In one case a precursor, progenitor of the drug is designed and synthesized which is called a <u>chemical delivery system</u> (CDS). This actually can be visualized as covalently binding to the drug's bioremovable moieties of two general types. The more important one is called the targetor (T) while the others are modifier (F) functions, which will optimize the physical-chemical properties of the molecule to achieve better delivery. By design, the CDS will undergo sequential metabolic conversions removing the modifiers after the targetor fulfills its site- or organ-targeting role. Important in the design processes is that the starting CDS and the intermediates leading to the active drug are all inactive and thus do not produce any unwanted side effects. In addition, the metabolic conversion should follow the designed route and no other metabolism should take place. This general design approach can be considered as one part of the retrometabolic design loop as shown in Figure 5. The concept of chemical delivery systems was successfully applied in a variety of drug targeting problems. Successful delivery to the brain, to the eye and other organs were accomplished (7,8,9). Similar approaches can or were used in targeted bioactivation of some environmentally important chemicals, in particular insecticides. These concepts should find even larger application in the development of safer commercial chemicals than they do in drug or pesticide design.

On the other side of the loop are newly designed drugs with desired intrinsic activity. These are designed based on isosteric-isoelectronic principles to provide intrinsic activity similar to the lead compound D. These so-called "soft drugs" are designed in such a way that they will be metabolized in a predictable and controllable way to an inactive metabolite (M_I) after they achieve their therapeutic role. As shown on the right side of the retrometabolic loop, the soft drug (SD) is designed in a retrometabolic fashion from the very metabolite it will be converted to. Thus, the design of the soft drug takes place with well-defined strategic modifications involving the drug metabolism.

The soft drug concept described here was introduced in 1980 (1-3) and it was emphasized that not all metabolic conversions are preferred in the design process. As a matter of fact, it was specifically emphasized to avoid as much as possible oxidative metabolic conversions, as these are the ones most responsible for formation of reactive intermediates and cell damage. As shown in Figure 6 the "active oxygen", representing various oxidative enzymes are responsible for a large number of metabolic processes, many of them involve intermediates with potential toxicity.

One of the most frequently involved classes of oxidative enzymes are the hepatic cytochrome P-450 and related enzymes. It is important to emphasize that in addition to the potential toxic intermediates, these enzymes are relatively slow and thus inefficient. In addition, they are involved in many endogenous metabolic processes and thus competition with foreign compounds will have indirect side effects as well. The relatively slow rates of these monooxygenase systems are illustrated in Table I, where one can see that they are several orders of magnitude slower than most other enzymes (4).

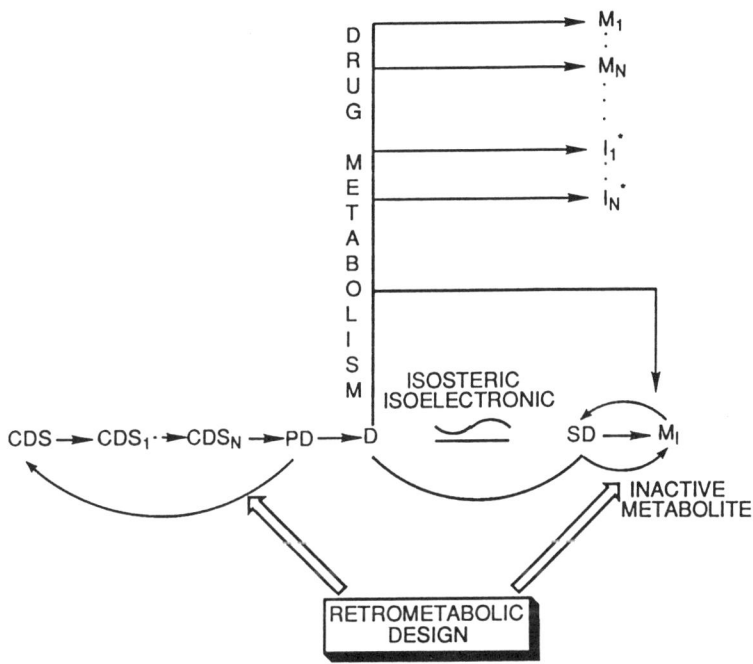

Figure 5. The Retrometabolic Drug Design Loop including Soft Drug and Chemical Delivery System Design. D = drug; PD = prodrug, i.e. the immediate precursor; CDS = the chemical delivery system (and its metabolites); SD = soft drug; A = active metabolites, inactive metabolite (MI) and reactive intermediates ($I1^*$-IN^*).

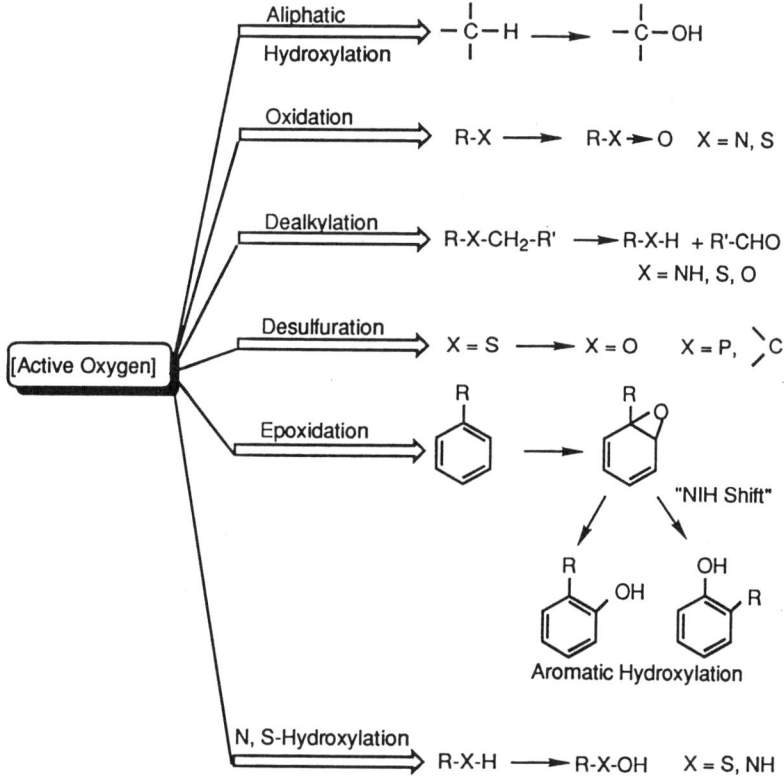

Figure 6. Characteristic Reactions Catalyzed by the Drug-Metabolizing Enzyme System through the Mediation of "Active Oxygen." The versatility of the system as depicted here generally does not extend to dehalogenation of aromatic chloro compounds.

Table I. Comparison of Reaction Rates of Some Hepatic Cytochrome P-450-linked Monooxygenase Reactions with the Turnover Numbers of Some Common Enzymes

Monooxygenase reactions (nmol substrate transformed/nmol P-450 per min)	
Aminopyrine N-demethylase	13
Ethylmorphine N-demethylase	10
p-Nitrophenetole O-deethylase	3
Hexobarbital hydroxylase	2.5
Aniline hydroxylase	1.2
Other reactions (nmol substrate transformed/ nmol of enzyme per min)	
Carbonic anhydrase c	36,000,000
Δ^3-ketosteroid isomerase	17,100,000
Catalase	5,600,000
β-amylase	1,100,000
β-galactosidase	12,500
Phosphoglucomutase	1,240
Succinatedehydrogenase	1,150

It is evident that the CDS and SD retrometabolic design concepts are opposite to each other. While the <u>CDS is inactive</u> and <u>metabolically activated</u> through a designed route to the drug, the <u>SD is active</u> itself which is then <u>metabolically deactivated</u> via a designed route.

A number of different classes of soft drugs were identified and classified (5,6) in five distinct retrometabolic-based groups. These are:

- activated soft compounds;
- soft analogs;
- active metabolite based drugs;
- controlled release endogenous agents; and
- inactive metabolites

Examples for practical use of each of the five classes for the design of drug products and pesticides are described in the literature (10-20). Four (excluding the Controlled Release Endogenous Agents class) of the five classes of the soft drug concept have the most potential application for the design of commercial chemicals. These four classes are briefly reviewed below.

Activated Soft Compounds. Using the activated soft compound approach in drug development, the design process starts with a known or designed nontoxic, inactive compound to which a metabolically labile substitiuent (the "activating" group) that bestows the desired biological effect without bestowing toxicity is added, and which is removed during metabolism to yield the original nontoxic, inactive compound. (The term "activated" should not be confused with bioactivation, which refers to metabolism that yields toxic metabolites.) The term "soft" refers to fact that the activated substance is metabolized readily to the original nontoxic substance from which it was designed. The important aspect of this approach is that metabolism of the activated substance yields the original nontoxic compound and perhaps other nontoxic substances subsequent to performing its desired pharmacological function.

An example of how the activated soft compound approach has been applied can be illustrated with locally (topically) acting antimicrobial agents. During the search for locally active antimicrobial agents of low toxicity, N-chloramines based on amino acids, amino alcohol esters, and related compounds were developed. These compounds represent a stable source of positive chlorine (Cl^+) which will be released before or after penetrating the microbial cell walls regenerating in this process the original amine, as shown in Figure 7. Here, the actual process of releasing the activating group (Cl^+) is responsible for the desired activity.

The antimicrobial activity of chloramines was known before but most of them could not be used because of their chemical instability, and generally high reactivity. Among these N-chloro derivatives of some amides and imides were of interest. After establishing the mechanism of their decomposition (21) it was found that if the α-carbon atoms lack hydrogen, stable N-chloramines can be obtained, however, they have a much lower "chlorine potential" and are much less corrosive. These low chlorine potential chloramines are represented by structures 1, 2, and 3 on Figure 7. Table II gives contact germicidal efficiencies (CGE) of a number of esters of the chlorinated 2-amino-2-methyl-1-propanol 3a-3l. Evidently the unchlorinated amino alcohol esters 3b and 3g are inactive and have no toxicity while the chlorinated derivatives at the relatively low concentration of 1-400 parts per million are very effective against a variety of bacteria. Thus the predictable and controllable metabolism of the soft chloramines implies a definite advantage of these compounds over the hard, lipophilic aromatic antimicrobials containing C-Cl bonds (22,23).

Unlike the design of drug substances, which is largely based on very strict structural requirements necessary for binding to a particular receptor

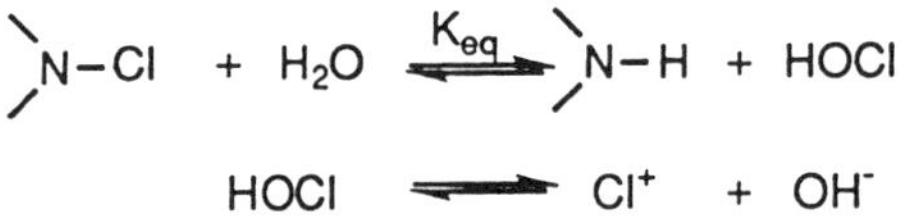

$$\text{\textbackslash}N-Cl + H_2O \underset{}{\overset{K_{eq}}{\rightleftharpoons}} \text{\textbackslash}N-H + HOCl$$

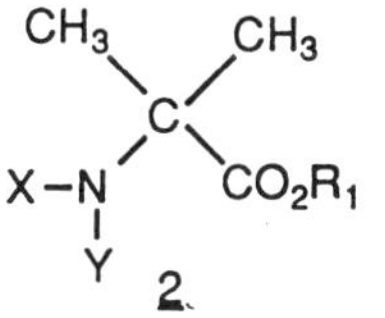

$$HOCl \rightleftharpoons Cl^+ + OH^-$$

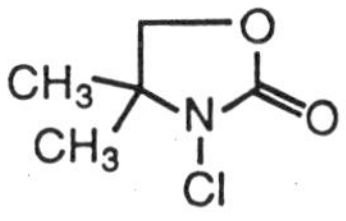

[Structure 1: oxazolidinone with CH₃, CH₃, N-Cl]

[Structure 2: C(CH₃)(CH₃)(CO₂R₁)-N(X)(Y)]   X = Cl; Y = H or Cl; R₁ = Alkyl, etc.

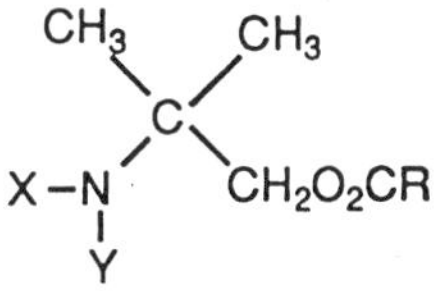

[Structure 3: C(CH₃)(CH₃)(CH₂O₂CR)-N(X)(Y)]

a: X = Cl, Y = Cl, R = CH₃

b: X = H, Y = H, R = (CH₂)₂CH₃·HCl

c: X = H, Y = Cl, R = (CH₂)₂CH₃

d: X = Cl, Y = Cl, R = (CH₂)₂CH₃

e: X = H, Y = Cl, R = C(CH₃)₃

f: X = Cl, Y = Cl, R = C(CH₃)₃

g: X = H, Y = H, R = (CH₂)₄CH₃

h: X = Cl, Y = Cl, R = (CH₂)₄CH₃

i: X - Cl, Y = Cl, R = (CH₂)₄CH₃

j: X = H, Y = H, R = (CH₂)₆CH₃·HCl

k: X = H, Y = H, R - (CH₂)₆CH₃

Figure 7. Application of the Activated Soft Compound Approach in the Design of Topical Antimicrobials.

TABLE II. Contact Germicidal Efficiency (CGE) of 2-Amino-2-methyl-1-propyl Carboxylates

Compound	Concentration[a] ppm	ppm, Cl⁻	Sterilization time[b] (min) S. aureus	S. pyogenes	E. coli	S. typhimurium	B. subtilis
3a	1292	458	5	2.5	0.5	2.5	2.5
3b	1070	—	<60	<60	<60	<60	<60
3c	2078	389	2.5	0.5	0.5	2.5	2.5
3d	979	304	2.5	0.5	0.5	2.5	0.5
3e	1886	323	5	0.5	0.5	2.5	2.5
3f	409	120	5	2.5	2.5	5	2.5
3g	1037	—	<60	<60	<60	<60	<60
3h	1319	211	2.5	0.5	2.5	5	2.5
3i	95	26	10	2.5	10	10	2.5
3j	1040	—	<60	30	15	30	5
3k	300	43	10	2.5	5	10	0.5

[a] 0.1 M sodium dihydrogen phosphate, pH 7.0.
[b] Time intervals screened were 0.5, 2.5, 5, 10, 45, and 60 min.

type, the design of commercial chemical substances is based on use function, and the structural requirements are therefore much less stringent. Factors such as stereochemistry, for example, are generally not as critical in the design of commercial substances as they are in drug design. This may allow the activated soft compound approach to have wider application in design of commercial chemicals than in the design of drugs. In the design of commercial chemicals, the chemist could incorporate into a nontoxic substance an activating group that bestows the desired use properties and is also easily removed metabolically to yield the original nontoxic substance if the activated substance happened to be absorbed in humans following exposure.

The Active Metabolite Principle. As shown in Figures 2 and 3, many drugs undergo stepwise metabolic degradation to yield intermediates and structural analogs that have similar activity as the original drug molecule. Examples of this include: oxyphenbutazone, an active metabolite of phenylbutazone; and 4-hydroxypropranolol, an active metabolite of propranolol. Many other active drug metabolites belong to this class. These transformations are in general oxidative in nature and will put a burden on the saturable and slow oxidative enzyme systems. But, even more importantly, these transformations will result in a mixture of active compounds that have different selectivity, pharmacokinetic properties, binding, distribution, and elimination properties. The overall result is that, depending on the presence of other compounds competing for the same enzyme system, as well as on the activity and regulation of specific enzymes, a variety of unpredictable combinations of the active species will be present in different individuals at different times. The full pharmacokinetic and pharmacodynamic evaluations of these cases becomes impossible. A good illustration of such a situation is provided by bufuralol (4) which has three active metabolites, 4a-4c, displaying different selectivities and rather different pharmacokinetic properties (24), as shown in Table III.

[Structure: benzofuran with R_2 at upper position, R_1 at lower position, and $-CHCH_2NHC(CH_3)_3$ with OH substituent]

4a-c

TABLE III. Metabolism and Pharmacokinetic Properties of Bufuralol 4 and Its Metabolites 4a-4c

Compound	R_1	R_2	Concentration[a] (μg ml)	$T_{1/2}$[b]
4	-CH$_2$CH$_3$	H	22	4
4a	-CH(OH)CH$_3$	H	19	7
4b	COCH$_3$	H	8	12
4c	CH$_2$CH$_3$	OH	>2	4

[a]Blood concentration at 9.5 h after administration to humans (20mg oral dose).
[b]Biological half-life.

The relative in vivo concentrations of compound 4 versus 4a-4c will vary with factors such as time, individual enzyme levels, and indirect drug interactions, to name a few. Such situations are too complex for a correlation between blood levels and pharmacological activity to be found. It is evident that this is a general problem. In most cases, it is not even clear which is the main active compound, the drug or its metabolites. In the latter case, the drug is only a prodrug provided it does not have a substantial intrinsic activity. One other complication is that in certain disease states involving renal insufficiency, edema, or significant alterations in protein binding, the elimination profile of the various metabolites will be substantially altered. It is very important to realize, however, that in a large number of cases, based on our present knowledge, many metabolites can be predicted, and one does not have to wait for the classic metabolism studies to synthesize, identify, and evaluate these candidates.

According to the fundamental principles of soft drug design, it is preferable to use as the drug of choice an active species that undergoes a one-step, singular, predictable metabolic deactivation. Thus, the active metabolite theorem of the soft drug design states that whenever oxidative metabolic transformations of a drug take place, going through possibly toxic, highly reactive intermediates or through pharmacologically active species, if activity and pharmacokinetic considerations permit it, the drug of choice should be the active metabolite that is in the highest oxidized state. The theorem can of course be expanded to other metabolic transformations besides the oxidative ones. According to these principles in the pharmaceutical field, one should look for those active metabolites that appear to have significantly different pharmacokinetic properties, particularly longer half-lives, and follow up on these leads to possibly replace the initial drug with these metabolites which are subject then to a much simplified metabolic deactivation process.

The very same principles can in general be extended to the design of safer commercial chemicals. Studying and/or predicting oxidative metabolites of various chemicals, and testing the ones in higher oxidized states for the very use the original compound is applied for, one could identify analog metabolites which would have a much simpler, more predictable and less toxic metabolic deactivation, while still performing the same way as the original compound.

Soft Analogs. Compounds classified as soft analogs are close structural analogs of known active drugs or commercial chemicals having a specific metabolically (preferably hydrolytic) sensitive spot built in their structure that provides their one-step, controllable detoxication. These sensitive structural parts in general are not oxidizable alkyl chains or functional groups subject to conjugation. The designed detoxication will take place as soon as possible after the desired role is achieved, not allowing other types of metabolic routes (e.g. bioactivation) to take place.

The basic principles of soft analog design are:

- the new soft compounds are close structural analogs of the lead compounds;

- a metabolically, preferentially hydrolytic, sensitive part is built into the molecule;

- the metabolically weak spot is located in a part of the molecule that the overall physical, physicochemical, steric, and complementary properties of the soft analog are very close to those of the lead compound;

- the built-in metabolism is the major or preferably the only metabolic route for deactivation of the substance;

- the rate of the predictable metabolism can be controlled by structural modifications;

- the products resulting from the metabolism are nontoxic and have no significant biological or other activities; and

- the predicted metabolism does not require enzymatic processes leading to highly reactive intermediates.

By designing the substance such that it, if absorbed, will undergo rapid metabolism to produce a profound structural change in the molecule that leads to complete loss of the characteristic properties of the substance and to a breakdown of any toxicophoric or toxicogenic sites of the molecule, one can avoid formation of the usual active metabolites that result in multicomponent active systems. Accordingly, facile metabolic deactivation should be the preferred or major metabolic route, and it also will make easier to control delivery, elimination, pharmacokinetic, and pharmacodynamic factors.

The first example of the above principles of the soft analog approach is represented by the "soft quaternary salts". These substances, represented by formula $\underline{5}$ undergo facile hydrolytic deactivation via the very short lived intermediate $\underline{6}$ to an aldehyde, an amine, and an acid.

$$R\text{-COOCH}(R_1)\text{-}\overset{|}{\underset{|}{N}}^{+}\!\!\!\!= \;\xrightarrow{\text{Hydrolysis}}\; R\text{-COOH} + \left[HO\text{-}CH(R_1)\text{-}\overset{|}{\underset{|}{N}}^{+}\!\!\!\!= \right]$$
$$\underline{5} \hspace{6cm} \underline{6}$$

$$R_1 CHO + H\text{-}\overset{|}{\underset{|}{N}}^{+}\!\!\!\!=$$

The simplest example (1) of a useful soft analog of this type is provided by the isosteric analog $\underline{7}$ of cetylpyridinium chloride $\underline{8}$. The latter substance can be considered a "hard" quaternary antimicrobial agent, which takes several oxidative (generally β-oxidation) steps to lose its surface active, antimicrobial properties. It is evident that the quaternary salts $\underline{7}$ and $\underline{8}$, are very similar, they are isosteric and accordingly their physical properties are also very similar, as measured by their critical micelle concentrations (1). Both compounds possess comparable antimicrobial activity, but $\underline{7}$ undergoes facile hydrolytic cleavage, leading to its deactivation. As a result of this, $\underline{7}$ is about 40 times less toxic than $\underline{8}$, as represented by their LD_{50} values. While cetylpyridinium chloride has a LD_{50} of about 100mg per kg when given orally to rats, the soft analog $\underline{7}$ requires over 4g of the molecule to produce similar toxicity.

[Structures 7 and 8: phenyl-N+ quaternary ammonium compounds with Cl⁻ counterions]

7: Phenyl-N⁺-CH₂-O-C(=O)-(CH₂)₁₂CH₃, Cl⁻

8: Phenyl-N⁺-CH₂CH₂CH₂(CH₂)₁₂CH₃, Cl⁻

Another example for the use of soft quaternary salts for the design of soft analogs is the class of soft anticholinergic agents. These soft analogs were designed to have high local but practically no systemic activity. It is well-known that anticholinergic compounds have many useful clinical effects, among which the local antisecretory activity was long thought to be beneficial for inhibiting eccrine sweating (perspiration). Indeed, a wide range of anticholinergic agents such as atropine, scopolamine, or their quaternary ammonium salts inhibit perspiration, but these substances produce a number of side-effects (e.g. dry mouth, urinary retention, drowsiness, mydriasis), and are therefore not useful as antiperspirants.

The structural difference between the "hard" and "soft" anticholinergics of this kind are relatively small but nonetheless profound. As represented by structures 9 and 10, the known quaternary ammonium type anticholinergics have two or three carbon atoms separating the quaternary head and the ester function. Actually, it was believed for a long time that this separation is critical for effective receptor binding. A corresponding soft analog 11 would contain just one carbon separation, which would allow facile hydrolytic deactivation as shown by the general structure 5. A number of compounds of type 11 were made and studied (3) and many of them found to be as or even more potent than atropine. For example, 11 (R_1 = -C_6H_5; R_2 = -cyclohexyl; R_3 = H; R_4 = H and N is N-methylpyrrolidine) is equipotent with atropine in various anticholinergic tests in vitro and in vivo, but it is very short-acting after i.v. injection, because of its facile hydrolytic deactivation. However, when applied topically to humans, it produced high local antisecretory activity without any systemic toxicity.

[Structure 9: R_1, R_2, R_3 attached to C-C(=O)-O-C-C-N⁺]

[Structure 10: R_1, R_2, R_3 attached to C-C(=O)-O-C-C-N⁺ (cyclic)]

[Structure 11: R_1, R_2, R_3 attached to C-C(=O)-O-C(H)(R_4)-N⁺]

A more recent application of a soft analog concept involves prostaglandins. The issues to be tackled here are two-fold. Prostaglandins are well-known to have multiple pharmacological effects, and it was hoped that a separation of these effects could be achieved by the soft analog design. On the other hand it is well recognized that prostaglandins undergo facile metabolism, sometimes too fast. Accordingly, the rationale objective could not be to further facilitate this metabolism but rather to control it. As shown in Figure 8a, oxidative metabolism takes place on both chains of the prostaglandins, leading to deactivation.

Isosteric analog design based on soft analog concepts could prevent some of this oxidative metabolism and could redirect metabolism hydrolytically in the desired direction with the desired rates. Using PGE_1 (12) as the lead compound, a variety of soft oxa-prostaglandin analogs were made and tested and interesting separations of activity were found (Figure 8b). Some of the soft PGE_1 analogs are represented by structures 13, 14, and 15 in Figure 8b.

These structures clearly illustrate one of the generally applicable isosteric types of soft analog design, whereby the neighboring methylene functions are replaced by ester or reversed ester functionalities. Compound 14 appears in a recent patent (25).

The Inactive Metabolite Approach. The inactive metabolite approach is one of the most promising and versatile methods for developing safe drugs and chemicals in general. The main objective of this type of design is again to include the metabolism of the substance into the design process. However, there is a basic variation in that the actual activity (in the case of drugs) and toxicity of the metabolite are known, and the design process also allows for the metabolic pathway to be operative once the drug has performed its role, or if the commercial substance has been absorbed. Accordingly, the principles of the inactive approach are:

- the design starts with a known inactive (i.e., nontoxic) metabolite of a drug or commercial substance that is used as the lead compound;

- starting with this inactive metabolite, isosteric and/or isoelectronic soft analogs of the lead active compound are designed. This is the activation stage;

- the new soft analogs are designed in such a way that their metabolism will yield the starting inactive metabolite in one step and without going through toxic intermediates (predictable metabolism);

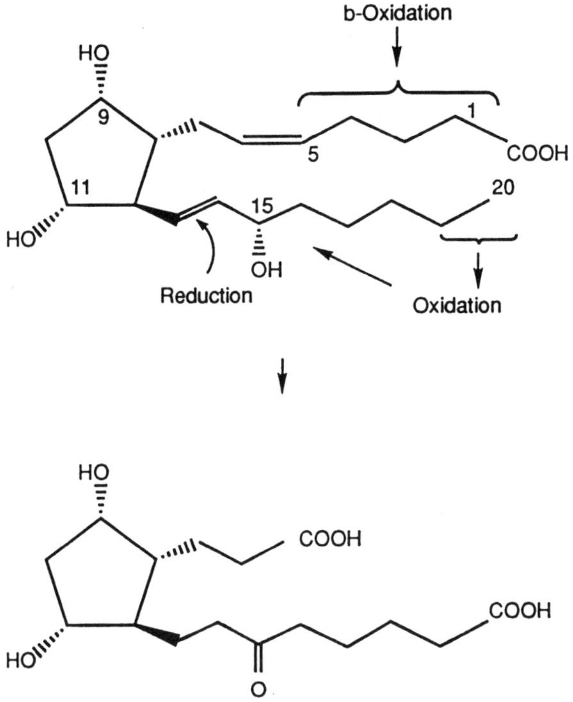

Figure 8a. Metabolism of Prostaglandins in Humans.

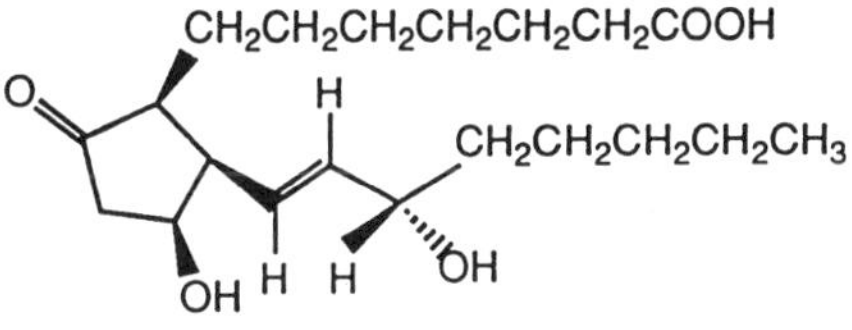

12

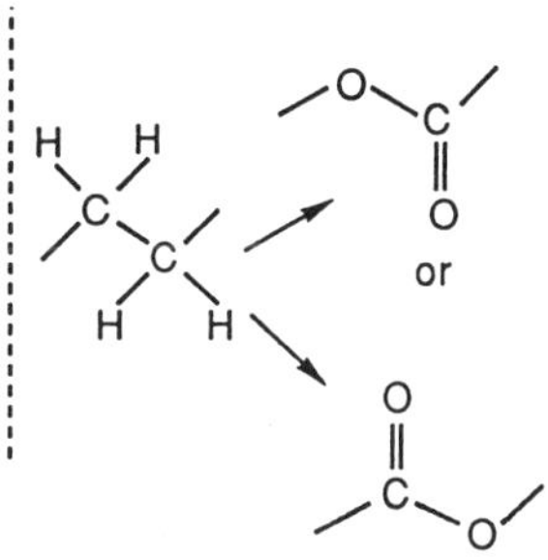

"Isosteric Substitution"

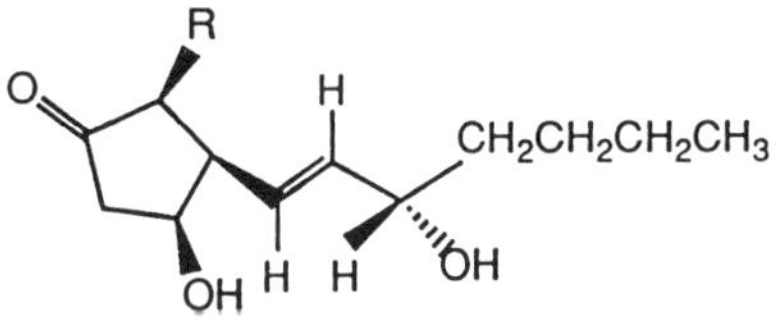

R = $CH_2CH_2O\underset{\underset{O}{\|}}{C}CH_2CH_2CO_2H$ 13

R = $CH_2\underset{\underset{O}{\|}}{C}OCH_2CH_2CH_2CO_2H$ 14

R = $CH_2CH_2\underset{\underset{O}{\|}}{C}OCH_2CH_2CO_2H$ 15

Figure 8b. Soft Oxa-Prostaglandin Analogs Designed Using the Soft Analog Approach.

- the transport and binding, rate of metabolism and pharmacokinetic properties of the soft molecules can be controlled by molecular manipulations in the first activation stage (controllable metabolism).

During the past 10 years or so, a number of classes of drug substances have been designed based on the inactive metabolite approach, and were used and advanced to human application. A few selected ones will be used here to illustrate the above general principles. For example anti-inflammatory corticosteroids are known to have multiple adverse effects that often limit their usefulness. As illustrated by hydrocortisone 16 in Figure 9, corticosteroids undergo a variety of metabolic conversions (oxidative and reductive) which lead to the formation of a large number of various steroidal metabolites. One of the major metabolic routes of 16 is oxidation of the dihydroxyacetone side chain, which ultimately leads to formation of cortienic acid (17) a major, pharmacologically inactive metabolite that does not have any of the adverse effects caused by 16 (*14,18*). Cortienic acid, therefore, is an ideal inactive metabolite to start the design process of a new drug substance using the inactive metabolite appproach.

The C-17 α-hydroxy keto and hydroxy substituents of 16 are critical for pharmacological activity, and hence are very important pharmacophoric components. Using 17 as basis for the design of new, pharmacologically active analogs of 16 was undertaken. It was felt that incorporating suitable isosteric\isoelectronic substitution of the α-hydroxy keto and hydroxy substituents in the form of esters should restore the original corticosteroid activity without restoring the potential to produce adverse effects. Indeed, a wide variety of ring substituted ester derivatives of cortienic acid have been synthesized, as represented by 18 (Figure 10), and found to be very potent corticosteroids. It was observed that the usual substitution in positions 6α, 9α, or 16α or β, together with 17α substitution (Figure 10) will provide the same type of activity enhancement or decrease as was seen in analogous "hard" corticosteroids series. The critical functions for the activity are clearly the ester in the 17β and also the 17α functions, where the novel carbonate (26) and ether (27) substitutions provided the best activity. The new soft analogs have outstanding activity as illustrated by some of the derivatives listed in (Table IV), where the intrinsic receptor binding activities compared to a standard are shown. It was demonstrated (*27*) that these active corticosteroids undergo singular, one-step hydrolytic deactivation to the corresponding inactive cortienic acid derivative 19.

Table IV lists some of the soft hydrocortisone-substitutes designed using the inactive metabolite approach. Many of these substances have high anti-inflammatory properties when applied topically to the skin, instilled in the eye, or taken orally. Due to their facile metabolism to inactive and nontoxic

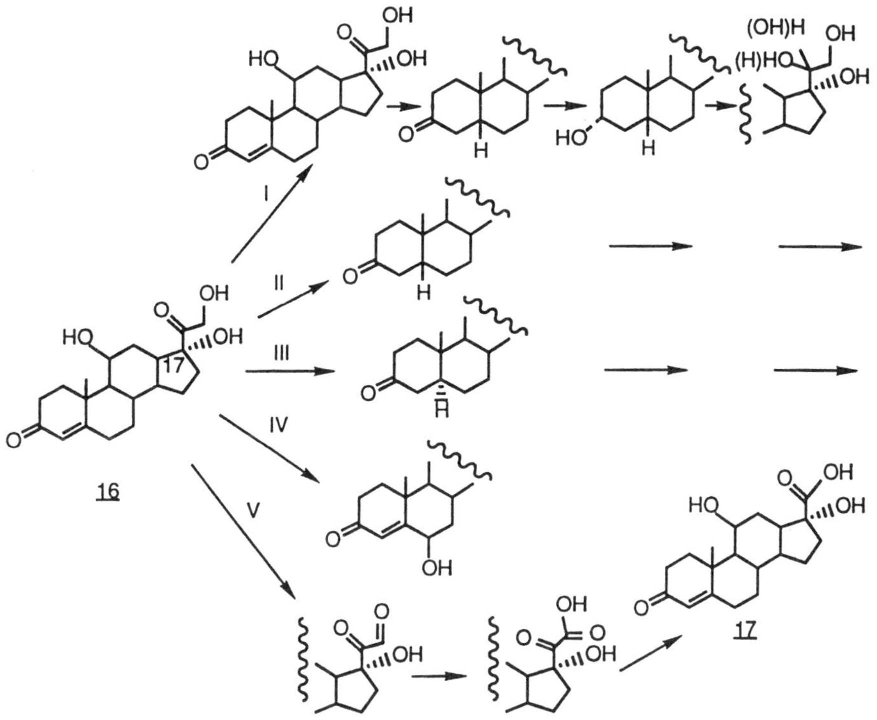

5. BODOR   *Biologically Safer Chemicals Based on Retrometabolic Concepts*   105

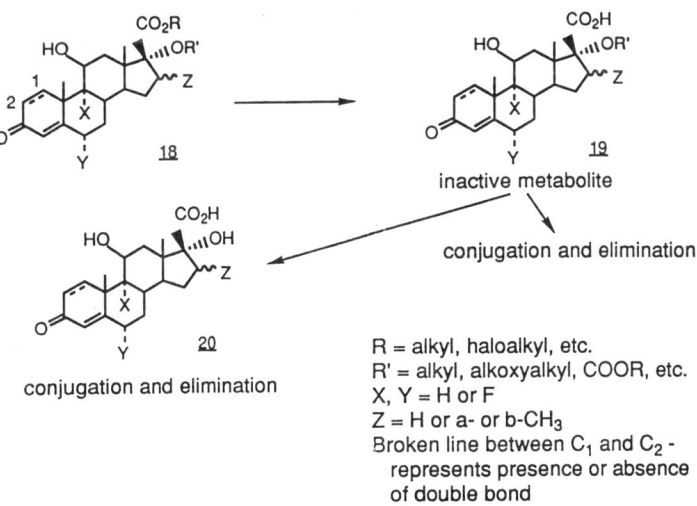

Figure 9. The Major Routes of Metabolism of Hydrocortisone.

R = alkyl, haloalkyl, etc.
R' = alkyl, alkoxyalkyl, COOR, etc.
X, Y = H or F
Z = H or a- or b-CH_3
Broken line between C_1 and C_2 - represents presence or absence of double bond

Figure 10. Design and metabolism of soft corticosteroids.

Table IV. Binding of Selected Soft Glucocorticoids to the Glucocorticoid Receptor of Rat Lung

Compound 20	R_1	R_2	X_1	X_2	X_3	RBA*
a	i-C_3H_7	α-CH_3	Cl	F	H	560
b	n-C_3H_7	α-CH_3	Cl	F	H	870
c		α-CH_3	Cl	F	H	840
d	i-C_3H_7	β-CH_3	-CH-Cl# $\|$ CH_3	H	F	11
e	C_2H_5	α-CH_3	CH_2Cl	F	H	19
f	C_2H_5	α-CH_3	Cl	F	H	740
g+	C_2H_5	α-CH_3	Cl	F	H	16
h	i-C_3H_7	α-CH_3	Cl	F	F	1,100
i	n-C_3H_7	α-CH_3	Cl	F	F	1,000
j	n-C_3H_7	α-CH_3	Cl	H	F	1,000
k	CH_3	α-CH_3	Cl	H	F	1,200
l	CH_3	β-CH_3	Cl	F	H	990
m	n-C_3H_7	β-CH_3	Cl	F	H	1,460
n	i-C_3H_7	α-CH_3	F	F	H	820
o	n-C_3H_7	α-CH_3	F	F	H	990
p	C_2H_5	H	F	H	H	200
r	i-C_3H_7	H	F	H	H	70
s	C_2H_5	α-CH_3	Cl	F	F	2,100
t	CH_3	H	Cl	H	H	180
u	C_2H_5	H	Cl	H	H	490
v	n-C_3H_7	H	Cl	H	H	540
w	17α-OH	β-CH_3	Cl	H	H	3
z	17α-OH	α-CH_3	Cl	F	H	7

* RBA dexamethasone = 100.
+ 11-keto.
# Note branching.

metabolites, however, they are essentially devoid of the well-known use-limiting adverse effects of anti-inflammatory steroids (such as adrenal suppression, immunosuppression, and growth inhibition). Substance 20u (Table IV), a non-fluorinated soft analog of prednisolone, was successfully developed for human use as a unique ophthalmic anti-inflammatory-antiallergic compound. This substance is highly active pharmacologically, but does not increase intraocular pressure: a debilitating side-effect of steriods used ophthalmically. This compound is now awaiting final approval by the FDA.

Soft analogs based on the inactive metabolite approach can also be obtained by introducing the hydrolytically sensitive function at a remote position from the pharmocophore. In this case, there is significantly more freedom in choosing the structural modification and thus the transport and rate of metabolism properties can better be controlled. For example the well-know β-blockers metoprolol 21 and atenolol 22 undergo metabolism to the corresponding inactive phenylacetic acid derivative, represented as 23 (Figure 11). The main pharmacophore in β-antagonists is the β-amino alcohol. Accordingly, substitution of the carboxylic acid in the inactive metabolite by esterification (represented by structure 24), will lead to various soft β-blockers, which will have different transport, and metabolism properties (28,29). For example, if membrane transport (lipophilicity) and relative stability is important for the desired pharmacological activity, the R group of the ester moiety of 24 should be relatively lipophilic and stable. The adamantane-ethanol moiety of 24c is an example of such an R group. This substance (24c) is known as adaprolol, and was shown to be very effective in humans in reducing intraocular pressure, without causing any of the well-known side effects of β-blockers (29). On the other hand, if systemic ultra short action is the objective, then R groups that make the ester moiety of 24 susceptible to rapid hydrolysis should be used. The recently developed methyl-thiomethyl and related esters represented by 24d, 24e, and 24f (Figure 11) were found to ultra short acting (30). These compounds hydrolyze extremely fast, much faster than simple alkyl esters, when injected intravenously.

Retrometabolic Design Principles for the Design of Safer Commercial Chemicals

The preceding sections have demonstrated how retrometabolic design principles have been used for the design of safer drugs. From these discussions it seems plausible that retrometabolic design principles could be equally or perhaps better applied to the design of safer commercial chemical substances. However, like many of the other drug design strategies discussed in Chapter 2 and other chapters in this book, the retrometabolic principles used for designing safer drugs have seldom been applied to the design of safer commercial chemicals. As alluded to in Chapter 1, this is probably due to the unfamiliarity of commercial chemists with the approaches used by medicinal chemists to design drugs.

One of the few instances in which a retrometabolic design principle has been used for the design of a nonpharmaceutical product was when the inactive metabolite approach was used for the design of safer chlorophenothane (DDT) substitutes (10). This is a classical early example that demonstrates the validity of the retrometabolic design concept, and was included in the very first conceptual and review papers on retrometabolic design (7,10). DDT (25) was widely used as a pesticide in the United States until it was banned as a result of suspicions that it causes a variety of health problems, including cancer. DDT undergoes cytochrome P450-mediated oxidative metabolism to kelthane (27) and dehydrohalogenation to bis([4-chlorophenyl] acetic acid (26), as shown in Figure 12. The latter metabolite (26) is nontoxic and inactive as a pesticide, and therefore was viewed as an ideal compound from which an active but safer DDT analog could be designed using the inactive metabolite approach of retrometabolic design. It was found that ethyl 4,4'-dichlorobenzilate (28), is reportedly highly active as a pesticide, but has much lower carcinogenicity when compared to DDT or kelthane (Table V) (7,10). The ethyl ester moiety of 28 apparently functions similarly to that of the trichloromethyl group of DDT in that it restores pesticidal activity. However, 28 is considerably less toxic than DDT because the ethyl ester group of 28 is labile and enables rapid metabolism of this substance to the free, nontoxic carboxylic acid in exposed subjects.

Table V. Toxicity of DDT and related pesticides

No	R_1	R_2	Name	Carcinogenicity[a] (ppm)
25	-H	-CCl_3	DDT	300
27	-OH	-CCl_3	Kelthane	300
28	-OH	-$COOC_2H_5$	ethyl 4,4'-dichlorobenzilate	6000

[a] For a summary of the toxicity data on these compounds, see the Registry of Toxic Effects of Chemical Substances (RTECs) on-line database.

The field of retrometabolic design has advanced significantly in the past several years, and this paper has reviewed only some of the latest achievements. With the development of reliable methods to calculate various molecular properties such as aqueous solubility, partition coefficient, molecular volume, surface area and others, more quantitative design of various soft analogs has become possible (31,32,33). The capabilities of

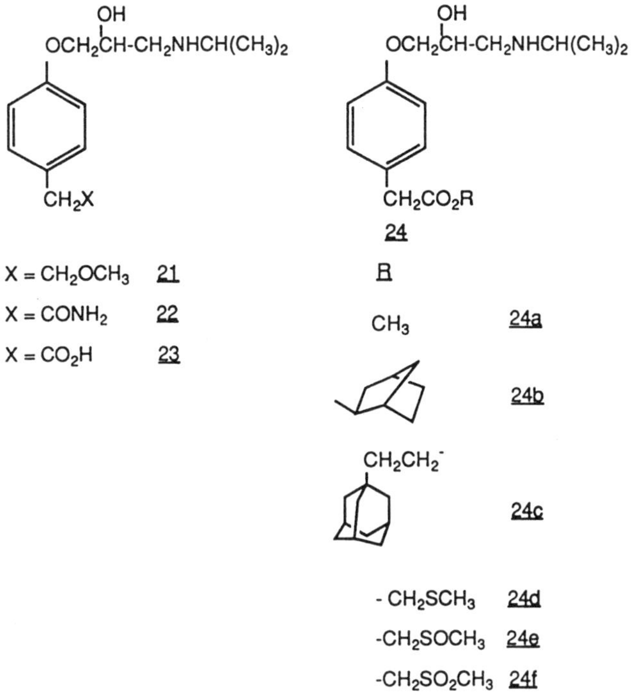

5. BODOR *Biologically Safer Chemicals Based on Retrometabolic Concepts* 109

OCH$_2$CH(OH)-CH$_2$NHCH(CH$_3$)$_2$ [p-substituted phenyl] CH$_2$X	OCH$_2$CH(OH)-CH$_2$NHCH(CH$_3$)$_2$ [p-substituted phenyl] CH$_2$CO$_2$R **24**

X = CH$_2$OCH$_3$ **21**
X = CONH$_2$ **22**
X = CO$_2$H   **23**

R
CH$_3$ **24a**
norbornyl   **24b**
CH$_2$CH$_2$-adamantyl   **24c**
-CH$_2$SCH$_3$ **24d**
-CH$_2$SOCH$_3$ **24e**
-CH$_2$SO$_2$CH$_3$ **24f**

Figure 11. β-Blockers Designed Using the Inactive Metabolite Approach.

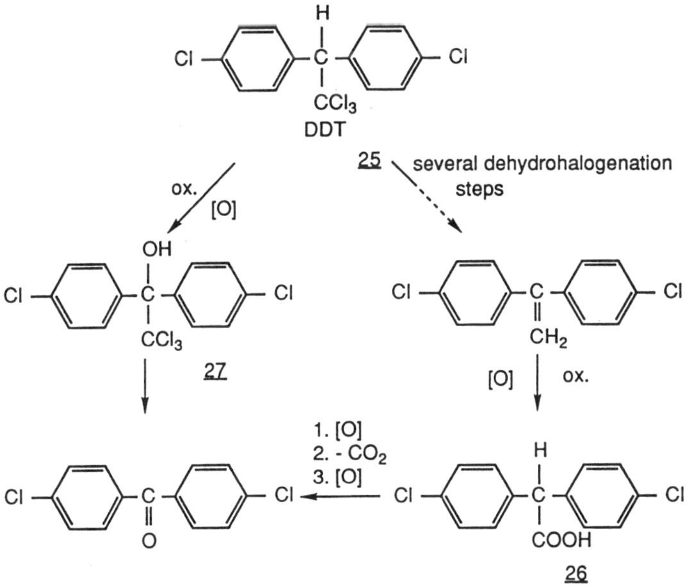

DDT: Cl-C$_6$H$_4$-CH(CCl$_3$)-C$_6$H$_4$-Cl

ox. [O] → 27: Cl-C$_6$H$_4$-C(OH)(CCl$_3$)-C$_6$H$_4$-Cl
  1. [O]
  2. -CO$_2$
  3. [O]
→ Cl-C$_6$H$_4$-C(=O)-C$_6$H$_4$-Cl

25 several dehydrohalogenation steps → Cl-C$_6$H$_4$-C(=CH$_2$)-C$_6$H$_4$-Cl
[O] ox. → **26**: Cl-C$_6$H$_4$-CH(COOH)-C$_6$H$_4$-Cl

Figure 12. Metabolism of DDT.

quantitative design has been further advanced by developing expert systems which predict soft analogs using the various approaches described in the preceding sections. These expert systems can select and rank the best soft analogs in order of a combination of the desired physical, chemical and activity (use) properties. The most important components of these expert systems are the application of the specific design rules and the ability to calculate and incorporate the properties that are used as the basis for ranking the soft compounds. For example some of the design rules involve computer identification of metabolizable groups like methyl, hydroxymethyl or other alkyl groups, or the presence of neighboring methylene or hydroxymethylene groups for which isosteric/isoelectronic groups such as esters, reversed esters or others are identified (34). This process can be exemplified by application of the ester type of soft analog and inactive metabolite derivatives on a cannabinoid analog HU-211 (29). Figure 13 shows the structure of HU-211 and its 20 computer-generated isosteric/isoelectronic soft analogs. Properties (i.e., partition coefficient, water solubility, volume, surface area, and dipole) of all these analogs are then calculated and, from these properties, the computer ranks the 20 soft analogs in order of closeness to HU-211 (the lead compound). and selects the soft analogs whose physicochemical properties are the closest to those of HU-211. Of the 20 soft analogs generated for HU-211, the closest ones are structures 30, 31 and 32 (Figure 13).

Similar expert systems could be developed for all kinds of soft chemical design, including soft chemical design of commercial chemicals. The fundamental architecture of these expert systems is represented in Figure 14, and a possible algorithm is shown in Figure 15. Accordingly, a given lead compound is selected for which a safer analog is desired, and the design project is then defined. The purpose, the problem to be solved, special features of the lead compound, and related information would be included in the database. The three dimensional structure is then generated and structurally and conformationally optimized. Next, the specific soft isosteric/isoelectronic transformations are selected and the soft chemical candidates are generated automatically based on the chosen transformations. Each of these soft analogs can be displayed and, if so decided, unreasonable or uninteresting ones can immediately be discarded. Next, detailed calculations of certain molecular descriptors are performed for each of the soft analogs. (Currently, the AM1 semiemperical molecular orbital method is used to calculate these descriptors). These descriptors are then used to calculate the important physicochemical properties (e.g., octanol/water partition coefficient (Log P), water solubility Log W), molecular surface area, molecular volume, ovality, etc.) and electronic properties such as charge distribution and dipole (31,32,33). A separate subroutine based on experimental metabolism of analogous structures stored and used to predict metabolic rates, would then estimate the rate of metabolism and corresponding metabolites. All this information is then stored and used to rank the soft analogs in accordance to closeness to the lead compound, from which the program will provide an order of preferred new soft analogs.

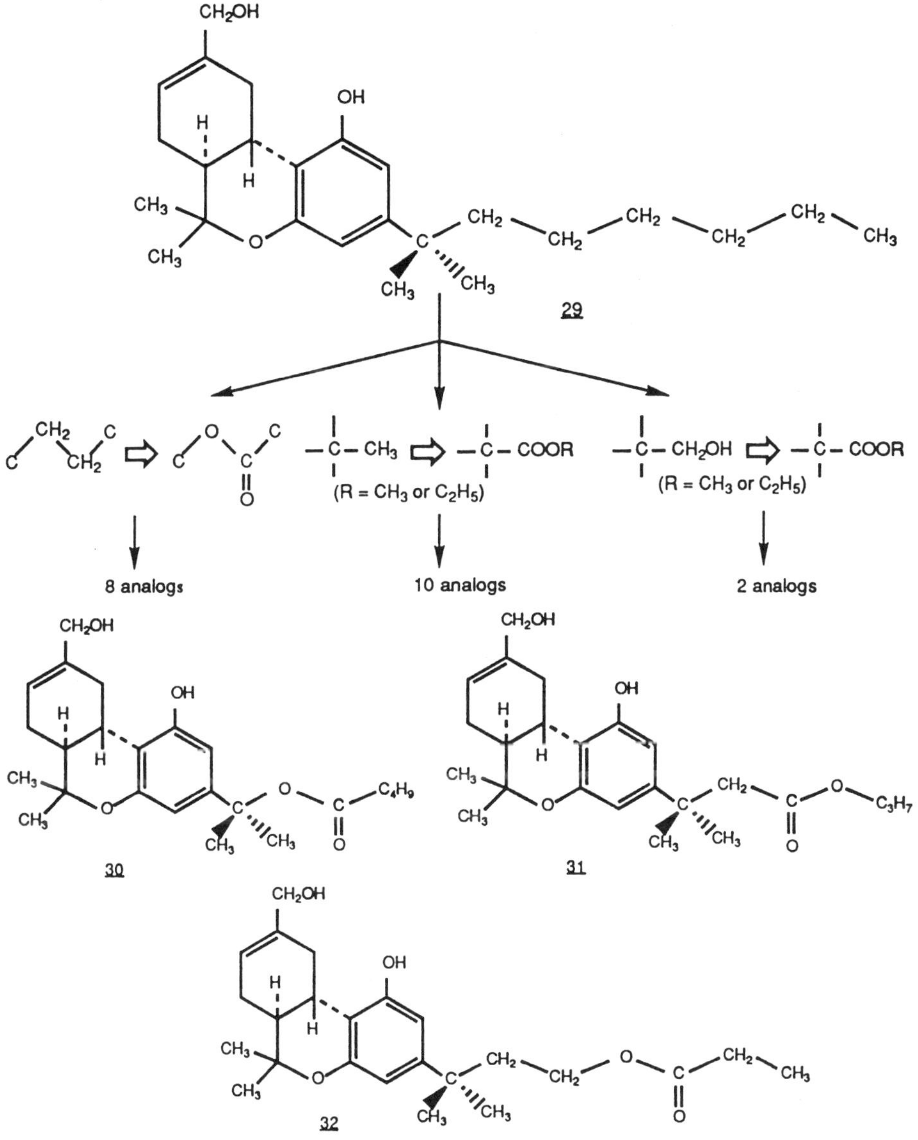

Figure 13. Expert System Designs 20 Soft Analogs of 29 using Three Transformation Rules. Equally weighted calculated contributions of logP, logW, volume, surface and dipole lead to structures 30, 31 and 32 as the closest overall analogs.

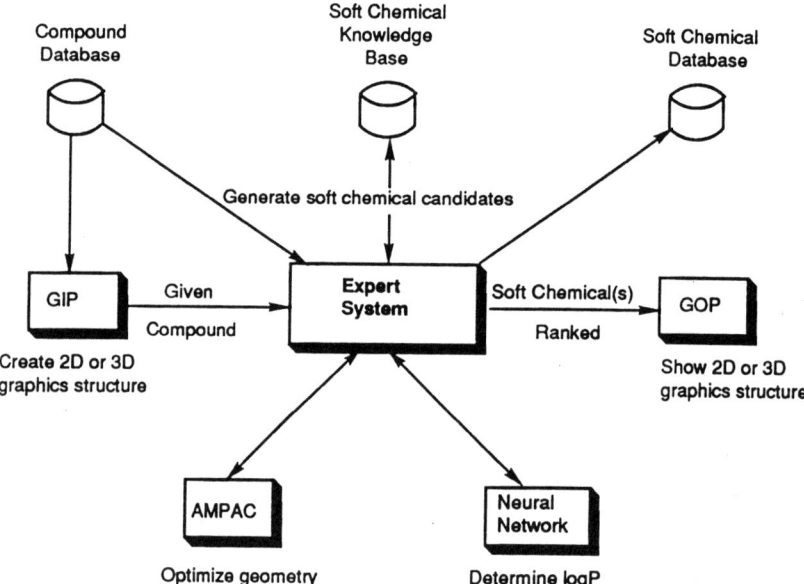

Figure 14. Architecture of the Safe Chemical Design System.

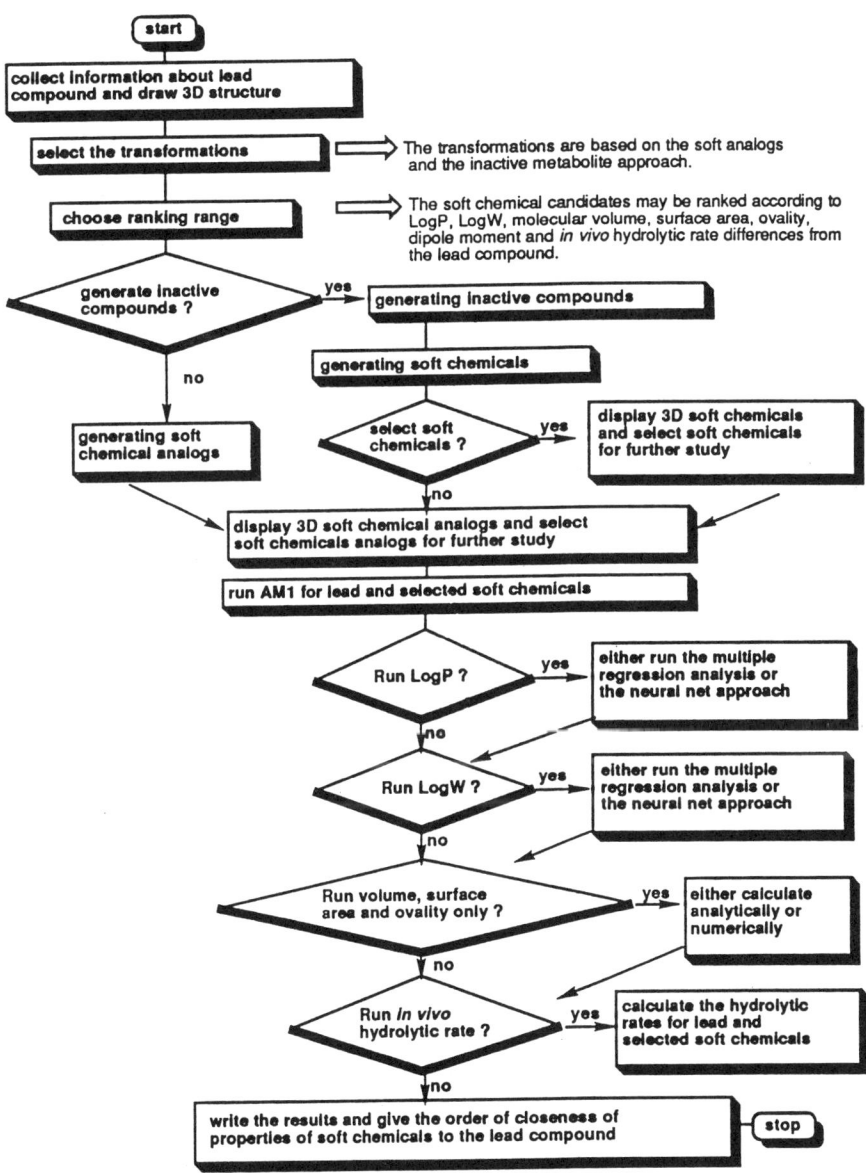

Figure 15. Flow Chart of Soft Chemical Design Expert System.

Finally, the preferred new soft analogs can be synthesized in the order of preference and the best (safest) new soft analog of the lead compound can be used to replace the lead compound.

Conclusions

The concept of retrometabolic-based soft drug design and the approaches therewith were reviewed. Several specific examples were given to provide an understanding of the basis for using retrometabolic drug design for the design of safer chemicals in general. Essentially, the <u>soft chemical design strategies outlined above combines structure-activity relationships with structure metabolism relationships to ultimately design the safety index</u>. Our knowledge of metabolic pathways is now sufficient to predict or evaluate the metabolism of a variety of chemical substances from the point of view of their likelihood of undergoing toxicification or detoxification metabolism. The rules in transformations (e.g., isosterism) applied for designing soft drug analogs could then be extended to the design of soft commercial chemical analogs that are less toxic than the existing commercial substances from which they were designed. One of the main features in the above design process is that there is no need for additional experimental data. The properties of the substances can be reliably calculated and the ranking could provide a very good basis for selecting the least toxic (safest) substances.

The strong emphasis on pollution prevention as a means of controlling the toxics that enter our environment and jeopardize our health has placed new challenges on commercial chemists to design products that are both useful from a commercial standpoint but also safe to human health and the environment. The concept of retrometabolic-based soft drug design needs to be applied to the design of safe commercial chemicals. In doing so, it is quite conceivable that many commercial chemicals currently implicated in environmental and human toxicity could be replaced with equally useful but safer substitutes.

Literature Cited

1. Bodor, N.; Kaminski, J. *J. Med. Chem.* **1980**, *23*, 469-474.
2. Bodor, N.; Kaminski, J. *J. Med. Chem.* **1980**, *23*, 566-569.
3. Bodor, N.; Wood, R.; Raper, C.; Kearney, P.; Kaminski, J. *J. Med. Chem.* **1980**, *23*, 474-480.
4. Mannering, G.J., In *Concepts in Drug Metabolism*; Jenner, P.; Testa, B., Eds.; Marcel Dekker, Inc: New York, 1981.
5. Bodor, N. In *Strategy in Drug Research*; Buisman, J.A.K., Ed.; Elsevier Scientific Publishing Company: Amsterdam, 1982.
6. Bodor, N. *Chemtech* **1984**, 28-38.
7. Bodor, N.; Brewster, M.E. *Pharm. Therapy* **1981**, *19*, 337-386.
8. Bodor, N.; Prokai, L.; Wu, W.; Farag, H.; Jonnalagadda, S.; Kawamura, M.; Simpkins,J. *Science* **1992**, *257*, 1698-1700.

9. Bodor, N. In *Trends in Medicinal Chemistry 90*; Sarel, S.; Mechoulam, R.; Agranat, I., Eds.; Blackwell Scientific Publications: 1992; pp 35-44.
10. Bodor, N. *Med. Res. Reviews* **1984**, 3(4), 449-469.
11. Bodor, N.; Oshiro, Y.; Loftsson, T.; Katovich, M.; Caldwell, W. *Pharm. Res.* **1984**, *3*, 120-125.
12. Bodor, N.; Sloan K.; Little, R.; Selk, S.; Caldwell, L. *Int. J. Pharm.* **1982**, *10*, 307-321.
13. Bodor, N.; Sloan, K. *J. Pharm. Sci.* **1982**, *71(5)*, 514-520.
14. Bodor, N. In *Topical Corticosteroid Therapy: A Novel Approach to Safer Drugs*; Christopers, E., et al., Eds.; Raven Press Ltd.: New York, 1988.
15. Bodor, N.; El-Koussi, A.; Kano, M.; Khalifa, M. *J. Med. Chem.* **1988**, *31*, 1651-1656.
16. Hammer, R.; Amin, K.; Gunes, Z.; Brouillette, G.; Bodor, N. *Drug Des. & Del.* **1988**, *2*, 207-219.
17. Bodor, N. In *Enc. of Human Biol.*: Academic Press: San Diego, 1991.
18. Bodor, N. In *Topical Glucocorticoids with Increased Benefit-Risk Ratio*; Korting, H., Ed.: A. G. Karger: Basel, 1993.
19. Bodor, N. and El-Koussi, A. *Cur. Eye Res.* **1988**, *7(4)*, 369-374.
20. Bodor, N.; Gabanyi, Z.; Wong, C. *J. Am. Chem. Soc.* **1989**, *111*, 3783-3787.
21. Kaminski, J.; Bodor, N.; Higuchi, T. *J. Pharm. Sci.* **1976**, *65*, 553-557.
22. Kaminski, J.; Bodor, N.; Higuchi, T. *J. Pharm. Sci.* **1976**, *65*, 1733-1737.
23. Kaminski, J. Huycke, M.; Selk, S.; Bodor, N.; Higuchi, T. *J. Pharm. Sci.* **1976**, *65*, 1737-1742.
24. Francis, R.J.; East, P.B.; McLaren, S.J.; Harman, J. *Biomed. Mass Spectrom.* **1976**, *3*, 281-285.
25. EPO Pat. 386901/Dec. 12, 1990 by ONO Pharmaceutical Co., Japan.
26. Druzgala, P.; Hochhaus, G.; Bodor, N. *J. Steroid Biochem.* **1991**, *38(2)*, 149-154.
27. Druzgala, P.; Bodor, N. *Steroids* **1991**, *56*, 490-494.
28. Bodor, N.; Oshiro, Y.; Loftsson, T.; Katovich, M.; Caldwell, W. *Pharm. Res.* **1984**, *3*, 120-125.
29. Bodor, N.; Elkoussi, A.; Kano, M.; Khalifa, M. *J. Med. Chem.* **1988**, *31*, 1651-1656.
30. Yang, H.; Wu, W.; and Bodor, N. *Pharm. Res.* **1995**, *12(3)*, 329-336.
31. Bodor, N.; Gabanyi, Z.; Wong. C. *J. Am. Chem. Soc.* **1989**, *111*, 3783-3786.
32. Bodor, N.; Huang, M. *J. Pharm. Sci.* **1992**, *81(3)*, 272-281.
33. Bodor, N.; Harget, A.; Huang, M. *J. Am. Chem. Soc.* **1991**, *113*, 9480-9483.
34. Bodor, N.; Huang, M. In *Computer-Aided Molecular Design*; Reynolds, C.H.; Holloway, M.K.; Cox, H.K., Eds.; ACS Symposium Series 589; American Chemical Society: Washington, DC, 1994; pp. 98-113.

Chapter 6

Predicting Rates of Cytochrome-P450-Mediated Bioactivation and Its Application to the Design of Safer Chemicals

Jeffrey P. Jones

Department of Pharmacology, University of Rochester, Rochester, NY 14642

A computational model is presented that can be used as a tool for predicting rates of cytochrome P450 mediated hydrogen atom abstraction. This model has been used to develop other models that predict the toxicity of nitriles and of hydrochlorofluorocarbons (HCFCs). Of particular note are the excellent correlations found for the *in vitro* metabolism of six halogenated alkanes by both rat and human hepatic microsomal enzyme preparations. Good correlations were also obtained for a set of inhalation anesthetics for both *in vivo* and *in vitro* metabolism by humans. These are the first *in vivo* human metabolic rates to be quantitatively predicted. This paper discusses how an understanding of toxic mechanisms involving cytochrome P450 bioactivation and the above computational model can be used for the design of safer chemical substances.

As a society we have learned from experience that many useful commercial (industrial) chemical substances are toxic to humans and the environment. The toxicity of these substances and their contribution to overall pollution is at least partly attributable to the fact that their potential for producing adverse health and environmental effects was not considered during their design. In cases where toxicity was realized during the design phase, the possibility of incorporating structural modifications that reduce toxicity was generally not recognized. There has emerged in recent years a new environmental paradigm known as pollution prevention. The premise of the pollution prevention paradigm is that the best way to control pollution (and therewith the hazards and risks associated with it) is to avoid creating it in the first place. Chapter 1 of this book discusses the importance of designing chemicals such that they are commercially useful but not toxic as a fundamental means of preventing pollution. Chapter 2 provides an overview of several different approaches that chemists can use for the rational design of safer chemical substances, and stresses that an understanding of toxic mechanisms, as a basis for deducing and incorporating structural modifications that reduce toxicity, is the most effective approach. The present chapter exemplifies how an understanding of toxic mechanisms involving cytochrome P450 (CYP) enzymatic bioactivation can be used for designing safer chemicals, by describing a computational model that can predict the relative likelihood of hydrogen atom abstraction (metabolism) by the CYP enzymes.

Bioactivation Reactions

Humans and most other mammalian life-forms have enzyme-mediated processes for converting exogenous substances into more water soluble, easier to excrete substances. These processes are generally referred to as metabolism or biotransformation. The overall purpose of metabolism is detoxication: a defense mechanism to convert potentially toxic chemical substances to other substances (metabolites) that are readily excreted. The chemical reactions involved in the metabolism of chemical substances foreign to the body can be classified as either phase-I or phase-II reactions. Phase-I reactions introduce or unveil a polar functional group such as hydroxyl (-OH) or carboxyl (-COOH) groups to increase water solubility. Phase-II reactions involve coupling (conjugation) of the chemical substance or a metabolite thereof with either glucuronate, sulfate, acetate, or an amino acid, which further increases water solubility and promotes rapid excretion.

Metabolism of certain chemical substances does not result in detoxication. For certain chemical substances, metabolism yields products that are toxic. Over the past several decades research efforts have elucidated biochemical mechanisms of toxicity of specific chemicals or chemical classes. It is becoming increasingly apparent that the toxicity of many substances is not due to the substances themselves, but rather as a result of their metabolism to toxic metabolites. This phenomenon, the conversion of a chemical substance into a toxic metabolite, is known as bioactivation *(1)*. While a number of enzymes systems such as glutathione transferase *(1,2)* sulfotransferase *(3)* and UDP-glucuronosyltransferase *(4)* can mediate bioactivation reactions, the CYP superfamily of enzymes *(5)* has been postulated to be responsible for the majority of bioactivation reactions of chemical substances *(6)*. The CYP superfamily of enzymes can activate chemical substances to toxic compounds by a plethora of reactions including epoxidation, N-hydroxylation, S-oxidation and aliphatic hydroxylation. A few such reactions are shown in Scheme 1.

Reaction 1 in Scheme 1, depicts the oxidation of an aromatic compound (e.g. benzene) to an epoxide. This reaction occurs through initial formation of a tetrahedral intermediate as was shown by Korzekwa et al. *(7)* in an elegant study using chlorobenzene as the substrate. The tetrahedral intermediate may then form an epoxide. The epoxide can react covalently with endogenous nucleophiles such as DNA or proteins, which results in toxicity. It should be noted that epoxides are not obligatory intermediates and the tetrahedral intermediate can form phenol directly.

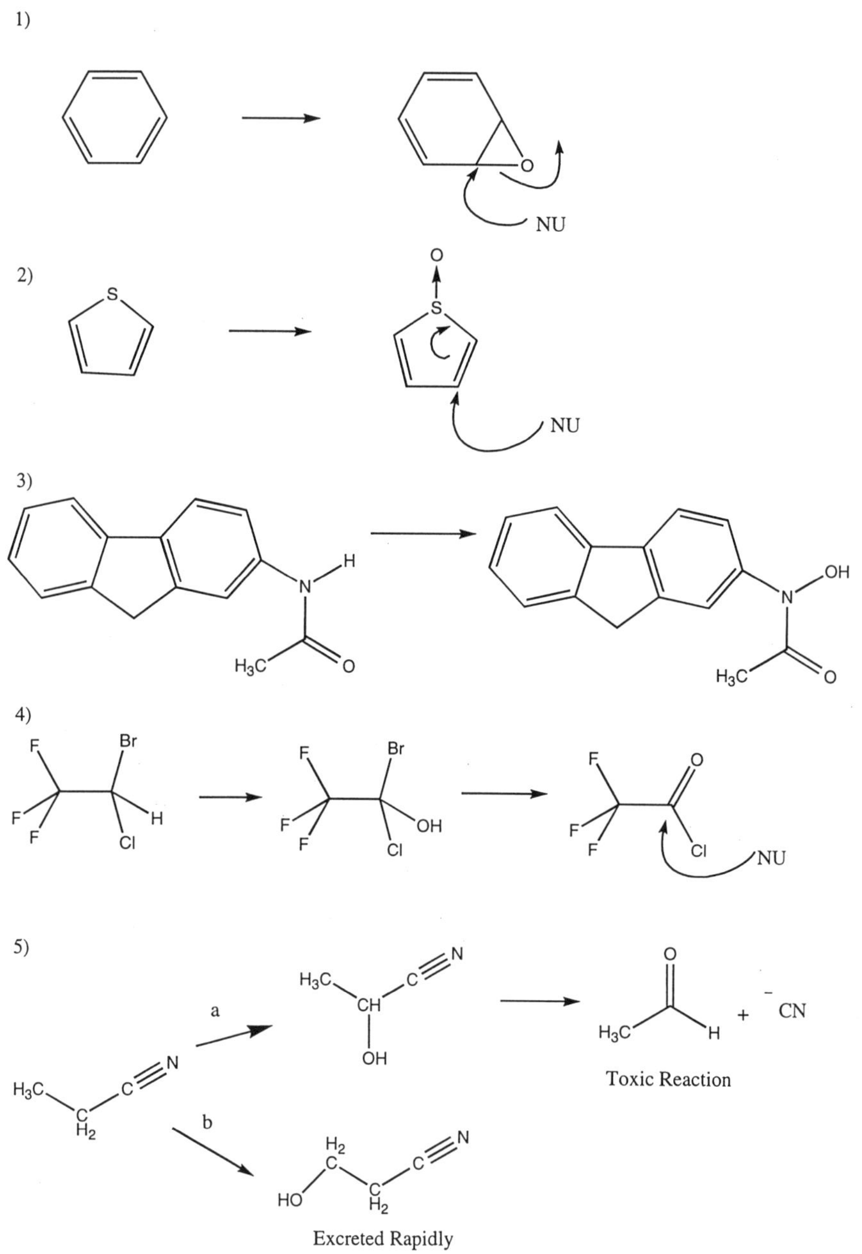

Scheme 1. Selected Examples of Cytochrome P450-Mediated Bioactivation Reactions (Nu = endogenous nucleophile).

The products of aromatic oxidation can also undergo further oxidation to form quinones. Quinones can also alkylate tissue nucleophiles or they can damage the cell through oxidative stress *(8)*.

Reaction 2 in Scheme 1 depicts the oxidation of thiophene to an alpha,beta-unsaturated sulfoxide metabolite. This metabolite can act as a Michael-acceptor, and undergo addition reactions with tissue nucleophiles. In fact, thiophene derivatives have been shown to bind to proteins and inactivate CYP enzymes *(9)*. An alternative bioactivation pathway for thiophene not shown in Scheme 1 is epoxidation of either double bond. The epoxide metabolite can also bind covalently to tissue nucleophiles *(9)*.

Acetylaminofluorene has been found to be carcinogenic *(10,11)*. The carcinogenicity of this substance is believed to involve N-hydroxylation (Scheme 1, reaction 3) *(10)*. After N-hydroxylation, a conjugation reaction occurs to form sulfate or acetyl group conjugates with the N-hydroxy group *(11)*. These conjugates are believed to be the ultimate carcinogenic metabolite, in that the sulfate or acetyl moieties make the nitrogen highly electrophilic and capable of binding to tissue nucleophiles such as DNA.

The oxidation of halothane to a halohydrin is shown in reaction 4 of Scheme 1. The halohydrin spontaneously decomposes to give trifluoroacetylchloride (an acyl halide) which then partitions between a reaction with water to form trifluoroacetic acid, and tissue nucleophiles to form acylated adducts. Halothane has been shown to induce hepatitis in humans through an antibody mediated response to trifluoroacetylated proteins *(12)*.

The acute toxicity of nitriles is due to their CYP-mediated hydroxylation at the carbon atom alpha (α) to the cyano group to form a cyanohydrin intermediate, which spontaneously decomposes to release cyanide and the corresponding carbonyl product (Scheme 1, reaction 5a) *(13)*. Hydroxylation at positions other than the alpha carbon result in the formation of nontoxic metabolites (Scheme 1 reaction 5b), and represents detoxication.

The few examples provided in Scheme 1 demonstrate that the CYP enzymes are involved with the bioactivation of a wide variety of substances. Many more examples are available *(1)*. Of the reactions shown in Scheme 1, reactions 3, 4 and 5 are likely to occur by hydrogen atom abstraction. This is an important point in that in that the relative toxicity of substances analogous to those shown in reactions 3, 4 and 5 (Scheme 1) is largely dependent on the relative rates of the CYP-mediated hydrogen atom abstraction critical for toxicity. With nitriles, for example, the hydrogen atom most critical for acute toxicity is the hydrogen(s) alpha to the cyano group. The easier (or faster) this hydrogen atom is abstracted, the more quickly cyanide is released and, consequently, the more toxic the nitrile will be. If, on the other hand, other hydrogen atoms are present that are easier for CYP to abstract, these hydrogen atoms are expected to be removed preferentially over the alpha hydrogens and, thus, this nitrile will be less acutely toxic. Thus, for a given class of substances whose toxicity is dependent on CYP-mediated abstraction of a particular hydrogen atom, the relative toxicity of the substances comprising the class is a function of the relative ease in which this hydrogen is abstracted. For such substances, the ability to quantitatively estimate the relative rates of CYP-mediated hydrogen atom abstraction should enable one to estimate the relative toxicity of an untested substance belonging to the class, such as a new or planned substance. In addition, the ability to quantitatively predict rates of CYP-mediated hydrogen atom abstraction better enables the chemist to design less toxic substances of the class in that molecular modifications that redirect metabolism away from bioactivation and direct it towards detoxication can be more easily deduced.

In this chapter a model for predicting the rates of hydrogen atom abstraction will be reviewed with respect to two classes of compounds, haloalkanes and nitriles (reactions 4 and 5, respectively, in Scheme 1). Results will be presented which validate this model for predicting the rates of metabolism and potential toxicity of these two classes of compounds. The kinetics and enzymology that make such predictions possible will be briefly discussed, and discussion of how this model can help in the design of safer haloalkanes and nitriles will be presented.

CYP Enzymology

The CYP superfamily is diverse in terms of subcellular location, as well as location in the organism and the type of organisms that express the enzymes. One of the most studied CYP enzymes is CYP101 (P450cam) which is isolated from the soil bacterium *psuedomonas putida (14,15)*. This enzyme allows this bacterium to utilize camphor as the sole carbon source. CYP101 is cytosolic, and many crystal structures have been obtained for this enzyme with various ligands bound in the active site *(14,16)*. Many other bacteria also express at least one CYP enzyme, and CYP enzymes have also been isolated from plants *(5)*. Wackett and co-workers have suggested that nonmammalian sources of CYP enzymes such as those mentioned above may be potentially useful in bioremediation of hazardous waste sites *(17)*. The mammalian CYP enzymes, unlike most bacterial enzymes, are membrane bound and are isolated either from the endoplasmic reticulum or the mitochondria. However, they are not integral membrane proteins, but instead appear to be tethered to the membrane by the amino terminus of the protein *(18)*. In most mammalian organisms, CYP enzymes are expressed at different levels in different organs, with the liver being the most abundant source of many CYP proteins. Other sites of high concentrations of CYP correspond to other sites of entry of xenobiotics (foreign chemical substances) into the body, including the lung, skin and nasal passages. While CYP is responsible for a large part of mammalian xenobiotic metabolism, it is also an important enzyme in the biosynthesis and degradation of endogenous compounds such as steroids and eicosanoids *(19)*. Most CYP enzymes require a second enzyme to function. The second enzyme provides reducing equivalents and usually is P450 reductase, however other reductases can be important for the different CYP enzymes, particularly the bacterial and mitochondrial enzymes.

Cytochrome P450 enzymes have a complex catalytic cycle that is shown diagramatically in Scheme 2. In this Scheme the metabolism of an alkane (RH) is demonstrated *(20)*. The initial event (Scheme 2, step 1) is binding of substrate (RH) to the enzyme. Binding is nonspecific in the sense that most P450 substrates are nonpolar and are often metabolized at several apparently rapidly interchangable sites. The binding of substrate changes the reduction potential of CYP so that step 2, which is a one-electron reduction of the substrate catalyzed by cytochrome P450 reductase, can proceed. Step 3 is the binding of the CYP-substrate complex with molecular oxygen. At this step carbon monoxide can bind which gives the characteristic 450 nM absorption spectra for which CYP was named. Step 4 is a second one-electron

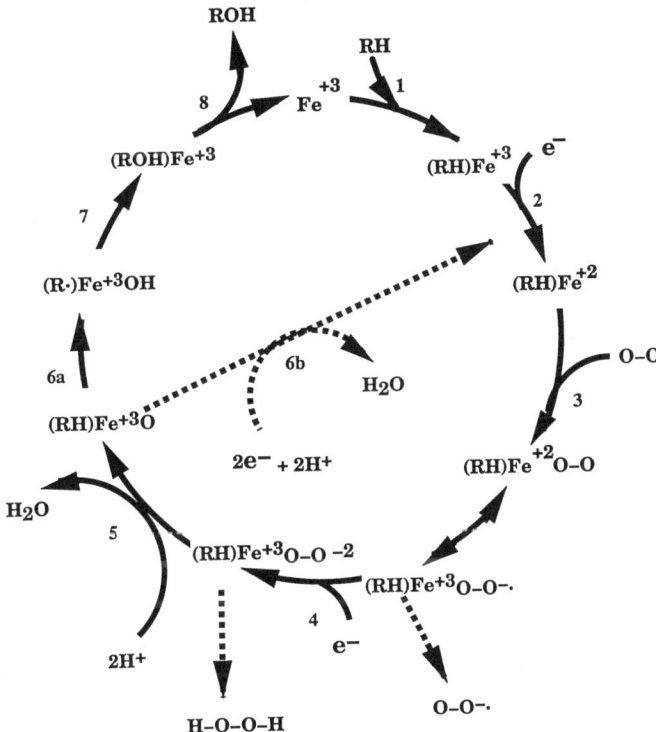

Scheme 2. Catalytic Cycle of Cytochrome P450-Mediated Hydrogen Atom Abstraction.

oxidation by either P450 reductase or cytochrome b5. At this stage hydrogen peroxide can be released, which uncouples electron flow from substrate oxidation. The catalytic steps (steps 1-5 in Scheme 2) result in the heterolytic cleavage of the hydroperoxide intermediate to produce the reactive monooxygen species. The transition state for this catalytic step is likely to be stabilized by proton donation from a hydroxy-containing amino acid (either threonine or serine) in the active site. The heterolytic cleavage is thought to be mediated by a proton relay mechanism from a glutamate to threonine to the peroxy intermediate based on solvent isotope effect experiments *(21)*.

The next step (step 6a in Scheme 2) in the catalytic cycle for alkanes (and also haloalkanes, ethers, and amino compounds) is hydrogen atom abstraction *(22,23,24)*. In step 6a the electron deficient active oxygen species can abstract a hydrogen atom from a hydrocarbon to form a carbon radical. However, if the hydrocarbon is difficult to oxidize, another two electron reduction of the active oxygen species can occur to generate water (step 6b). Step 6a is the step that must be modeled to predict the effect of structure on the rate of CYP mediated reactions. This step is not catalyzed by the enzyme since no evidence exists for transition state stabilization of the hydrogen atom abstraction step, and in fact, this process can be mimicked exactly with a chemical system *(22)*. Studies by Karki et al., *(22)* indicate that *t*-butoxy radical is a good model for CYP mediated reaction of substituted N,N-dimethylanilines. These studies reveal that the isotope effects observed for CYP and *t*-butoxy radical are identical, indicating that the potential energy surfaces for these two reactions are extremely similar. This is to be expected since it was predicted by the results of Jones et al., *(23)* that the active oxygen species is isoenergetic with the primary carbon radical of octane, and octane has a bond dissociation energy close to *t*-butanol *(25)*. The close biomimetic nature of *t*-butoxy radical argues against any transition state stabilization by the enzyme. Furthermore this step has also been shown to have extremely similar potential energy surfaces for all of the enzymes that have been studied. This includes human, rodent and even bacterial enzymes *(22,23)*. If the hydrogen atom abstraction step was catalyzed by the enzyme, one would expect to observe differences in potential energy surfaces due to differences in protein-substrate interactions for these proteins because of their very different primary sequence. Thus, it is more appropriate to discuss CYP *mediated* hydrogen atom abstractions, as opposed to CYP catalyzed hydrogen atom abstractions.

Step 7 involves the rapid recombination of the hydroxy radical equivalent with the carbon based radical to give an alcohol. Step 8 is release of the metabolite (hydroxylated substrate) and return of the enzyme to its resting state.

The fact that the CYP family has a conserved mechanism and can be modeled as a simple chemical process has important implications in the computational prediction of the rates of CYP mediated reactions. This means that: 1) apoprotein stabilization of the transition state need not be modeled; 2) any apoprotein effects will only be steric in nature, and thus easier to model; and 3) a single computational model can predict rates for all isoenzymes of the CYP superfamily.

Development of a CYP Model

To model the CYP active oxygen species it was decided to find a small molecule surrogate for the iron-protoporphyrin IX oxygen complex *(26)*. The heme-prosthetic group of CYP was not modeled due to its complex electronic nature. The following three criteria were established to guide model development: 1) the model must contain an oxygen and be isoenergetic with the C-1 radical of octane; 2) a linear free energy relationship must exist, so that rates can be predicted based on ground state properties alone; and 3) the model must reproduce known experimental results.

The first criterion, that the active oxygen surrogate mimic the reactivity of a primary carbon based radical, led to the exploration of the reactivity of a number of different oxygen containing compounds. All of the calculations were performed with the AM1 semiempirical Hamiltonian. Of the compounds studied para-nitrosophenoxy radical (PNR) proved to be the closest in energy to the primary carbon radical. Furthermore, this compound mimicked the isotope effects observed for hydrogen atom abstraction by CYP. This criteria is believed to be important since a symmetrical reaction coordinate should be the most sensitive to changes in electronic structure, based on the Hammond postulate *(27)*. If the reaction modeled was very exothermic, small differences in activation energies may be obscured.

The second criterion was invoked since the isolation and characterization of a transition state by computational means can be very tedious. It was believed that a model would be more useful if a linear free energy relationship could be established. To determine if the activation energy could be predicted from ground state properties a plot of the activation energy versus the enthalpy of reaction was made, and a respectable correlation was observed: r^2 (the square of the regression coeficient) was found to be 0.73 *(26)*. It was noticed that the outliers were mostly compounds that could resonance stabilize or destabilize the transition state. This led to the use of modified Swain-Lupton resonance parameters in the correlation *(26)*. While the correlation improved ($r^2 = 0.97$), these parameters were not available for every compound of interest and certainly would not be available for new chemicals. Thus, a search was undertaken for computational methods that would correct for resonance effects. After fruitless efforts to correlate a number of parameters it was observed that ionization potentials of the intermediate radical, as determined by Koopman's theorem, gave a very good correlation ($r^2 = 0.94$) when included in the model. This PNR computational model can be used to predict activation energies from ground state energies using the established free energy relationship shown in equation 1, where $\Delta H_{act.}$ is the activation energy, $\Delta H_{reac.}$ is the heat of reaction and IP is the ionization potential of the intermediate carbon radical. Thus, this model depends only on very simple ground state energies to predict the rates of reaction *(26)*.

$$\Delta H_{act.} = 2.60 + 0.22\,(\Delta H_{reac.}) + 2.38\,(IP) \qquad (1)$$

With this model prediction of activation energies can be based on ground state properties alone. This means that activation energies can be calculated very rapidly for even relatively large molecules. Furthermore, it eliminated some of the uncertainty associated with the characterization of transition states.

Criteria 3 proved difficult to explore. A search of the literature revealed that no systematic study had been conducted on the differences in reactivity of different function groups across the broad spectrum of compounds oxidized by CYP. While computational results *(26)* paralleled the rank order reactivity assumed by most researchers in the field, no experimental data to confirm these assumptions were found. At present, controlled experiments on the regioselectivty of reaction of a number of compounds are now underway to rectify this situation.

One study, however, was published on the reactivity of a small group of closely related para substituted toluenes *(28)*. In this study, the reaction rates for CYP2B4 (a rabbit liver enzyme) mediated hydroxylations of 8 toluenes were determined. The investigators of this study describe the rate of reaction by using Hammett sigma values and the log of the octanol/water partition coefficient (expressed as log P) values of these substances. The correlation with sigma indicates that the reaction rate is partially under control of electronic factors. Thus, the enthalpy of activation for each of the toluenes using the PNR model (equation 1) were determined. A plot of the natural logarithm of the reaction rates versus the enthalpy of reaction gave a weak correlation. However, when log P is included as descriptor in the correlation better agreement between the predicted and experimental results were obtained (Higgins and Jones, unpublished results). A plot of the predicted versus actual rate is given in Figure 1. Thus, this provides evidence that the electronic factors that control the rate of reaction can be modeled by the PNR model for CYP mediated reactions.

Predicting Rates of CYP-Mediated Oxidation with the PNR Model

Nitriles. Nitriles are important in the manufacture of pharmaceuticals, polymers, plastics, dyestuffs, vitamins, resins, and fibers. Hence, occupational exposure to these compound can be significant. Many nitriles are quite acutely toxic (lethal). The acute toxicity of nitriles is dependent upon the extent of CYP-mediated hydroxylation at the alpha carbon, which results in cyanide liberation (Scheme 1, reaction 5a) and death *(13,29)*. Nitriles which contain structural features that favor CYP-mediated hydroxylation at the alpha carbon are generally highly toxic, whereas nitriles that contain structural features (e.g., hydroxy groups at non-alpha positions, non-alpha tertiary carbons) that favor hydroxylation at other positions tend to be considerably less toxic *(29)*. A model to predict the acute lethality (expressed as the acute median lethal dose or LD_{50}) of nitriles was reported *(30)*. This model uses log P as the only descriptor of acute lethality, and was found to poorly correlate ($r^2 = 0.35$) acute lethality with log P when applied to a larger dataset of nitriles *(29)*.

Using the same dataset used in ref. *29*, another model to predict the acute lethality of nitriles was published by Korzekwa and coworkers *(31)*. This model, also uses log P as a descriptor but includes an additional descriptor ($k_{\alpha corr}$) that represents CYP-mediated hydrogen atom abstraction (oxidation) of the alpha carbon, statistically corrected for hydrogen atom abstraction at other carbon positions, and gave a significantly better correlation. (The $k_{\alpha corr}$ descriptor was based on the PNR model.) In developing this model, it was assumed (based on the results of the publications cited in refs. *29*) that oxidation (hydroxylation) of the carbon alpha (adjacent) to the nitrile moiety represents bioactivation and leads to toxicity, and that oxidation at other carbon atoms would lead to detoxification of the parent nitrile.

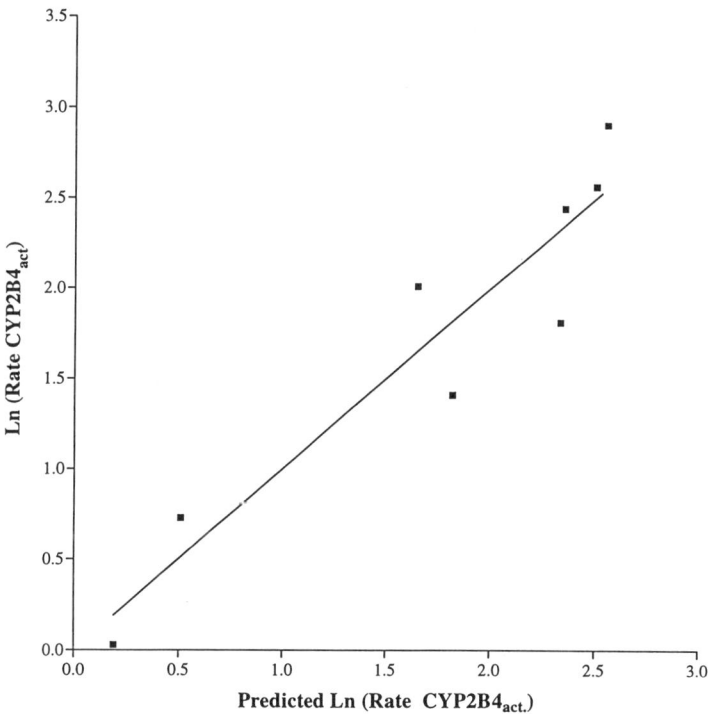

Figure 1. Regression of actual rates of metabolism of eight substituted toluenes to rates predicted using log P values and activation energies (ΔH_{act}) predicted by equation 1.

(These two pathways are respectively shown in reaction 5a and 5b of Scheme 1). Thus, the restrictions on this model are: 1) that the acute lethality of a particular nitrile is a function of its ability to release cyanide; and 2) nitriles contain at least one hydrogen atom on the carbon atom alpha to the nitrile moiety. Calculation of CYP-mediated hydrogen atom abstraction at the alpha carbon pseudo rate constant ($k_{\alpha corr}$) was accomplished by first calculating the activation energy, using equation 1. A rate constant for cyanide release was then obtained using the Arrhenius equation. This rate constant was corrected based on the relative rates of reaction at other sites in the molecule and also corrected for the number of hydrogen atoms present at each position in the molecule, to a give $k_{\alpha corr}$ value. This procedure was performed for each nitrile in the dataset. The LD_{50} values of the 26 nitriles comprising the dataset were correlated with the corresponding $k_{\alpha corr}$ and estimated log P values *(32)* using multivariate regression analysis *(31)*. The model is represented by equation 2, where n is the number of compounds used and r^2 is the square of the correlation coefficient. This model gave a much better correlation ($r^2 = 0.72$) than the model that uses only log P as a descriptor.

$$\log (1/LD_{50}) = - 0.16 \ (\log P)^2 + 0.22 \log P + 0.11(\ln k_{\alpha corr}) + 5.67 \qquad (2)$$

$$n = 26, \ r^2 = 0.72$$

The improved predictive quality of this model is clearly attributable to the $k_{\alpha corr}$ term, which is a descriptor of the likelihood of CYP oxidation of the alpha carbon of a nitrile relative to the other carbon atoms of the same nitrile. This model can be used as a tool to estimate the toxicities of nitriles under consideration for synthesis and development, and to evaluate the effects of structural features deliberately added to reduce toxicity *(29)*. The design of safer nitriles based on structure-activity relationships and mechanisms of toxicity is covered in great detail elsewhere in this book (see the chapter entitled Designing Safer Nitriles, by DeVito) and a more detailed discussion of the development of this model and its application to the design of safer nitriles is included.

Halogenated Hydrocarbons. Halogenated hydrocarbons of the type shown in reaction 4 of Scheme 1 are commercially important substances used for a variety of purposes that include refrigerants, blowing agents, solvents, and inhalation anesthetics. The general population as well as workers in industrial settings may be exposed to these substances. As shown in reaction 4 of Scheme 1, halogenated hydrocarbons that contain a hydrogen atom and two halogen atoms on a terminal carbon can undergo CYP metabolism to yield acid halide metabolites, which react with water (to form a carboxylic acid) or a host of endogenous nucleophiles, and cause a variety of toxic effects *(1)*. To form the reactive species shown in reaction 4 of Scheme 1, CYP enzymes must first abstract a hydrogen atom from the halogenated hydrocarbon. The toxicity of halogenated hydrocarbons from this mechanism is a function of how quickly the acid halide metabolites are formed. Knowledge of how quickly this bioactivation occurs for untested halogenated hydrocarbons can provide insight into the relative toxicity of the substances. In addition, knowledge of how quickly this bioactivation occurs can enable one to design halogenated hydrocarbons that will not be metabolized to acid halides, or will metabolize to acid halides extremely slowly and thus will be less toxic than those that are bioactivated more quickly. The PNR model has been used to predict the rate of hydrogen atom abstraction from a number of halogenated hydrocarbons, and the model has proven to successfully predict the rates of biotransformation *in vivo* and *in vitro* in both rats and humans *(33a)*. The results of these efforts are outlined briefly below.

To determine if the rates of CYP mediated biotransformations of the halogenated hydrocarbons can be predicted using the PNR model, the methodology outlined above (equation 1) was used to predict the activation energy for each compound. Trihaloacetic acid metabolites were measured using NMR, GC/MS or by using a fluoride specific electrode. *In vitro* rates were predicted in rat microsomes and in expressed human enzymes preparations (CYP2E1). The natural logarithm (ln) of the rates of trihaloacetic acid formation *in vitro* in each enzyme system was correlated with the PNR model predicted activation energies for six substances (HCFC-121, 122, 123, 124, 125 and halothane). For the rat microsomal enzyme preparations an excellent linear correlation ($r^2 = 0.86$) with predicted activation energies (ΔH_{act}) was found, and is represented by equation 3 and shown in Figure 2. The line obtained for the correlation of the expressed human CYP2E1 enzyme with the PNR model predicted activation energies of the same six substances is shown in Figure 3 ($r^2 = 0.97$) and represented by equation 4.

$$\ln (\text{rate microsomes}_{act}) = 44.99 - 1.79 \ (\Delta H_{act}) \qquad (3)$$

$$n = 6, \quad r^2 = 0.86$$

$$\ln (\text{rate CYP2E1}_{act}) = 46.99 - 1.77 \ (\Delta H_{act}) \qquad (4)$$

$$n = 6, \quad r^2 = 0.97$$

These results indicate that the PNR model is an excellent predictive model for CYP mediated *in vitro* oxidation of these halogenated hydrocarbons. Also, the slopes of the lines for the two enzymes systems are very similar, and are consistent with a similar kinetic mechanism and reaction energetics for the biotransformation reactions mediated by the rat and human CYP enzymes.

The rates of *in vivo* trifluoroacetic acid formation were determined in rats *(33)*. The amount of acid formed was determined for four compounds by comparison of the ^{19}F NMR signal with internal standard. A very good correlation ($r^2 = 0.9$) was observed for these four compounds when the natural logarithm (ln) of the rate was correlated with the activation energy.

In addition to the *in vivo* rat data collected for the halogenated hydrocarbons, a number of *in vivo* data points for halogenated hydrocarbon metabolism have been collected in humans. This dataset is comprised of the peak plasma fluoride concentrations for five inhalation anesthetics (enflurane, sevoflurane, desflurane, methoxyflurane, and isoflurane) *(33a)*. In humans, these inhalation anesthetics are intially biotransformed by CYP to give fluoride ion *(33a)*. An excessive amount of fluoride ion causes kidney damage *(33a)*. The activation energies of these anesthetics were calculated as described above using the PNR model. The direct

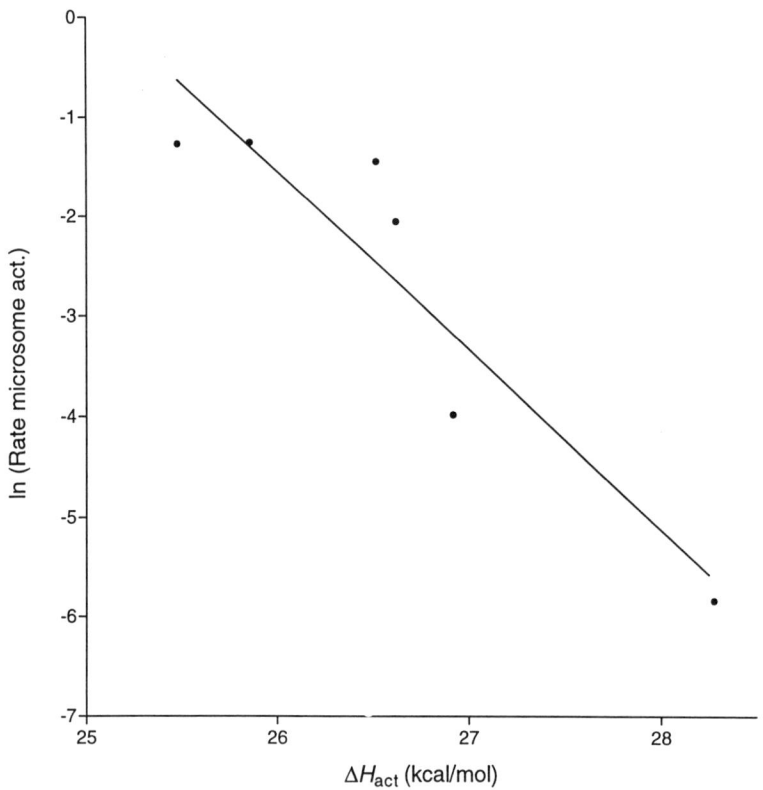

Figure 2. Regression of *in vitro* rates of metabolism (microsomal activation) in rat liver microsomes of six HCFCs using activation energies (ΔH_{act}) predicted by equation 1.

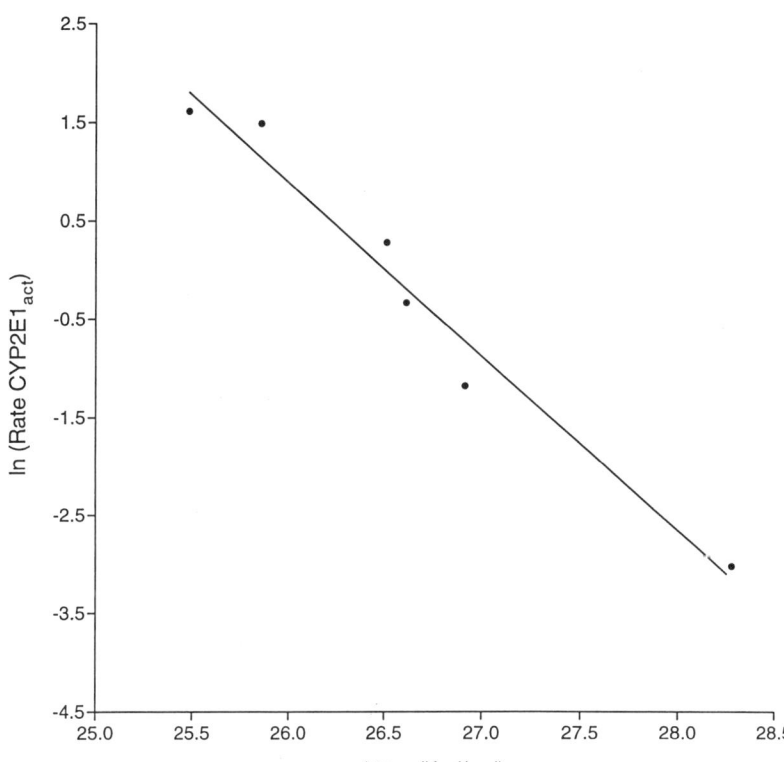

Figure 3. Regression of *in vitro* rates of metabolism in human expressed CYP2E1 of six HCFCs using activation energies ($\Delta H_{act.}$) predicted by equation 1.

correlation of activation energy with peak plasma fluoride concentrations is represented by equation 5 ($r^2 = 0.86$) and is shown in Figure 4. Because the peak plasma fluoride levels correlate to a high degree with the amount of fluoride released from *in vitro* incubations with human liver microsomes *(34)*, the computational model can also predict the in vitro metabolic rates of this group of five compounds.

$$\ln [F]peak = 42.87 - 1.57(\Delta H_{act.}) \qquad (5)$$

$$n = 5, \quad r^2 = 0.86$$

This appears to be the first *in vivo* metabolic rate to be quantitatively predicted for humans. Furthermore, this is one of the first examples where *in computero, in vivo* and *in vitro* data have been shown to agree. It should be noted that the doses of inhalation halogenated hydrocarbon anesthetic administered to patients are high enough to saturate the CYP enzymes responsible for metabolism, and as such the peak plasma fluoride levels are closely related to maximum velocity (V_{max}) rates.

Although the sample sizes used in these datasets are small, the excellent correlations obtained using the PNR model provide some validation that the PNR model is useful as a predictive method for establishing rates of biotransformation of halogenated hydrocarbons and provide a rigorous quantitative-structure-metabolism relationship. This relationship can be used to help in the design of safer halogenated hydrocarbons, based on the predicted amount of metabolism to potentially toxic intermediates. For example if two compounds are under consideration for development as a new solvent, chlorofluorocarbon replacement, or inhalation anesthetic, and have similar physical properties such as boiling point or lipophilicity (log P), the compound with the lowest level of predicted metabolism would be the logical choice for development.

When the PNR Model will Fail

While the PNR model has been validated to be successful in predicting the rates of halocarbon and nitrile metabolism, one of the most important pieces of knowledge that can be obtained about any predictive model is when it will fail to predict the correct result. While a number of reasons can be proposed for a potential failure of the PNR model (such as those described below), the most likely reason is when steric factors play a major role in the determination of the rates of reactions. Because the PNR model only predicts the electronic components of cytochrome P450 hydrogen atom abstraction, any steric interactions that affect regioselectivity can also affect the predictive quality of the model. The best examples of this phenomenon are when stereoselective product formation occurs from a prochiral substrate. The electronic factors in an achiral environment are by definition the same, however numerous examples of prochiral stereoselectivty product formation have been documented for P450 mediated reactions. For example, the products formed from hydroxylation of the prochiral compound cumene are chiral and the product distribution differs for different CYP enzymes *(35)*.

While electronic models such as the PNR model will not predict enzyme mediated steric effects, some progress has been made in predicting these type of effects using molecular dynamics. Stereoselectivity is dependent on the tertiary

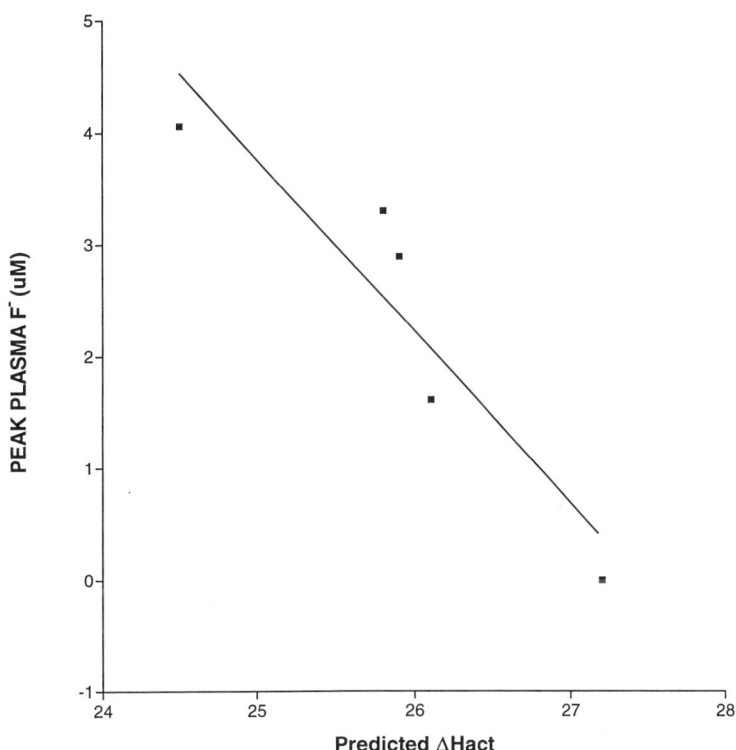

Figure 4. Correlation of peak plasma fluoride levels from in vivo metabolism (in humans) of five inhalation anesthetics with activation energies ($\Delta H_{act.}$) predicted by equation 1.

structure of the enzyme. This means that in order to be able to predict the binding of a chemical substance to the active site of an enzyme we must have knowledge of the structure of the active site of the enzyme. For the CYP family the only known tertiary structures are for the bacterial CYPs *(16,36,37)*. Thus, any structural factors important in mammalian CYPs can only be obtained by comparison of mammalian and bacterial enzymes. One can compare the metabolic profiles of the bacterial and mammalian isoforms to determine if they give the same results. This approach allows us to establish which bacterial crystal structure should be used to model the tertiary interactions of the mammalian isoforms. Until very recently it was believed that these enzymes only metabolized the natural substrates camphor (P450cam), fatty acids (P450BM3) and terpine (P450terp) and some closely related substrate analogs. The first study to be published on a non-natural substrate-P450cam interaction was a study of styrene epoxidation *(38)*. Since then a group of diverse compounds that include styrene *(38)*, ethylbenzene *(39)*, [R]- and [S]-nicotine *(40)*, benzo[a]pyrene *(41)*, para-substituted thioanisoles *(42)* and valproic acid *(43)* have been identified as P450cam substrates. Thus, while these models are all based on bacterial systems, it has been established that steric factors can be successfully predicted using existing computational methodology. Therefore, while at this time it is impossible to predict with certainty the steric effects on rates of CYP-mediated metabolism it would appear that a PNR model that considered the steric and electronic effects involved in CYP metabolism could be developed in the future. This will require knowledge of the structure of each of the important human CYP isoforms.

The current PNR modeling method will also fail to predict rates of reaction if the oxidation step is fast relative to other steps in the overall disposition pathway unless branched pathways allow for the oxidation step to affect product ratios. Thus, pharmacokinetics and enzyme kinetics play an important role in whether the model will predict rates. This is described briefly below. Furthermore, if the model is used to predict toxicity the model will fail if the toxicity does not arise from the believed bioactivation pathway. For example, if the mechanism of toxicity of a substance or class of substances is independent of CYP metabolism, the toxicity of these substances cannot be predicted using the PNR model. In fact, this may also be one of the important functions of the model, that is, when a compound is shown to be a significant outlier other mechanisms of toxicity should be considered. Thus, PNR-based models may help to verify if the presumed mechanism of toxicity is correct.

An example of the failure of the PNR model was recently published by Testa and coworkers *(44)*. These workers found that the genotoxicity of allylbenzenes and propenylbenzenes did not correlate with the PNR derived enthalpy of activation. Instead it was found that the relative stability of the benzylic carbonium ion could predict the genotoxicity. It is likely that the stability of the carbonium ion controls whether the metabolite will be excreted or bind to DNA. Thus, it controls a branch point after the CYP-mediated event.

Pharmacokinetics, Enzyme Kinetics and the PNR Model

The role of pharmacokinetics in predicting the rates of CYP mediated reactions is not well understood. In our nitrile model, log P is used as a descriptor of the relative absorption, distribution and the ability of these substances to reach the active site of the enzyme. To this author's knowledge, little is known about the volumes of distribution of these substances, and no data are available about the lethal concentrations of the various nitriles at the site of the enzymes. This information is important since the PNR model should predict zero order elimination (V_{max} effects) for saturated enzyme concentrations but may be less reliable at subsaturating first order elimination concentrations. In the model for halocarbon metabolism (equation 4) we know that all the concentrations used of the HCFCs, HFCs and inhalation anesthetics are saturating CYP2E1, the major isoform involved in halocarbon metabolism. It is also likely that since all the halocarbons have similar structure and lipophilic properties that the enzyme binding constants for the different halocarbons are very similar. Thus, it is possible that even at subsaturating conditions the rates can be predicted. This hypothesis remains to be tested. It should be noted that the regioselective effect, which plays a major role in predicting nitrile toxicity, is intramolecular and should be accurate even at subsaturating conditions.

The importance of understanding the enzyme kinetics cannot be overemphasized. It is difficult to reconcile the fact that the rate limiting step in a CYP-mediated reaction is not hydrogen atom abstraction and the results obtained with the predictive model, which predicts the ease of hydrogen atom abstraction, are based on predicting the rate of this step. Some light can be shed on this contradiction by the observation that while no isotope effects can be observed for electron consumption, large isotope effects are usually observed for product formation. This difference has been interpreted in terms of branched pathways and their ability to unmask the intrinsic differences in rates inherent in isotopic substitution (45,46). For the nitriles, branched pathway are the likely explanation for our ability to predict LD_{50} values and in fact are inherent in the method. The $k_{\alpha corr}$ value is obtained by predicting the amount of branching to alternate positions in the molecule. This branching probably allows for the intrinsic differences in reactivity at the position adjacent to the nitrile moiety to be expressed. For the halohydrocarbons the predictive ability of the method is more difficult to understand. For most of these compounds only one carbon-hydrogen bond is available for oxidation. Thus, no intramolecular alternative products are available from the substrate. At this time we can only speculate that the alternative product is water. Branching to water has been shown to be important in unmasking the isotope effects on norbornane metabolism mediated by P450cam (46). To fully understand the limitations of our modeling method, further studies must be conducted on the stoichiometry of these CYP mediated reactions.

Using the PNR Model to Design Safer Chemicals

As discussed earlier in this chapter, CYP enzymes are known to metabolize many chemical substances. Depending upon the substance, CYP metabolism can result in either detoxication or toxication (bioactivation). Because hydrogen atom abstraction by CYP is often a key step in both of these metabolic pathways, *a priori* knowledge of which hydrogen atom(s) in a given substance will be or are likely to be abstracted by the CYP enzymes provides insight as to whether metabolism will result in detoxication or bioactivation and, thus, helps to establish the potential toxicity of the substance. Such knowledge is useful for the design of safer analogous substances because it provides a rational basis for deciding what structural modifications can be made that direct CYP metabolism of the substance to detoxication. The PNR model can be a very useful tool for designing safer analogs of substances whose toxicity is known to result from bioactivation caused by CYP-mediated hydrogen atom abstraction. Using the PNR model, one can predict the relative rate of CYP-bioactivation of the toxic substance, and design new, analogous substances such that they contain structural modifications that are expected to redirect metabolism away from bioactivation to detoxication, or at least lower the rate of bioactivation relative to the toxic substance.

To design a less acutely toxic nitrile, for example, one should incorporate structural modifications that are expected to divert CYP-mediated hydroxylation away from the alpha carbon. Such structural modifications may include making or keeping the alpha carbon a secondary carbon, and make another carbon (e.g., beta, gamma, etc.) atom of the nitrile tertiary by adding to it an alkyl substitutent such as a methyl or ethyl group. The tertiary carbon of such a nitrile is expected to be hydroxylated more readily than the secondary alpha carbon, because the initial CYP-mediated radical formation (a requisite step of CYP hydroxylation) is expected to occur more readily at the tertiary carbon due to stabilization by hyperconjugation. The predicted $k_{\alpha corr}$ of such a nitrile should be much lower than that of an analogous nitrile that lacks the added alkyl substituent, which means it will be less acutely toxic (will have a higher LD_{50} value) according to equation 2 (assuming no substantial difference in log P). If a number of new nitriles are under consideration for development, the ones with the lower predicted $k_{\alpha corr}$ are likely to be the least acutely toxic. One should then use equation 2 to quantitatively estimate the LD_{50} of these nitriles to delineate more precisely their relative acute toxicity. The design of safer nitriles is covered more extensively elsewhere in this book (see chapter by DeVito entitled Designing Safer Nitriles).

The connection between predicting rates of metabolism and the toxicity of a halogenated hydrocarbon is less certain. Although for the inhalation anesthetics presented in this study such a correlation appears to exist *(34)*, this is not necessarily the case for other halogenated hydrocarbons. For example, 1,2-difluoroethane is predicted to have a relatively low metabolic rate and, thus, low toxicity. This substance, however, is metabolized (albeit slowly) to fluoroacetic acid which is highly toxic *(47)*, and accounts for the greater than expected toxicity of 1,2-difluoroethane. For the HCFCs presented in this chapter, however, the toxicity is likely to correlate directly with metabolism, because these specific HCFCs produce trihaloacyl halides as metabolites. For HCFCs in general the potential for toxicity may not be as easily predicted because of the different types of products that can be formed. These include aldehydes *(48)*, alcohols *(49)* as well as acylhalides *(49,50)*. Information about the toxicity of each particular class of metabolite has yet to be gathered. For example, while it is known that chlorodifluoroacetaldehyde, a metabolite of HCFC-132b, inhibits serine hydrolases *(48)*, the physiological significance of this finding from a toxicity standpoint is not known. Thus, at this stage we can only make the reasonably logical assumption that the compound with the lowest level of metabolism will be the least toxic.

This is likely to be the case for most classes of compounds since the toxicity of the entire class of compounds will not be mediated by the same pathway. Therefore, while the computational model can help in the design of safer chemicals, knowledge of the underlying mechanism of toxicity of a compound and its metabolites is very important in interpreting the results. The PNR model should not be considered to be a black box that accepts a structure and predicts the toxicity. While many such models exist, they are bound to fail unless the results are interpreted in terms of the potential toxic pathways for a given compounds or class of compounds. In this sense, the PNR model is a true design model which allows the user to ask how important a given pathway will be relative to another pathway or relative to another compound. This is what the PNR model does for nitriles. In this case it is assumed that all other pathways are detoxication, and toxicity is predicted. However, a psuedo rate constant could be generated for other pathways and we could predict the relative flux through a number of pathways. The goal is to maximize the flux through nontoxic pathways.

Future Direction for PNR Model Validation and Development

As described above a number of studies need to be done to help us understand the kinetics of the CYP mediated reactions, both *in vivo* and *in vitro*. This information could be very important in understanding when the model will work. These studies are currently underway for the nitriles, and halocarbons. One of the major limitations to the PNR model is that it can only predict the rate of hydrogen atom abstractions, and not other metabolic transformations. Attempts to develop new computational models that can predict the rates of other metabolic reactions such as S-oxidation, olefinic oxidation and aromatic hydroxylation (Scheme 1, reactions 1 and 2) are underway. For the majority of these CYP-mediated oxidative reactions the precise chemical mechanisms need to be further elucidated before we can hope to develop models that predict their reaction rates. Thus, it is important that basic mechanistic studies be performed on the mechanism of oxidation of each functional group that is not thought to proceed via hydrogen atom abstraction. Once a better understanding of CYP-mediated oxidative transformations are known, a unified predictive model of these CYP-mediated transformations that includes empirical corrections for each mechanism of oxidation can be developed. Finally, to predict bioactivation rates we must understand and have models for each of the metabolic enzymes that are important in the metabolism of a given compound. Thus, models of other bioactivating enzymes such a glutathione tranferase must be developed.

Acknowledgments

The author would like to acknowledge Professor Kenneth Korzekwa for his collaboration on this project, and Dr. Hequn Yin, LeeAnn Higgins and Grace Bennett for their contributions to the work presented in this chapter. This work was funded in part by NIEHS grant 06062. Computer time for this project was obtained from the U.S. Environmental Protection Agency's National Environmental Supercomputing Center located in Bay City, Michigan.

Literature Cited

(1) *Bioactivation of Foreign Compounds*; Anders, M. W., Ed.; Academic Press: New York, 1985.

(2) Lash, L. H.; Dekant, W.; Anders, M. W. In *Nephrotoxicity. In Vitro to In Vivo, Animals to Man*; P. H. Bach and E. A. Lock, Ed.; Plenum Press: London, 1989; pp 595-600.
(3) Falany, C. N.; Wilborn, T. W. *Adv. Pharmacol.* **1994**, *27*, pp 301-329.
(4) Spahn-Langguth, H.; Benet, L. Z. *Drug Metab. Rev.* **1992**, *24*, pp 5-48.
(5) Nelson, D. R.; Kamataki, T.; Waxman, D. J.; Guengerich, F. P.; Estabrook, R. W.; Feyereisen, R.; Gonzalez, F. J.; Coon, M. J.; Gunsalus, I. C.; Gotoh, S.; Okuda, K.; Nebert, D. W. *DNA and Cell Biology* **1993**, *12*, pp 1-51.
(6) Guengerich, F. P.; Shimada, T. *Chem. Res. Toxicol.* **1991**, *4*, pp 391-407.
(7) Korzekwa, K. R.; Swinney, D. C.; Trager, W. F. *Biochemistry* **1989**, *28*, pp 9019-9027.
(8) Monks, T. J.; Lau, S. S. *Crit. Rev. Toxicol.* **1992**, *22*, pp 243-270.
(9) Mansuy, D.; Valadon, P.; Erdelmeier, I.; Lopez-Garcia, P.; Amar, C.; Girault, J.-P.; Dansette, P. M. *J. Am. Chem. Soc.* **1991**, *113*, pp 7825-7826.
(10) Flammang, T. J.; Yamazoe, Y.; Guengerich, F. P.; Kadlubar, F. F. *Carcinogenesis* **1987**, *8*, pp 1967-1970.
(11) DeBaum, J. R.; Miller, E. C.; Miller, J. A. *Cancer Research* **1970**, *30*, pp 577-584.
(12) Pohl, L. R.; Kenna, J. G.; Satoh, H.; Christ, D.; Martin, J. L. *Drug Metab. Rev.* **1989**, *20*, pp 203-217.
(13) Hartung, R. In *Patty's Industrial Hygiene and Toxicology*; 3rd revised ed. Wiley-Interscience: New York, 1982; pp 4845-4900.
(14) Poulos, T. L. In *Methods in Enzymology*; W. &. Johnson, Ed.; Academic Press: San Diego, 1991; Vol. 206; pp 11-30.
(15) *Bacterial P-450cam Methylene Monooxygenase Components: Cytochrome m, Putidaredoxin, and Putidaredoxin Reductase*; Gunsalus, I. C.; Wagner, G.C., Ed.; 1978; Vol. 52, pp 166-188.
(16) Poulos, T. L.; Finzel, B. C.; Howard, A. J. *J. Mol. Biol.* **1987**, *195*, pp 687-700.
(17) Wackett, L. P.; Sadowsky, M. J.; Newman, L. M.; Hur, H.; Shuying, L. *Nature* **1994**, *368*, pp 627-629.
(18) Korzekwa, K. R.; Jones, J. P. *Pharmacogen* **1993**, *3*, pp 1-18.
(19) Coon, M. J.; Ding, X. X.; Pernecky, S. J.; Vaz, A. D. *FASEB Journal* **1992**, *6*, pp 669-673.
(20) Raag, R.; Poulos, T. L. *Biochemistry* **1989**, *28*, pp 917-922.
(21) Aikens, J.; Sligar, S. G. *J. Amer. Chem. Soc.* **1994**, *116*, pp 1143-1144.
(22) Karki, S. B.; Dinnocenzo, J. P.; Jones, J. P.; Korzekwa, K. R. *J. Amer. Chem. Soc.* **1995**, *117*, pp 3657-3664.
(23) Jones, J. P.; Rettie, A. E.; Trager, W. F. *J. Med. Chem.* **1990**, *33*, pp 1242-1246.
(24) Guengerich, F. P.; Macdonald, T. L. *Acct. Chem. Res.* **1984**, *17*, pp 9-16.
(25) Dinnocenzo, J. P.; Karki, S. B.; Jones, J. P. *J. Amer. Chem. Soc.* **1993**, *115*, pp 7111-7116.
(26) Korzekwa, K. R.; Jones, J. P.; Gillette, J. R. *J. Am. Chem. Soc.* **1990**, *112*, pp 7042-7046.
(27) Isaacs, N. S. *Physical Organic Chemistry*; John Wiley & Sons: New York, 1987, pp 104-106.

(28) White, R. E.; McCarthy, M. *Arch. Biochem. and Biophysics* **1986**, *246*, pp 19-32.
(29) DeVito, S.C.; Pearlman, R.S. *Med. Chem. Res.* **1991**, *1*, pp 461-465.
(30) Tanii, H.; Hashimoto, K. *Arch. Toxicol.* **1984**, *55*, pp 47-54.
(31) Grogan, J.; DeVito, S. C.; Pearlman, R. S.; Korzekwa, K. R. *Chem. Res. Toxicol.* **1992**, *5*, pp 548-552.
(32) CLOGP computer program for estimating octanol/water partition coefficient. Available through the Pomona College Medicinal Chemistry Project, Claremont, California, 91711.
(33) Harris, J. W.; Jones, J. P.; Martin, J. L.; LaRosa, A. C.; Olson, M. J.; Pohl, L.R.; Anders, M. W. *Chem. Res. Toxicol.* **1992**, *5*, pp 720-725.
(33a) Yin, H.; Anders, M.W.; Korzekwa, K.R.; Higgins, L.; Thummel, K.E.; Kharasch, E.D.; Jones, J.P. *Proc. Natl. Acad. Sci.* **1995**, *92*, pp 11076-11080.
(34) Kharasch, E. D.; Thummel, K. E. *Anesthesiology* **1993**, *79*, pp 795-807.
(35) Sugiyama, K.; Trager, W. F. *Biochemistry* **1986**, *25*, pp 7336-7343.
(36) Ravichandran, K. G.; Boddupalli, S. S.; Hasserman, C. A.; Peterson, J. A.; Deisenhofer, J. *Science* **1993**, *261*, pp 731-736.
(37) Li, H.; Poulos, T. L. *Acta Crystallogr.* Sect. D. **1995**, *51*, pp 21-32.
(38) Ortiz de Montellano, P. R.; Fruetel, J. A.; Collins, J. R.; Camper, D. L.; Loew, G. H. *J. Amer. Chem. Soc.* **1991**, *113*, pp 3195-3196.
(39) Filipovic, D.; Paulsen, M. D.; Loida, P. J.; Sligar, S. G.; Ornstein, R. L. *Biochem. Biophys. Res. Comm.* **1992**, *189*, pp 488-495.
(40) Jones, J. P.; Trager, W. F.; Carlson, T. J. *J. Am. Chem. Soc.* **1993**, *115*, pp 381-387.
(41) Jones, J. P.; Shou, M.; Korzekwa, K. R. *Biochemistry* **1995**, *34*, pp 6956-6961.
(42) Fruetel, J.; Chang, Y.; Collins, J.; Loew, G.; Ortiz de Montellano, P. R. *J. Amer. Chem. Soc.* **1994**, *116*, pp 11643-11648.
(43) Chang, Y.; Loew, G. H.; Rettie, A. E.; Baillie, T. A.; Sheffels, P. R.; Ortiz de Montellano, P. R. *Int. J. Quant. Chem. Quant. Biol. Sym.* **1993**, *20*, pp 161-180.
(44) Tsai, R.-S.; Carrupt, P.-A.; Testa, B.; Caldwell, *J. Chem. Res. Toxicol.* **1994**, *7*, pp 73-76.
(45) Jones, J. P.; Korzekwa, K. R.; Rettie, A. E.; Trager, W. F. *J. Am. Chem. Soc.* **1986**, *108*, pp 7074-7078.
(46) Atkins, W. M.; Sligar, S. G. *J. Am. Chem. Soc.* **1987**, *109*, pp 3754-3760.
(47) Brady, R. O. *J. Biol. Chem.* **1955**, *217*, pp 213-224.
(48) Yin, H.; Jones, J. P.; Anders, M. W. *Chem. Res. Toxicol.* **1993**, *6*, pp 630-634.
(49) Yin, H.; Jones, J. P.; Anders, M. W. *Toxicologist* **1994**, *14*, p 279.
(50) Yin, H.; Jones, J. P.; Anders, M. W. *Chem. Res. Toxicol.* **1994**, *8*, pp 262-268.

Chapter 7

Use of Computers in Toxicology and Chemical Design

G. W. A. Milne, S. Wang, and V. Fung

Laboratory of Medicinal Chemistry, National Cancer Institute, National Institutes of Health, Building 37, Room 5B29, Bethesda, MD 20892

Experimental determination of the acute or chronic toxicity of chemicals has traditionally been carried out by treating animals with different doses of the compound and determining the effects. This procedure requires that a sample of the material be available. The testing is expensive, particularly if chronic toxicity is to be examined and as the testing is done in animals, the extrapolation of the results to humans is often of dubious reliability. For these reasons, efforts have been expended in recent years to develop computer programs which can predict toxicity from chemical composition. The accuracy of such predictions is improving steadily and currently they offer an alternative to traditional toxicological methods. Further, they can be used to predict the toxicities of chemicals which do not exist and thus can be used as a guide to ecologically sensitive chemical development.

The toxicity of a chemical compound is an important property which can place significant restrictions upon the acceptability of the compound for use in the normal environment. Because of this, the measurement of toxicity is a well established task and, depending upon the specific type of toxicological information that is sought, there are standard procedures for making these measurements. Until relatively recently, these methods invariably involved exposure of animals to different doses of the compound and recording of the resulting symptoms, up to and including death. Testing of this sort has been done for many years to determine both acute and chronic toxicities.

The measurement of chronic toxicity, primarily carcinogenicity, is particularly problematical, for various reasons. Such testing must continue for as long as two years and the expense of maintaining test animals in a controlled environment throughout such a long test can be prohibitive. The different endpoints associated with cancer are often difficult to observe: death is the only unequivocal endpoint, but even there, cause may be obscure. Finally, numerous cases are now known of chemicals which, while carcinogenic to rats or mice are not to humans and this casts doubt upon the relevance of the animal data.

This chapter not subject to U.S. copyright
Published 1996 American Chemical Society

Beginning in the 1970s, the value of mutagenicity, measured *in vitro*, as a marker for possible carcinogenicity was examined (*1*). After much work on this subject, the current consensus is that mutagenicity is indeed a possible marker for carcinogenicity. Mutagenesis testing is relatively inexpensive but the imperfect relationship between mutagenicity and carcinogenicity in animals has allowed animal testing for chronic toxicity to survive, its expense notwithstanding.

An alternative to either of these approaches is the estimation, from the chemical structure, of the toxicity of the compound. Development of this technique began in the 1980s and considerable improvements have been made in the methods during the last ten years. Because they are almost without cost and require no physical sample of the chemical, these approaches have presented themselves as serious alternatives to either *in vivo* or *in vitro* testing. The mechanics of estimation methods and their role, present and future, in toxicology are the subject of this chapter.

Cost of Toxicity Testing in Animals

The cost of conducting toxicity testing in animals is considerable and increases steadily. Obviously, acute testing is much less expensive than chronic testing but even here, the cost of toxicity measurement conducted to the standards used in the U.S. is in the thousands of dollars, ranging from $3,500 for a primary eye irritation study to $10,000 for a dermal study or an acute oral study which provides a toxicity limit and an LD_{50}. A ninety day sub-chronic study ranges from $375,000 (oral dosage) to $950,000 (inhalation) and a chronic, two year study of the sort used in carcinogenicity testing costs about $1,125,000 for an oral dosage and $3,900,000 for an inhalation dosage. These latter costs are so huge that they are in and of themselves a powerful incentive to the development of alternative methods.

Methods of Toxicity Estimation

Methods for the calculation of the values of biological effects, such as toxicity, from chemical structure all pay great allegiance to statistics. Some are outright products of statistical analysis of a large dataset; others are so-called "expert systems" in which the dependence upon statistics is much less explicit. There are many expert systems for toxicity prediction and many more continue to be developed; so far, there is little consensus within the toxicology community regarding the predictive quality of these systems. The methods described here represent only a sampling and begin with those most heavily dependent upon statistical analysis progressing to the expert systems.

1. TOPKAT

Beginning in the late 1970s, Enslein and co-workers (*2-14*) conducted a major effort to relate the biological properties of molecules to their structure. Their methodology, which is summarized in Figure 1, was based upon a statistical analysis of a large dataset using stepwise regression techniques to establish a relationship between various structural features and the toxicity data. A "learning database" was established from a subset of the large database and each chemical structure in this subset was entered as its linear SMILES code (*15*) in which structures are represented by a string of characters. Thus C1CC1C(=O)O denotes cyclopropane carboxylic acid, while c1ccccc1C(=O)O represents benzoic acid. Each chemical structure in the learning database is analyzed by a program which generates for that structure a set of descriptors which encode substructural features and other properties such as charge, shape and connectivity - all properties which can potentially exert an influence upon toxicity. Stepwise regression and discriminant analysis are next used to search out mathematical relationships between the structural descriptors and the toxicity. The result is an equation which relates the two and which can be used with other structures, not in the learning database, to predict their toxicity. In an early use of this method (*2*), 549 compounds were taken from the NIOSH Toxic Substances List (*16*). Substructural descriptors as well as descriptors encoding logP (n-octanol/water partition coefficient) and molecular weight were developed and these parameters were related to the oral LD_{50} of the compound in rats. The resulting equation had the form:

$$(1000*MW)/LD_{50} = (0.111*FG51R) + (0.0624*FG120) + ... + (0.000225*MW) + 0.426$$

in which the parameters FGnnn (FG = functional group) assume the value 0 or 1 depending upon whether the specific substructural feature (FG51R represents a carbonate ester attached to a ring, FG120 an olefinic bond, and so on) is present in the structure. In the early attempt to correlate structure with acute toxicity, a correlation coefficient (R^2) of 0.493 was obtained. This method was thus used as the basis of a program for prediction of approximate acute toxicity from chemical structure. Subsequent work by the Enslein group led to refinement of the method and exploration of its utility in ranking in terms of acute toxicity of chemicals regulated under the Toxic Substances Control Act.

The method used by TOPKAT may in principle be used to attempt to correlate any biological activity with chemical structure. It is reasonable to expect some sort of correlation and, provided the appropriate structural features are examined, a useful regression equation should be obtained. The method was studied for its utility in the prediction of chronic toxicity, expressed as carcinogenicity (*3-5, 7*), mutagenicity (*8*), teratogenicity (*9*), skin irritation (*10,11*) and biodegradability (*14*). In the case of carcinogenicity, a retrospective examination of the carcinogenic potential of chemicals classified by the International Agency for Research in Cancer (IARC) was carried out (*2*). A total of 343 "definite" carcinogens or non-carcinogens were taken from Volumes 1-17 of the IARC Monographs (*17*) and used as a learning database. The resulting regression equation was validated with data from IARC Volumes 18-24 and for the 38 compounds in those volumes, a correct classification was obtained for 28 (74%). There were 12 steroids in this group and 7 of these were mis-classified. Removal of the 12 steroids from the dataset gave a correct classification rate of 87%.

Results of this sort have established TOPKAT as one of several methods which could be used as an alternative to animal testing for chronic toxicity. This issue is explored further in a later section.

2. ADAPT

The group headed by Jurs at Pennsylvania State University has, over a period of several years, developed statistically based methodology for the correlation of chemical structure with molecular properties, including toxicity. Their approach uses newer statistical tools such as pattern recognition and neural networks and thus differs from that of Enslein, who relies upon the traditional statistical analysis methods.

In the ADAPT software developed by Jurs (*18*), all the chemical structures in the database are entered in the form of connection tables. The program generates structural descriptors using these 2-dimensional structures, physicochemical parameters, elemental composition, and molecular shape. Then, using pattern recognition techniques, classifiers based upon these descriptors are developed which can discriminate between carcinogens and non-carcinogens. Finally, the number of descriptors used is reduced as far as possible without degrading the classification accuracy. This approach was applied to a dataset of 209 chemicals (*19*) containing 130 carcinogens and 79 non-carcinogens. It was not possible to classify all 209 compounds accurately; the best results, which used about 30 or fewer descriptors, had a classification accuracy of between 90% and 95%. The best set of descriptors had a predictive accuracy of about 85%, producing false positives (non-carcinogen predicted to be carcinogenic) more often than false negatives.

Further development of the ADAPT method has been directed towards quantitative estimation of toxicity, particularly acute toxicity (*20*) and has been quite successful. This method, a variant of the Enslein technique is a competing method of toxicity estimation.

3. SIMCA

A variation on the approach embodied in ADAPT is found in SIMCA (*21*). This program treats the toxicity prediction question as one of classification of compounds as toxic or non-toxic and uses cluster analysis (*22*), another pattern recognition technique, to achieve this classification. Application of this technique allows classification of unknowns as to type and degree of toxicity, but the method apparently has not been used by other workers in the field. SIMCA use partial least squares (PLS) analysis, a powerful statistical technique which has more recently seen successful application to Comparative Molecular Field Analysis (CoMFA), a form of 3D QSAR.

3. CASE

CASE was developed by Klopman at Case Western Reserve University during the 1980s (*23-28*) and is a statistical analysis method closely related to those already described. This program, shown schematically in Figure 2 is based upon statistical analysis of a database and as such as similar to TOPKAT. It requires a learning set of compounds whose structure and biological activities are known. Structures can be entered in a variety of formats and are then decomposed into molecular fragments containing between 2 and 5 linearly connected non-hydrogen atoms. These fragments, together with other structure-based parameters are evaluated vis-à-vis the activity associated with them, using primarily discriminant analysis. After removal of outliers

Toxicity Prediction by Komputer-assisted Technology (TOPKAT)

Structure-Activity model development scheme. After assembly of the database and generation of parameters, statistical techniques are used to find a correlation between the toxicity and some or all of the parameters defined in step 3. Step 8 is used to determine the predictive power of the model.

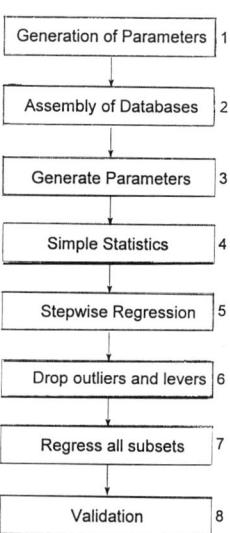

Figure 1. Estimation of Toxicities by Stepwise Regression.

The CASE program. The Database is composed of structures and their activities and the statistical evaluation finds the most discriminatory of the descriptors.

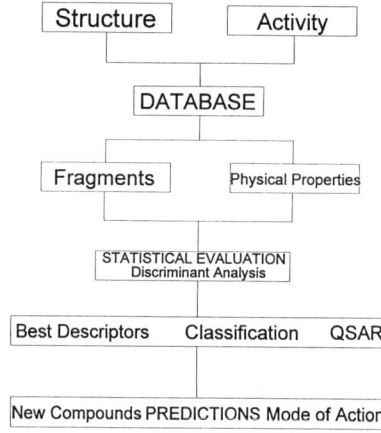

Figure 2. The CASE Program.

and levers, the descriptors which relate best to activity or lack of activity are retained. For continuous properties, such as LD_{50} values, linear regression is used to provide a quantitative relationship between the presence of the fragments and other molecular properties and the toxicity value.

When chronic toxicity is considered, CASE can be used to determine the structural basis of the carcinogenicity of chemical compounds (*28*). Analysis of the structures of

A   B   C

189 chemicals tested in the NCI/NTP Bioassay revealed 23 molecular fragments that could be linked to carcinogenicity, or the lack thereof, in the parent compounds. These fragments include aromatic amino, nitro and methoxyl, and various aliphatic fragments, such as ether oxygen.

Armed with this knowledge base, CASE is a practical means of predicting carcinogenic potential. CASE analysis of 4-chloro-*p*-phenylenediamine shows it to contain three crucial substructural fragments, A, B and C. The presence of A in a compound associated with a 79% likelihood that the compound will be a carcinogen. The figures for B and C are 79% and 66%, respectively. The probability that any compound containing all three fragment will be carcinogenic is calculated to be 96.8%. This suggests that in this structure at least, the two aromatic amino groups reinforce one another and the chlorine has no effect upon the carcinogenicity of the compound.

A valuable enhancement to CASE, and its successor MULTICASE, is the program META, also developed by Klopman's group. This program provides the potential for CASE to predict the toxicity not only of a chemical, but also of its expected metabolites.

META (*29, 30*) is a knowledge-based expert system, similar to those discussed below, which is designed to simulate the biotransformations of chemicals. It does this with reference to dictionaries of known biotransformations.

4. Structural Alerts

In 1988, Ashby and Tennant conducted a detailed analysis (*31*) of the structure-activity relationships in the voluminous data from the NCI/NTP Bioassay, in which over 300 chemicals had been tested in animals for carcinogenicity. The central conclusion of the analysis was that mutagenicity to *Salmonella* and presence in the structure of an electrophilic site were both criteria that are consistent with carcinogenicity. The correlation of either of these properties with carcinogenicity is high. They both appear however, to be necessary, but insufficient criteria. Bioavailability for example, is clearly important, but is neglected in this study.

This painstaking analysis is in effect, a qualitative statistical analysis and lies somewhere between the statistical analyses discussed above and the expert systems described below. The analytical aspect of the work is augmented by the expertise of the authors and the outcome is a set of what have become known as "structural alerts" - items such as specific substructures or mutagenesis data which are found to correlate well with carcinogenicity. The structural alerts that emerge from the analysis by Ashby and Tennant are observation of mutagenic activity in the *Salmonella* assay, and presence in the structure of any of several critical substructures, such as aromatic amine, epoxide, nitrosamine, and so on. To these critical factors, other more mechanical items are often added and the usual list of structural alerts is:

1. The structure or its metabolites are DNA-reactive (electrophilic).
2. The compound is mutagenic to *Salmonella*.
3. The compound can be administered at doses which do not lead to acute toxicity.
4. Organ damage is observed at sub-chronic doses.
5. Ancillary factors such as solubility, evidence of carcinogenicity are present.

The use of structural alerts involves a simple examination of each of these categories for the chemical in question. If any of the 5 alerts is triggered, the estimate of potential carcinogenicity may be made. If more than one item is checked, the potential increases. This method for determining carcinogenic potential is disarmingly simple: in place of factors derived from a statistical analysis of the database, one use structural alerts which evolve from the experience of experts. This is in fact a rudimentary "expert system" which has much in common with the expert systems that will be described below.

5. COMPACT

It has been proposed that specificity towards one of the P450 family of enzymes is a surrogate for toxicity, particularly chronic toxicity and this is the basis of a toxicity prediction system called COMPACT (*32*). This system, shown in Figure 3, relies on prior data but focuses upon the dimensions and electronic structure of molecules which it regards as predictors of toxicity. Two key parameters are the molecular shape, expressed as length/width or and electronic activation energy ΔE, defined as:

$$\Delta E = E_{lumo} - E_{homo}$$

where *lumo* denotes the lowest unoccupied, and *homo* the highest occupied molecular orbital. When a shape parameter such as area/depth2 is plotted against ΔE, the scatterplot that that results (Figure 4) shows carcinogens and non-carcinogens to be fairly well distinguished from one another.

6. DEREK

DEREK is an expert system (*33*) which can assimilate a chemical structure, decompose it into substructural fragments and assess the toxic potential of each fragment, and hence of the complete molecule with reference to a set of rules to which it has access.

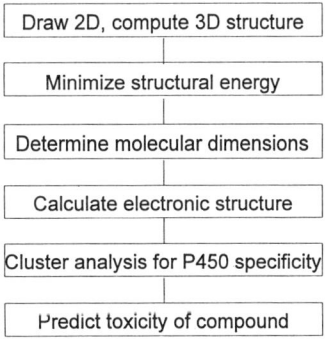

Figure 3. COMPACT Flowchart.

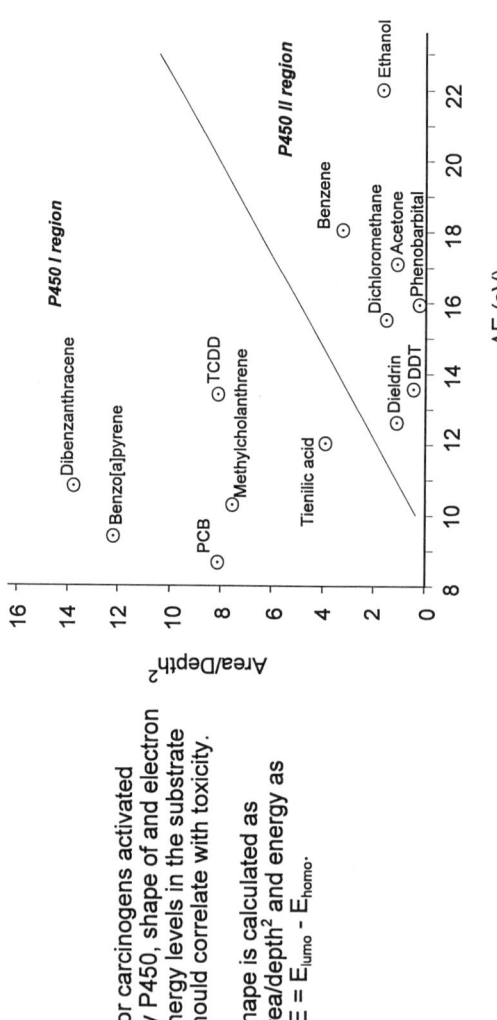

Figure 4. COMPACT and Cytochrome P450 Substrates.

A strength of the system is its facility in dealing with structures. One can enter a structure in graphical form, or retrieve a structure from a database and submit it to DEREK. Generation of fragments is automatic as is consultation with the rulebase. The result is a listing of the possible risks associated with the entered structure. For each risk the apposite rule is cited so the perceived structure-activity association can be seen.

The rulebase is open-ended; many of the rules represent a distillation of much toxicological experience, others may be more arbitrary. As an example, the FDA structural alerts for carcinogenicity have been built into a rulebase. Examination by the program of *p*-hydroxyacetanilide (1) led to its categorization as an irritant because it is a phenol with no counter base (rule 41) and as a carcinogen because it is an aromatic amido compound and as such triggers FDA structural alert #1 (rule 107). Similarly, *o*-benzyl-*p*-chlorophenol (2) is classified as an irritant by rule 41 but it is ruled to be non-carcinogenic because it triggers none of the alerts. When DEREK encounters a

structure which is not adequately covered by its rules, it demurs. Thus 3,4-dihydrocoumarin (3) is dismissed with "no comment".

7. HAZARDEXPERT

This system, which is produced by the CompuDrug Corporation of Budapest and is shown in Figure 5, uses physical properties and molecular fragments to predict the toxicity of a compound. It is thus similar to TOPKAT except that it derives all the parameters it uses, including logP and pK from the entered molecular structure. In the same way that CASE and DEREK attempt to account for toxic metabolites, HAZARDEXPERT uses a related program METABOLEXPERT to examine the possibility that a compound could give rise to toxic metabolites. This appears to be a sophisticated estimation program but performance data are not readily available.

8. ONCOLOGIC

During the last several years, the Environmental Protection Agency (EPA) has supported the development of a toxicity estimation system called OncoLogic. This is a frankly expert system which relies only indirectly upon statistics. The OncoLogic code was written by computer programmers in close consultation with experts in different areas of toxicology and is ambitious in that it attempts to deal with all manner of materials, including organic and inorganic chemicals, metals and metalloids, and complex materials such as fibers and polymers. Each of these requires a quite distinct approach and the program goes some way towards accommodation of these eclectic requirements. The code is in the form of a large tree structure. At each level of the tree, the material in question is categorised into a more precise area. Thus at the top level, as shown in Figure 6, chemicals compounds are classified as fibers, polymers, organic compounds or metals and metalloids. Next, as shown in Figure 7, organics for example, are further subdivided. Compounds in one of the new subgroups, aromatic hydrocarbons, are then examined in terms of a number of expert rules. The result of this examination will be a carcinogenicity rating which is based upon the most current knowledge available and appropriate to that case.

The National Cancer Institute/National Toxicology Program Bioassay

The National Toxicology Program (NTP), operated jointly by the National Cancer Institute and the National Institute for Environmental Health Sciences has, as one of its major activities, the task of testing chemicals for carcinogenicity towards mammals, primarily mice and rats. This testing is slow; a complete test requires over two years and the NTP has completed work on between 300 and 400 chemicals.

In 1990, a list was compiled (*36*) by the NTP of 44 chemicals for which carcinogenicity testing was to be completed by 1994. The publication of the list was accompanied by the suggestion that interested scientists might use it for prospective estimation of the carcinogenicity of the chemicals and, in this way, obtain an objective measure of the efficacy of the various programs. Several groups responded to this challenge and the computer programs described in this chapter that were tested in this way were TOPKAT, COMPACT and CASE. The list was also examined in terms of structural alerts and a comparison of these four, and other methods was thus possible.

To date testing has been completed on 37 of the compounds; 11 were established by testing as non-carcinogens and 3 were classified as "unknown" (Table I) and of the remaining 30 (Table II) 22 were determined to possess carcinogenicity while 8 were found to have "equivocal" carcinogenicity. The performance of the different prediction systems is given in the Tables and summarized in Figure 8. This summary ignores all "unknown" or "equivocal" results and thus covers 31 compounds. As can be seen from Figure 8, COMPACT and the structural alerts performed equally well with 58% correct predictions.

Both CASE and TOPKAT performed less well, although a mitigating factor for TOPKAT is that it declined to make a prediction for one third of all the compounds, on the grounds that it had too little relevant data. TOPKAT made 21 predictions and of these 12 (57%) were correct.

The mixing of "equivocal" bioassay results with firm data reflects the uncertainties in chronic toxicological testing referred to earlier, but in spite of this, it seems fair to conclude from this experiment that all the methods have 50-60% accuracy with COMPACT and structural alerts out-performing TOPKAT and CASE. This accuracy is not adequate, but as research on these methods proceeds, it will presumably improve.

Summary

We are at an interesting stage in the development of toxicity estimation capabilities. There is available a large amount of experimental data, including acute toxicity data for

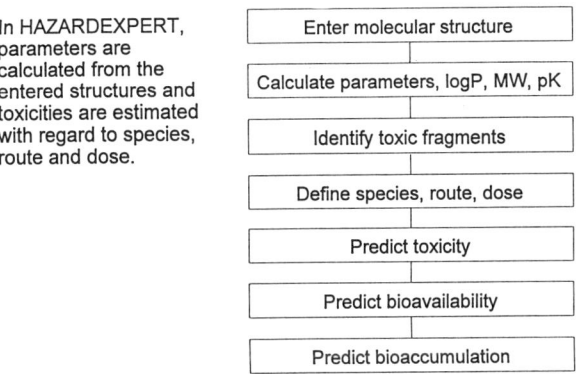

Figure 5. The HAZARDEXPERT System.

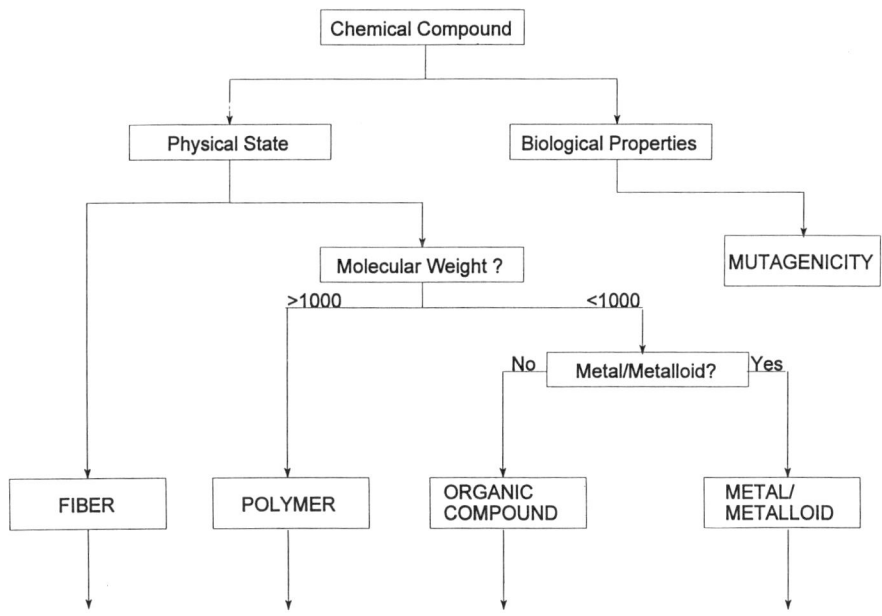

Figure 6. OncoLogic-Cancer Expert System.

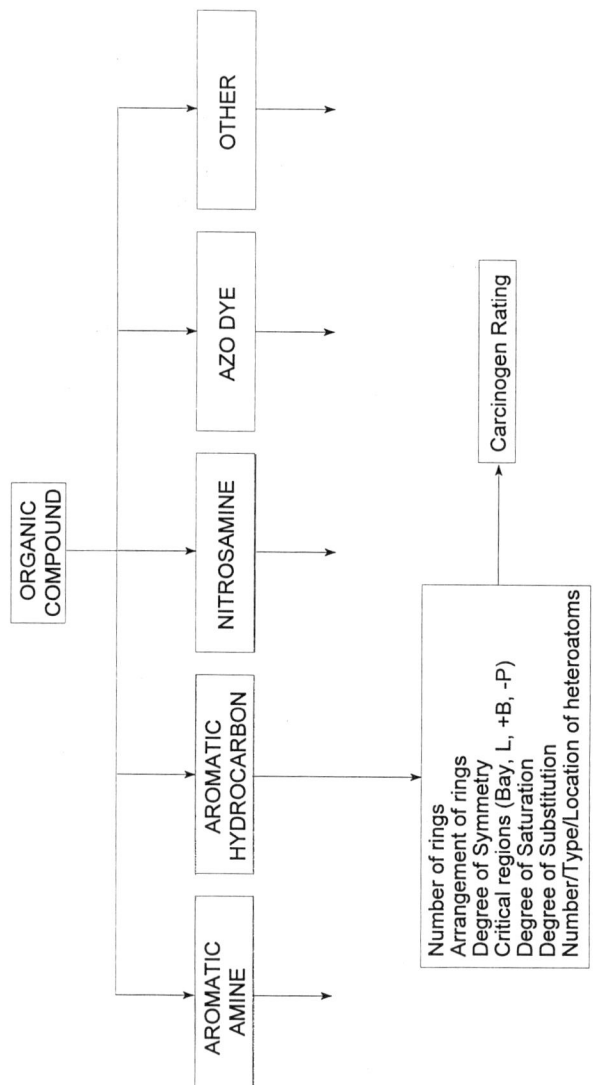

Figure 7. OncoLogic Cancer Organic Subsystem.

Table I. Carcinogenicity Estimation for Non-Carcinogens (N) and Unknowns (U)

CAS RN	COMPOUND	Structural Alerts	Topkat	Compact	CASE	Bioassay
58-33-3	Promethazine HCL	-	?	+	+	N
58-55-9	Theophylline	-	-	+	-	U
60-13-9	Amphetamine sulfate	-	?	-	-	N
74-83-9	Methyl bromide	+	?	-	+	N
79-11-8	Monochloracetic acid	+	-	-	-	N
81-11-8	Diaminostilbene disulfonic acid	+	+	+	+	N
96-69-5	4,4'-Thiobis(*t*-butyl-*m*-cresol)	-	-	?	+	N
100-02-7	*p*-Nitrophenol	+	+	+	-	N
107-21-1	Ethylene glycol	-	-	-	-	N
108-46-3	Resorcinol	-	-	+	+	N
599-79-1	Salicylazosulfapyridine	+	+	+	-	U
1330-78-5	Tricresyl phosphate	-	?	+	?	N
10599-90-3	Chloramine	+	?	-	?	U
26628-22-8	Sodium azide	-	?	-	?	N

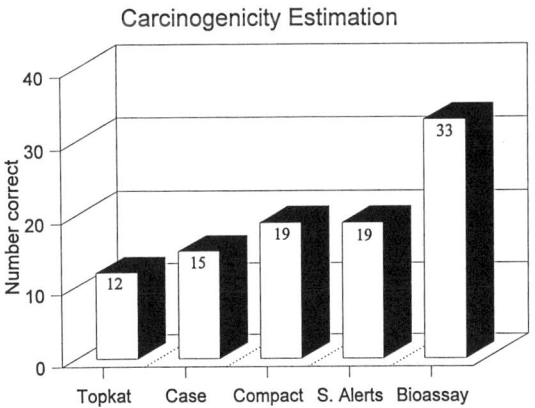

Figure 8. Performance of Carcinogenicity Estimators.

Table II. Carcinogenicity Estimation for Confirmed (C) and Equivocal (E) Carcinogens

CAS RN	COMPOUND	Structural Alerts	Topkat	Compact	CASE	Bioassay
57-41-0	Diphenylhydantoin	-	?	-	-	C
62-23-7	p-Nitrobenzoic acid	+	+	+	-	C
75-65-0	t-Butyl alcohol	-	-	-	-	C
91-20-3	Naphthalene	-	?	+	?	C
91-23-6	o-Nitroanisole	+	+	+	+	C
91-64-5	Coumarin	+	?	+	-	C
96-13-9	2,3-Dibromo-1-propanol	+	?	-	+	C
96-18-4	1,2,3-Trichloropropane	+	+	-	+	C
96-48-0	γ-Butyrolactone	-	-	-	+	E
100-01-6	p-Nitroaniline	+	+	+	+	E
103-90-2	P-Hydroxyacetanilide	+	+	+	-	E
115-96-8	Tris-(2-chloroethyl) phosphate	+	+	-	+	C
119-84-6	3,4-Dihydrocoumarin	-	-	+	-	C
119-93-7	3,3'-Dimethylbenzidine	+	+	+	+	C
120-32-1	o-Benzyl-p-chlorophenol	-	-	-	-	C
137-09-7	2,4-Diaminophenol 2HCl	+	+	+	+	C
298-59-9	Methyl phenidate Hcl	-	-	-	-	C
396-01-0	Triamterene	-	-	+	+	C
1271-19-8	Titanocene dichloride	-	?	?	?	E
1825-21-4	Pentachloroanisole	-	-	+	+	C
2425-85-6	Pigment Red 3	+	+	+	+	C
2429-74-5	Direct Blue 15	+	+	+	-	C
3296-90-0	2,2-bis(BrCH$_2$)-1,3-propanediol	-	-	+	+	C
6459-94-5	Acid Red 114	+	?	+	-	C
6471-49-4	Pigment Red 123	+	+	+	+	E
7487-94-7	Mercuric chloride	-	?	?	?	C
9005-65-7	Polysorbate 80	-	?	?	+	E
10034-96-5	Manganese sulfate	-	?	?	?	E
28407-37-6	Direct Blye 218	+	?	+	+	C
52551-67-4	HC Yellow 4	+	+	+	-	E

hundreds of thousands of compounds, mutagenicity and teratogenicity data for thousands of compounds and data on carcinogenicity for hundreds of compounds. These data provide a fairly adequate resource upon which to build statistically-based toxicity estimation methods and indeed, such methods have been shown to work quite well. The expert systems on the other hand, seem to work even better and it seems clear that they are taking into account information which, because it is sparse, is discounted by the statistical analyses. Space constraints have precluded a detailed description of the numerous expert systems that have been developed and for further information the reader is referred to the "suggested readings" provided at the end of this chapter. Dependency upon experts, even to build a computer system, is not a long term solution however and statistical methods will probably supervene in the long term.

The more important issue is whether estimation methods can ever displace the traditional animal testing. The answer to this it that they must, because the financial and social costs of toxicity testing in animals are becoming prohibitive. At present we are faced with a choice between fairly reliable toxicity data at a huge cost and less reliable data at a greatly reduced cost. So, what price reliability? If, for around $10 one can say that there is a 70% chance that a compound will be carcinogenic in humans, is that better than spending over a million dollars to be 95% sure that it will be so? This depends upon the circumstances. If the issue is to protect an established industry based upon the chemical, the greater expense may be justifiable. On the other hand, if one is merely contemplating the manufacture and sale of a chemical, the inexpensive and less reliable estimated data may be adequate to direct one's attention to different chemicals.

Literature Cited

1. Maron, D. M.; Ames, B. N. *Mutat. Res.*, **1983**, 113, 173-215.
2. Enslein, K ; Craig, P. N. *J. Environ. Pathol. Toxicol.* **1978**, *2*, 115-121. Enslein, K.; Lander, T. R.; Tomb, M. E.; Landis, W. G. *Bench. Pap. Toxicol.*, **1983**, *1*, 1-123. Enslein, K. *Pharmacol. Rev.* **1984**, *36*, 131S-135S.
3. Enslein, K.; Borgstedt, H. H.; Tomb, M. E.; Blake, B. W.; Hart, J. *Toxicol. Ind. Health*, **1987**, *3*, 267-287.
4. Enslein, K. *Toxicol. Ind. Health*, **1988**, *4*, 479-498.
5. Enslein, K.; Borgstedt, H. H. *Toxicol. Lett.* **1989**, *49*, 107-121.
6. Blake, B. W.; Enslein, K.; Gombar, V. K.; Borgstedt, H. H. *Mut. Res.*, **1990**, *241*, 261-271.
7. Enslein, K.; Blake, B. W.; Borgstedt, H. H. *Mutagenesis*, **1990**, *5*, 305-306.
8. Enslein, K.; Blake, B. W.; Tomb, M. E.; Borgstedt, H. H. *In Vitro Toxicol.* **1986**, *1*, 33-44.
9. Gombar, V. K.; Borgstedt, H. H.; Enslein, K.; Hart, J. B.; Blake B. W. *Quant. Struct.-Act. Relat.* **1991**, *10*, 306-332.
10. Enslein, K.; Borgstedt, H. H.; Blake, B. W.; Hart, J. B. *In Vitro Toxicol.* **1987**, *1*, 129-147.
11. Enslein, K.; Blake, B. W.; Tuzzeo, T. M.; Borgstedt, H. H.; Hart, J. B. *In Vitro Toxicol.* **1988**, *2*, 1-14.

12. Enslein, K.; Tuzzeo, T.M.; Borgstedt, H. H.; Blake, B. W.; Hart, J. B. Prediction of Rat Oral LD_{50} from *Daphnia magna* LC, and Structure. In *Proceedings of the 2nd International Workshop on QSAR in Environmental Toxicology* (Kaiser, K. L. E., Ed.) D. Reidel Publishing Co., Dordrecht, Holland.
13. Enslein, K.; Tuzzeo, T. M.; Blake, B. W.; Hart, J. B.; Landis, W. G. Prediction of *Daphnia magna* EC_{50} Values from Rat Oral LD_{50} and Structural Parameters. ASTM Special Technical Publication 1007, American Society for Testing & Materials, Philadelphia, PA 19103.
14. Gombar, V. K.; Enslein, K. A Structure-Biodegradability Relationship Model by Discriminant Analysis. In *Applied Multivariate Analysis in SAR and Environmental Studies* (Devillers, J.; Karchner, W., Eds.) pp 377-414, Kluwer Academic Publishers, Dordrecht, Holland (**1987**).
15. Weininger, D. *J. Chem. Inf. Comput. Sci.*, **1988**, *28*, 31-36.
16. Christensen, H. E. (Editor), *The Toxic Substance List, 1974 Edition*. HEW Publication Number (NIOSH) 74-134, **1974**.
17. IARC Monograph: Evaluation of Carcinogenic Risk of Chemicals to Humans, **1972-1978**, Vols. *1-17*.
18. Stuper, A. J.; Brugger, W. E.; Jurs, P. C. In *Chemometrics: Theory and Application*; Kowalski, B. R. Ed.; American Chemical Society: Washington, D.C. **1977**, p. 165.
19. McCann, J.; Choi, E.; Yamasaki, E.; Ames, B. N. Determination of Carcinogens and Mutagens in the *Salmonella*/microsome Test: Assay of 300 Chemicals. *Proc. Natl. Acad. Sci. U.S.*, **1975**, *72*, 5135-5139.
20. Xu, L.; Ball, J. W.; Dixon, S. L.; Jurs, P. C. *Env. Tox. & Chem.*, **1994**, *13*, 841-851.
21. Dunn III, W. J. *Tox. Let.*, **1988**, *43*, 277-283.
22. Dunn III, W. J.; Wold, S. *J. Med. Chem.*, **1980**, *23*, 595-602.
23. Klopman, G. *J. Amer. Chem. Soc.*, **1984**, *106*, 7315-7321.
24. Klopman, G.; Rosenkranz, H. S. *Mutat. Res.*, **1984**, *126*, 227-238.
25. Klopman, G.; Contreras, R.; Rosenkranz, H. S.; Waters, M. D. *Mutat. Res.* **1985**, *147*, 343-356.
26. Kloprnan, G.; Frierson, M. R.; Rosenkranz, H. S. *Environ. Mutagen.*, **1985**, *7*, 625-644.
27. Klopman, G.; Frierson, M. R.; Rosenkranz, H. S. *Mutat. Res.*, **1990**, *228*, 1-50.
28. Rosencranz, H. S.; Klopman, G. *Mutat. Res.*, **1990**, *228*, 105-124.
29. Klopman, G.; Dimayuga, M.; Talafous, J. *J. Chem. Inf. Comput. Sci.*, **1994**, *34*, 1320-1325.
30. Talafous, J.; Sayre, L. M.; Mieyal, J. J.; Klopman, G. *J. Chem. Inf. Comput. Sci.*, **1994**, *34*, 1326-1333.
31. Ashby, J.; Tennant, R. W. *Mut. Res.*, **1988**, *204*, 17-115.
32. Lewis, D. F. V. *Rev. Comp. Chem.*, **1992**, *3*, 173-222.
33. Lewis, D. F..; Ioannides, C.; Parke, D. V. *Poly. Arom. Compd.*, **1993**, *3*, 719-724.
34. Ioannides, C.; Ayrton, A. D.; Lewis, D. F. V.; Walker, R. Modulation of Cytochromes P450 and Chemical Toxicity by Food Constituents in *Food, Nutritional Chemistry and Toxicology* (Parke, D. V.; Ioannides, C.; Walker, R. Eds.) pp 301-310, **1993**.

35. Sanderson, D. M.; Earnshaw, C. *Hum. Exp. Tox.*, **1991**, *10*, 261-273.
36. Woo, Y.-T.; Lai, D. Y.; Argus, M. F.; Arcos, J. C. Development of SAR Rules for Assessing/Predicting Carcinogenic Potential of Chemical Compounds. *Proc. Amer. Assocn. Cancer Res.*, **1994**, *35*, 160-183.
37. Woo, Y.-T.; Lai, D. Y.; Argus, M. F.; Arcos, J. C. Development of Structure Activity Relationship Rules for Predicting Carcinogenic Potential of Chemicals. *Tox. Lett., in press*, **1995**.
38. Tennant, R. W.; Spalding, J.; Stasiewicz, S.; Ashby, J. *Mutagenesis*, **1990**, *5*, 3-14.

Suggested Readings

Auer, C. M.; Nabholz, J. V.; Baetcke, K. P. *Envir. Health Perspect.,* **1990**, *87*, 183-197.
Benigni, R.; Andreoli, C.; Giuliani, A. *Mut. Res.,* **1989**, *221*, 197-216.
Benigni, R.; Giuliani, A. *Arch. Toxicol. Suppl.*, **1992**, *15*, 228-237.
Frierson, M. G.; Klopman, G.; Rosenkranz, H. S. *Environ. Mutagenesis*, **1986**, *8*, 283-327.
Hileman, B. *Chem. Eng. News*, **1993**, *71*, 35-38.
Klopman, G. *Quant. Struct.-Act. Relat.*, **1992**, *11*, 176-184.
Klopman, G.; Rosenkranz, H. S. *Mutation Res.*, **1992**, *272*, 59-71.
McKinney, J. D.; Waller, C. L. *Envir. Health Perspect.*, **1994**, *102*, 290-297.
Lewis, D. F. V. *Prog. Drug Metab.*, **1990**, *12*, 205-255.
Lewis, D. F. V. Computer-Assisted Methods in the Evaluation of Chemical Toxicity, in: *Reviews in Computational Chemistry*, Lipkowitz, K. B.; Boyd, D. B. (Eds), VCH Publishers Inc., New York, pp. 173-221.
Phillips, J. C.; Anderson, D. *Occ. Health Rev.*, **1993**, 27-30.
Richard, A. M. *Mut. Res.*, **1994**, *305*, 73-95.
Waller, C. L. *J. Med. Chem.*, **1992**, *35*, 3660-3666.

Chapter 8

Designing Biodegradable Chemicals

R. S. Boethling

Office of Pollution Prevention and Toxics, U.S. Environmental Protection Agency, Mail Code 7406, 401 M Street, SW, Washington, DC 20460

One way to reduce pollution at the source is to design safer chemicals. Chemicals that persist in the environment remain available to exert toxic effects and may bioaccumulate. Since microbial degradation is the major loss mechanism for most organic chemicals in soil, water and sewage treatment, biodegradability should be included as a factor in product design along with function and economics. Biodegradability has been an important design consideration for down-the-drain consumer products like laundry detergents for more than 40 years, but not for chemicals used mainly in commerce. The relationship between molecular structure and biodegradability is generally well enough understood to make its application feasible in product design. We must extend the benign by design concept to commercial chemicals because (i) production volume and thus release may increase with time; (ii) new uses may develop and with them, the potential for greater release; (iii) we cannot know in advance all of the possible toxic effects of released chemicals. By designing a chemical such that it will biodegrade more readily to nontoxic products, the risks posed by the chemical to human health and the environment are reduced.

Under Section 6602(b) of the Pollution Prevention Act (PPA) of 1990, Congress declared it to be the policy of the United States that pollution should be prevented or reduced at the source whenever possible. Further, the Act defines source reduction as any practice that:

"...(i) reduces the amount of any hazardous substance...prior to recycling, treatment or disposal; and (ii) reduces the hazards to public health and the environment associated with the release of such substances..."

Product reformulation and redesign are specifically included in source reduction.
The implementation of PPA and relocation of EPA's Pollution Prevention Division to the Office of Pollution Prevention and Toxics (OPPT) have given OPPT a new sense of purpose in which pollution prevention is considered the first principle in achieving environmental stewardship of chemical products. One outcome of these changes in OPPT's mission has been a flurry of activity in alternative synthetic design. EPA has supported a variety of extramural projects aimed at developing safer

This chapter not subject to U.S. copyright
Published 1996 American Chemical Society

technologies for synthesizing industrial chemicals, has organized symposia, and published the proceedings (*1*).

Yet scant attention has been given to the notion that the "benign by design" concept might also be pursued in the **molecular** design of the target substance, and specifically, by designing into the molecular structure lower toxicity and enhanced biodegradability to nontoxic products. It is not the design of the synthetic sequence but rather the design of the molecule itself that is the earliest design phase of chemical manufacture. Molecular design to enhance biodegradability to nontoxic products is pollution prevention if the safer chemical eliminates the production and release of more persistent and potentially hazardous substitutes.

Enhanced biodegradability is also a worthy goal precisely because pollution cannot always be prevented at the source. The PPA acknowledged this in establishing a hierarchy of options such that

> "...pollution that cannot be prevented or recycled should be treated in an environmentally safe manner whenever feasible; and disposal or other release into the environment should be conducted in an environmentally safe manner."

Chemicals that resist biodegradation remain available to biota to exert toxic effects, not all of which may be known or predictable at the time of release to the environment. Moreover, persistent chemicals that are also bioaccumulative are of even greater concern because levels may be achieved in organisms that appear safe on the basis of acute toxicity criteria, but which ultimately result in chronic or other unforeseen toxic effects. Microbial degradation is the major loss mechanism for most organic chemicals in aquatic and terrestrial environments, and is the cornerstone of the modern wastewater treatment plant. Thus, both the treatability of generated wastes and safety of materials that ultimately must be released to the environment normally are enhanced by responsible molecular design.

In this paper I set forth the principles involved in molecular design to enhance biodegradability, and I show by means of examples from actual existing and Premanufacture Notification (PMN) chemicals how this can be accomplished. Before proceeding, however, it is necessary to review briefly the biochemistry of biodegradation, its relationship to toxicity, and what is known about the effects of chemical structure on biodegradability.

The Microbial Basis of Biodegradation

Biodegradation is not a process confined to the microbial world. Nevertheless, microorganisms (primarily bacteria and fungi) are by far the most important agents of biodegradation in nature, in terms of the mass of material transformed as well as the extent to which it is degraded. An abundance of evidence exists to show that microorganisms are responsible for the degradation of many organic chemicals that cannot be altered significantly by higher organisms. For the most part, animals excrete chemicals that they cannot metabolize, and plants tend to convert chemicals into water-insoluble forms that can be easily stored. In contrast, microbial populations are characterized by catabolic versatility, rapid growth in the presence of food, high metabolic activity and species diversity. The eventual mineralization of organic compounds (their conversion to inorganic substances such as CO_2 and water) can be attributed predominantly to microbial degradation.

Knowledge of how microorganisms bring about the degradation of organic substrates derives largely from studies of pure cultures (single strains or species) able to grow at the expense of the selected compound. Through such studies, biodegradation pathways have been defined by characterizing intermediate products and the enzymes that catalyze successive steps (2). Ordinarily, an organic compound must first enter the microbial cell by passing through the cell wall and cytoplasmic membrane. This may occur by passive diffusion or with the assistance of specific transport systems, the latter being the more common situation in aquatic and terrestrial environments, which are typically characterized by low levels of organic substrates and other nutrients. In some cases, such as with large polymeric substrates like proteins and polysaccharides, biodegradation is initiated by extracellular enzymes, the action of which yields smaller compounds that can be transported into the cell.

Once inside the cell, the reactions that a compound may undergo are determined by its molecular structure. Hundreds of transformations have been described in the literature, but almost all can be classified broadly as oxidative, reductive, hydrolytic or conjugative. The catabolic pathways employed by microbial populations are also diverse and vary with the environmental conditions. But despite the immense structural variety of naturally occurring as well as anthropogenic compounds, their utilization by microorganisms always involves the same basic strategy. That strategy is stepwise degradation to yield one or more intermediate products capable of entering the central pathways of metabolism. The overall objective is always to produce carbon and energy for growth (3). Persistent and toxic intermediates occasionally arise from partial biodegradation of a compound, but this is the exception rather than the rule.

Naturally occurring organic compounds are degraded via pathways that represent evolutionary adaptations to prevailing conditions. Of these substances, Dagley (4) has said
"...it is reasonable to believe at the present time that every biochemically synthesized organic compound is biodegradable." Many manmade chemicals are identical or similar to naturally occurring substances, but human activities have also produced structures never before seen or at least infrequently encountered in nature. Many of these, nonetheless, can be attacked by microorganisms (some very readily) by virtue of a phenomenon referred to as "fortuitous" or "gratuitous" metabolism. This is attributable to the fact that degradative enzymes are generally not absolutely specific for their natural substrates.

Chemical Structure and Biodegradability. Studies in the chemical industry, research in universities, and the results of environmental monitoring conducted over the past 40 years have shown that relatively small changes in molecular structure can appreciably alter a chemical's susceptibility to biodegradation. These studies have resulted in several "rules of thumb" (5) about the effects of chemical structure on biodegradability. The following molecular features generally increase resistance to aerobic biodegradation:
- Halogens; especially chlorine and fluorine;
- Chain branching, especially quaternary carbon and tertiary nitrogen, or extensive branching such as in surfactants derived from tri- or tetrapropylene;
- Nitro, nitroso, azo, arylamino groups;
- Polycyclic residues (such as in polycyclic aromatic hydrocarbons or PAHs), especially with more than 3 fused rings;
- Heterocyclic residues; e.g., pyridine rings; and
- Aliphatic ether (C-O-C) bonds.

This list is not exhaustive, nor should it be inferred that the presence of even a single atom or group from the list necessarily renders a compound recalcitrant. Moreover, in most cases the mechanism by which increased resistance to biodegradation is conferred is not known in detail. But this should not blind us to the fact that sufficient information is available to allow application of these principles in chemical **design**. For the most part, the features listed above affect the ability of the compound to serve as an inducer or substrate, or both, of degradative enzymes and cellular transport systems. For example, addition of a chlorine atom to a phenyl ring makes the ring less susceptible to attack by oxygenase enzymes, which utilize a form of electrophilic oxygen as a cosubstrate. Strongly electron-withdrawing substituents such as halogens are therefore to be avoided in chemical design if possible.

In contrast, biodegradability is usually enhanced by the presence of potential sites of enzymatic hydrolysis (e.g., esters, amides); by the introduction of oxygen in the form of hydroxyl, aldehydic or carboxylic acid groups; and by the presence of unsubstituted linear alkyl chains (especially ≥4 carbons) and phenyl rings, which represent possible sites for attack by oxygenases. The second of these three factors is particularly important because the first step in the biodegradation of many compounds (e.g., hydrocarbons) is the enzymatic insertion of oxygen into the structure, and this step is almost always rate limiting. More generally, if the first biodegradative step is some form of oxidation, it seems logical to expect that biodegrability will be enhanced if the synthetic chemist has in effect already carried it out during molecular design.

The number of substituent groups appended to a base structure (such as a phenyl ring) and a compound's water solubility also seem to have some bearing on biodegradability, but it is more difficult to apply these generalizations to specific substances. For some polymers such as modified cellulosics (e.g., methyl cellulose), degree of substitution is a relatively precise concept and has predictive value. But for most nonpolymeric structures this is not true. The effect of solubility probably involves one or more of the following:

(i) Microbial bioavailability. Insoluble chemicals tend to partition to the adsorbed state in activated sludge, sediments and soil. Most studies have shown that this tends to reduce the rate of biodegradation;

(ii) Rate of solubilization. Most studies have shown that for solids with very low solubility, only the dissolved or dispersed phase is available to microorganisms. Therefore, the rate of dissolution of a solid in water may control the rate of biodegradation. Many microorganisms secrete biosurfactants (e.g., rhamnolipids) that enhance the rate of solubilization;

(iii) Low aqueous concentration. Some studies have shown that for chemicals soluble to the extent of only a few micrograms per liter or less, this concentration may be too low for optimal function of cellular enzymes or transport systems.

At the present it can be stated that (i) highly substituted structures are likely to be less rapidly biodegraded than much simpler compounds; and (ii) for very insoluble chemicals, replacement of a given functional group with one that increases solubility may also result in enhanced biodegradability, all other things being equal (they never are).

Group Contribution Method for Predicting Biodegradability

An ability to predict relative rates of biodegradation from chemical structure alone would greatly facilitate the design of safer chemicals. In this section I describe the approach we used to develop mathematical models capable of such predictions. These models utilize the molecular features listed in the previous section as the basis for prediction. Fragment contribution methods such as those used in our models have been used for many years in chemical engineering, but only more recently in environmental chemistry. The basic premise is that the activity of interest is a function of the contribution of one or more molecular substructures or fragments of which the molecule is composed, and that the contribution of each fragment does not vary from compound to compound (i.e., there is no interaction between fragments). The ideal situation is that each fragment in a model has a clear mechanistic relationship to the activity of interest, which is understood at the molecular level. This situation is rarely if ever realized. Fortunately, it doesn't really matter as long as (i) there exists for model development a set of measured values ("training set") of adequate size, and (ii) a reasonably comprehensive set of structural fragments associated in some way with activity can be identified.

We used this approach to develop a set of 4 models for predicting biodegradability. Two of the models (based on linear and nonlinear algorithms) classify chemicals as easily or not easily biodegradable, and the other two (for primary and ultimate biodegradation) make semi-quantitative estimates of aquatic biodegradation rates (*6,7*). For purposes of model development the positive and negative molecular features listed in the previous section were formally defined and constituted the independent (predictor) variables. A training set of biodegradability data for the two classification models was developed using data from the literature that are well documented and widely available. To take advantage of all available data for the largest possible universe of chemicals we instituted a data evaluation procedure (*8*) that utilized biodegradation data from all types of studies other than pure cultures. The objective of this "weight of evidence" approach was to increase confidence that model predictions reflect chemical structure rather than experimental conditions.

Data were retrieved from BIODEG (*8*), a component of the Environmental Fate Data Base (*9*) that contains extracted biodegradation data on more than 800 organic chemicals. The records for each chemical constitute a comprehensive assessment of existing biodegradation data for the chemical. Each chemical is assigned a qualitative descriptor such as BR (Biodegrades Rapidly) or BSA (Biodegrades Slowly even with Acclimation) and a reliability code, reflecting the amount and consistency of available data, for each of several endpoints (e.g., screening studies, grab sample studies, etc.). Biodegradability and reliability codes are also assigned for an overall assessment of aerobic biodegradability based on all of the data, and it was this biodegradability endpoint that was used in modeling.

What the two classification-type models predict is the probability that a chemical is in the BR category. The models predicted biodegradation category correctly for approximately 90 % of the 295 chemicals in the training set (*7*); an earlier study showed similar results for an independent validation set (*6*). This level of accuracy is on par with other published models for predicting biodegradability that used different training sets and statistical methods (e.g., *10-13*), but we consider it an advantage of our approach that the predictor variables **explicitly** reflect generally accepted rules of thumb. That these rules are so accepted has been confirmed by carefully conducted surveys of expert knowledge in the field (*7,14*). Fragments and their coefficients are listed in Table I. Both the signs and relative magnitudes of the coefficients are generally consistent with expectation.

There is no need to use these particular models, or any other model, to make an educated guess about biodegradability of an untested compound. But the fragment contribution models do provide a rapid, convenient, and systematic way to accomplish this, in a way that is consistent with knowledge in the field. Molecular structure is the only input needed and is entered from the PC keyboard via the chemical's SMILES (Simplified Molecular Information and Line Entry System; *15*) notation. The models clearly identify and list for the user relevant fragments and their predicted contributions. Assumptions about the potential utility of these models have focused mainly on scoring exercises aimed at prioritizing long lists of chemicals for more

8. BOETHLING Designing Biodegradable Chemicals

Table I. Structural Fragments and Coefficients

fragment or parameter	BIODEG models				survey models		
	freq[a]	linear coeff	nonlinear coeff	freq[a]	primary coeff	ultimate coeff
equation constant		0.748	3.01		3.848	3.199
M_w	295	-0.000476	-0.0142	200	-0.0144	-0.00221
unsubstituted aromatic (≤3 rings)	2	0.319	7.191	1	-0.343	-0.586
phosphate ester	5	0.314	44.409	6	0.465	0.154
cyanide/nitrile (C≡N)	5	0.307	4.644	11	-0.065	-0.082
aldehyde (CHO)	4	0.285	7.180	5	0.197	0.022
amide (C(=O)N or C(=S)N)	9	0.210	2.691	13	0.205	-0.054
aromatic (C(=O)OH)	24	0.177	2.422	6	0.0078	0.088
ester (C(=O)OC)	23	0.174	4.080	25	0.229	0.140
aliphatic OH	34	0.159	1.118	18	0.129	0.160
aliphatic NH_2 or NH	13	0.154	1.110	7	0.043	0.024
aromatic ether	11	0.132	2.248	11	0.077	-0.058
unsubstituted phenyl group (C_6H_5)	25	0.128	1.799	22	0.0049	0.022
aromatic OH	46	0.116	0.909	21	0.040	0.056
linear C4 terminal alkyl ($CH_2CH_2CH_2CH_3$)	44	0.108	1.844	26	0.269	0.298
aliphatic sulfonic acid or salt	4	0.108	6.833	4	0.177	0.193
carbamate	4	0.080	1.009	6	0.194	-0.047
aliphatic (C(=O)OH)	33	0.073	0.643	10	0.386	0.365
alkyl substituent on aromatic ring	36	0.055	0.577	36	-0.069	-0.075
triazine ring	5	0.0095	-5.725	4	-0.058	-0.246
ketone (CC(=O)C)	12	0.0068	-0.453	10	-0.022	-0.023
aromatic F	1	-0.810	-10.532	1	0.135	-0.407
aromatic I	2	-0.759	-10.003	2	-0.127	-0.045
polycyclic aromatic hydrocarbon (≥4 rings)	6	-0.657	-10.164	2	-0.702	-0.799
N-nitroso (NN=O)	4	-0.525	-3.259	1	0.019	-0.385
trifluoromethyl (CF_3)	1	-0.520	-5.670	2	-0.274	-0.513
aliphatic ether	11	-0.347	-3.429	16	-0.0097	-0.0087
aromatic NO_2	14	-0.305	-2.509	13	-0.108	-0.170
azo group (N=N)	2	-0.242	-8.219	3	-0.053	-0.300
aromatic NH_2 or NH	32	-0.234	-1.907	23	-0.108	-0.135
aromatic sulfonic acid or salt	11	-0.224	-1.028	8	0.022	0.142
tertiary amine	10	-0.205	-2.223	10	-0.288	-0.255
carbon with 4 single bonds and no H	9	-0.184	-1.723	32	-0.153	-0.212
aromatic Cl	40	-0.182	-2.016	27	-0.165	-0.207
pyridine ring	18	-0.155	-1.638	8	-0.019	-0.214
aliphatic Cl	12	-0.111	-1.853	14	-0.101	-0.173
aromatic Br	5	-0.110	-1.678	4	-0.154	-0.136
aliphatic Br	5	-0.046	-4.443	2	0.035	0.029

[a] Number of compounds in the training set containing the fragment.

detailed assessment. However, industrial screening of chemicals in the premanufacture phase of development (i.e., during product design) would seem to be an equally valid application.

Designing Biodegradable Chemicals: Examples

Linear Alkylbenzene Sulfonates. The development of laundry detergents based on linear alkylbenzene sulfonate (LAS) is a brilliant success story, and a case can be made that this is still the best illustration to date of molecular engineering to enhance biodegradability and thus environmental acceptability. The replacement of soap as the workhorse surfactant in household laundry products occurred as early as the 1940s (*16*), with the development of manmade alkylbenzene sulfonate (ABS) surfactants. At first the alkyl chains were derived from a kerosene fraction, but these products were soon replaced by ABS produced from propylene tetramer. Tetrapropylene alkylbenzene sulfonate (TPBS) (**1**) was a more efficacious and economical product, obtained via a one-step Friedel-Crafts process involving addition of benzene at the double bond of the olefin feedstock to yield the branched alkylbenzene, followed by sulfonation of the benzene ring. As manufactured TPBS is actually a complex mixture but a typical structure is

1

Environmental problems with these highly branched products appeared almost immediately, as they were found to be incompletely biodegraded in municipal sewage treatment systems. Painter (*17*) offers a colorful description of what this meant in real terms:

> "[TPBS] was degraded by only about 50 % in sewage treatment units and as a result excessive foaming occurred in activated sludge aeration tanks, as well as in receiving rivers. The foaming was far worse than that caused by proteinaceous material in sewage prior to the introduction of synthetic surfactants and in extreme cases sewage-works operators were killed by asphyxiation after falling into foaming tanks from walkways made slippery by the foam...because of its incomplete biodegradation, the concentration of TPBS in river waters [was] as high as 2 mg L^{-1}, and water tended to foam when coming out of the tap."

Other results of the foaming were impaired efficiency of the treatment plants and increased dispersal of potentially pathogenic bacteria (*17*). Eventually methods were developed that permitted the economical manufacture of a more environmentally acceptable product LAS (**2**):

$$CH_3\!-\!\!\underset{\underset{n}{|}}{C}\!H\!-\!\!\underset{\underset{2}{|}}{(CH}\!-\!)_n\!\overset{\overset{CH_3}{|}}{C}\!H\!-\!\!\left\langle\!\!\!\begin{array}{c}\\\end{array}\!\!\!\right\rangle\!\!-\!SO_3H \qquad n = 7\text{-}11$$

2

This technology involved use of molecular sieves to obtain predominantly linear alkanes from petroleum, followed by any of several methods for producing the olefin.

LAS surfactants are almost completely biodegradable in sewage treatment, and this has been amply demonstrated in hundreds of studies, including numerous monitoring studies conducted at full-scale treatment plants (*18*). Voluntary changeover from TPBS to LAS was complete by the early 1960s in the U.S. (*16*). Of course we now take it for granted that the lower biodegradability of TPBS is due to its highly branched alkyl group, and that the greatly enhanced biodegradability of LAS is due to the absence of such branching.

Dialkyl Quaternaries. Surface-active quaternary ammonium compounds (generally referred to as QACs or "quats") first gained prominence more than 50 years ago, with Domagk's discovery that the biocidal properties of simple quaternary ammonium compounds were greatly enhanced by the presence of a long alkyl group (*19*). There are still many QAC-type biocides in use, but household fabric softeners presently constitute the largest market by far for QACs. Other applications are mainly industrial and include multiple uses in textile processing, road paving, oil well drilling and mineral flotation, to name just a few. According to Cross (*20*) 66 % of the market for QACs is dominated by three classes, all of which are dialkyl quaternaries, meaning that hydrophobicity is imparted to the molecule by two linear alkyl chains in the C_{10} to C_{18} range. The three classes are dialkyl dimethylammonium salts (**3**), imidazolium quaternary ammonium salts (**4**), and ethoxylated ethanaminium quaternary ammonium salts (**5**), with typical structures shown on the following page. Most uses of QACs, especially in high-volume fabric softeners, lead to their release to municipal wastewater treatment systems.

Until recently the fabric softener market was dominated by a QAC of the first type, dihydrogenated tallow dimethyl ammonium chloride (DHTDMAC). The long alkyl groups in DHTDMAC are derived, as the name implies, from purified animal fat (tallow), and consist of a mixture chiefly in the C_{16}-C_{18} (tallow fatty acids) range. The true aqueous solubility of DHTDMAC is exceedingly low, and the chemical sorbs strongly to solids in wastewater treatment and the environment. Removal in treatment is therefore high (> 95 %), unlike TPBS, but does not necessarily correspond to ultimate biodegradation (*21*). More importantly, the heavy use of DHTDMAC before 1990, its relatively low rate of biodegradation in aquatic sediments, and its high intrinsic ecotoxicity led to a public perception in some parts of the world, particularly Europe, that DHTDMAC was placing a critical load on surface waters. As a result DHTDMAC has been phased out of the European market, with consumption declining from a peak of about 50,000 metric tons per year before 1990 to 4,500-9,000 tons in 1993 (*22*). Similar changes are occurring in the U.S. market as manufacturers are voluntarily reformulating their products.

DHTDMAC is being replaced by dialkyl QACs in the other two classes listed above, the imidazolium and ethoxylated ethanaminium QACs. Although the database on environmental fate and removal in wastewater treatment of these latter compounds is less extensive than for DHTDMAC, the new dialkyl QACs seem to biodegrade more rapidly due to the linkage of the alkyl groups to the remainder of the molecule via hydrolyzable amide bonds (21). These new, more biodegradable fabric softening

$$CH_3-(CH_2)_n-\overset{\overset{CH_3}{|}}{\underset{\underset{CH_3}{|}}{N^+}}-(CH_2)_n-CH_3 \quad n = 9\text{-}17 \quad \mathbf{3}$$

$$n = 14\text{-}16 \quad \mathbf{4}$$

$$n = 14\text{-}16 \quad \mathbf{5}$$

agents thus represent another illustration of how safer surfactants can be developed by rational molecular design. The approach here may appear to be different from that for TPBS/LAS since it involves direct incorporation of a molecular feature known to favor biodegradability (Table I), not the deletion of an impediment (branching) as in TPBS. But this distinction is really not very meaningful since the replacement of tetrapropylene with a linear alkyl group in ABS/LAS also facilitates beta-oxidation by competent microorganisms. Attack on the alkyl group is in fact the principal degradation pathway for LAS (16).

Until now discussion has been limited to chemicals used in high-volume consumer products, but the same principles can be applied to products used chiefly in industry. A good example of appropriate technology is contained in a recent non-confidential PMN received by EPA. The notified chemical substance (6) proposed for use as a textile fiber finish, is essentially the ester analog of the amide-containing ethoxylated ethanaminium dialkyl QACs (5) used as fabric softening agents. Due to the incorporation of ester linkages which are known to enhance biodegradability (Table I) and in contrast to DHTDMAC, 6 is predicted to be readily biodegradable.

$$\text{CH}_3\text{-(CH}_2\text{)}_n\text{-C(=O)-O-CH}_2\text{CH}_2\text{-N}^+(\text{CH}_2\text{CH}_3)(\text{CH}_2\text{CH}_2\text{OH})\text{-CH}_2\text{CH}_2\text{-O-C(=O)-(CH}_2\text{)}_n\text{-CH}_3 \quad n = 10\text{-}16 \quad \mathbf{6}$$

Alkylphenol Ethoxylates. Alkylphenol ethoxylates (APEs), **7**, are one of two major classes of nonionic surfactants. APE uses are mainly industrial and cover a wide range, including applications in textile processing, emulsion polymerization, printing, metal cleaning, oil well drilling and papermaking. According to the Chemical Manufacturers Association (CMA) 450 million pounds of APEs were sold in the U.S. in 1988 (*23*). Nonylphenol ethoxylates (NPEs) represent the largest class of APEs, accounting for nearly 75 % of U.S. APE production in 1980. A good example of NPE use in industry is in printing. In screen reclamation NPEs are used in ink, emulsion and haze removal formulations, according to EPA's draft *Cleaner Technologies Substitutes Assessment* (CTSA) for the screen reclamation use cluster (*24*). And in a new CTSA being prepared for the lithographic blanket wash process, NPEs appear in 7 of 37 product formulations.

Unlike linear alcohol ethoxylates (the other major class of nonionic surfactants), APEs are mostly branched (*25*). The industrial synthesis of APEs and evolution of the manufacturing process are somewhat parallel to those of alkylbenzene sulfonates (*26*). As with alkylbenzene sulfonates the alkylphenyl portion of the molecule is synthesized by addition of an aromatic feedstock (phenol in this case) to the double bond of an olefin. At first the olefins were formed from polymerization of butene and isobutene, which resulted in highly branched alkyl groups with an abundance of quaternary carbons. These products long ago were replaced by APEs derived from propylene oligomers (represented by general structure **7**), which are structures without quaternary carbons but still containing branched alkyl groups as in TPBS.

$$\text{CH}_3\text{-CH(CH}_3\text{)-CH}_2\text{-CH(CH}_3\text{)-CH}_2\text{-CH(CH}_3\text{)-C}_6\text{H}_4\text{-(OCH}_2\text{CH}_2\text{)}_n\text{-OH} \quad \mathbf{7}$$

n = 12-14, typically

But the substitution of linear for branched olefins, which would logically constitute the next step in the evolution of safer APEs, has not occurred on a large scale. There may be various reasons for this, but the fact that the stringent biodegradability criteria adopted by the U.S. Soap and Detergent Association apply only to consumer products (*27*) would seem to be especially important, since APE uses are mainly commercial. This and the absence of obvious environmental problems such as foaming in waterways have translated into a lack of impetus for change.

The environmental risks associated with APEs and especially NPE are a complex and contentious issue. Most attention has focused on the mono- and diethoxylated nonylphenol adducts NP1EO (structure 7, n = 1) and NP2EO (structure 7, n = 2), which have been reported to be relatively stable intermediates in NPE biodegradation (26). NP1EO, NP2EO and nonylphenol itself are highly toxic to aquatic organisms, whereas the parent NPEs (the number of ethoxylate groups may be as high as 30-50, but 12-14 is more typical) are much less toxic. Recently added to this is a new controversy, as nonylphenol, NP2EO and related compounds have been reported to be estrogenic in fish (28).

The actual margins of safety under environmental conditions for the above effects are a subject of intense debate, but the current situation might not have arisen in the first place if APEs were manufactured mainly from **linear** olefins. Branched APEs biodegrade initially by stepwise shortening of the polyethoxylate hydrophile one ethoxylate group at a time (16), and attack on the **branched** alkyl group does not seem to be a significant pathway. Substitution of a linear alkyl group would eliminate an impediment to biodegradation and provide another site for microbial attack, by analogy to TPBS and LAS. This should lead to faster biodegradation and make transient accumulation of toxic intermediates unlikely. Experimental data support this contention (Table II). Indeed, studies (16) suggest that a fundamental shift in the principal route of breakdown occurs such that degradation by beta-oxidation of the alkyl group (not attack on the polyethoxylate chain) now predominates. Note that the enhanced biodegradability of linear APEs is easily predicted from information in Table I because the molecule now contains an additional fragment positive for biodegradability that is not present in branched APEs, the linear terminal alkyl with ≥4 carbons.

Table II. Biodegradability of Linear and Branched Alkylphenol Ethoxylates[a]

APE[b]	% Biodegradation Linear[c]	Branched	Method[d]	Analysis[d]
C_8APE_9	71	46	In, 28 d	Wt
	51	49	In, 20 d	Wt
C_9APE_9	65	25	RW, 15 d	CT
	65	30	SF, 5 d	CT
	88	55	CAS, 4 hr	CT
	57	33	In, 9 d	CT
	66	32	In, 9 d	ST
	75	0	In, 9 d	F
	62	10	SF, 7 d	CT
	60	18	SF, 7 d	ST
	0-50	0	SF, 7 d	F
	89	75	RW, 10 d	CT

[a] Adapted from Swisher (16). [b] APE = alkylphenol ethoxylate. C_8APE_9 signifies octylphenol with 9 ethoxy groups and C_9APE_9 signifies nonylphenol with 9 ethoxy groups. [c] Linear secondary. [d] Abbreviations: In=natural or synthetic medium inoculated with acclimated or unacclimated microorganisms; RW=river water die-away; SF=shake-flask culture; CAS=continuous flow activated sludge; Wt=weight of soluble organics in cell-free medium; CT=cobalt thiocyanate; ST=surface tension; F=foaming properties.

Other Pollution Prevention Opportunities

Other opportunities for pollution prevention by molecular design exist in industry. Textile processing seems especially amenable to this because surfactants have many uses as antistats, dye leveling agents, lubricants, etc., and substantial volumes of wastewater that must be treated before being released are generated. An ethoxylated ethanaminium dialkyl QAC proposed for use as a fiber finish that illustrates appropriate molecular design for enhanced biodegradability has already been mentioned. An example of molecular design that hinders rather than improves biodegradability is furnished by a class of anionic surfactants with general structure **8**, shown on the next page. In general structure **8**, R_1, R_2 and R_3 consist of one or two long alkyl groups and any other R groups are H (M is the counterion). This class of surfactants has been the subject of numerous PMNs over more than 10 years at EPA, and most but not all proposed uses have been in textile processing. In most of these PMNs and possibly all (it's not always possible to determine) the proposed source of alkyl groups has been tetrapropylene, despite Swisher's admonition (*16*) that "Polypropylene derivatives such as [TPBS] are obsolescent because of the exceptional biological stability of some of the components."

8

Another example from EPA's PMN program is a group of alkylphosphate surfactants (**9**) with similar uses as in the preceding example. Isooctadecanol phosphate (structure **9**, n = 10) is typical and contains a highly branched hydrophobe.

9

Here the specific uses proposed (which were expected to result in 100 % release to water) and hydrolyzable phosphate group suggest an additional concern for eutrophication potential. Whether this effect would ever be realized would of course depend on the production volume, circumstances of release and other factors. Nevertheless, Gleisberg (*29*) has strongly recommended that phosphate inputs to surface waters be reduced significantly in the interest of environmental protection.

The last example of molecular design that hinders rather than enhances biodegradability is the incorporation of propoxylate groups into polyethoxylated nonionic surfactants. EPA frequently receives PMNs for such compounds and no doubt many enter commerce eventually. Data on biodegradation of mixed ethylene oxide (EO)/propylene oxide (PO) surfactants are very sparse in comparison to EO-only nonionics, but there is clear evidence that PO groups markedly retard biodegradation (*16*). In designing such products the synthetic chemist is undoubtedly maximizing some functional property of the surfactant, since the incorporation of propoxylate groups allows control over the hydrophobicity of the final product. For example, low-foaming nonionics can be prepared by capping polyethoxylates with polyoxypropylene groups (*27*). The trouble with this is that since biodegradation would otherwise proceed stepwise from the free EO end as noted above, terminating the EO chain with a PO function effectively prevents biodegradation or at least greatly lowers the rate. Thus we can be sure that rational design from an environmental safety viewpoint is **not** foremost in the chemist's mind.

The Future

Absent specific knowledge of a chemical's environmental behavior, the way to design more biodegradable chemicals is to incorporate molecular features such as ester linkages and hydroxyl groups, and exclude halogens, quaternary carbons, nitro groups and the like. The positive and negative features highlighted earlier in the text are a good starting point. Table I contains a more extensive list of relevant fragments and coefficients from the BIODEG model (*7*). The BIODEG model provides a convenient way for chemists in research and development to quickly compare alternative designs.

Product performance and economics obviously are as closely linked as biodegradability is to molecular structure. This makes the task of following these recommendations potentially very difficult, and it further suggests that chemical design is properly a multidisciplinary process. It's easy to see the need to consider biodegradability for high-volume, down-the-drain chemicals used in consumer products. But manufacturers, processors and users of chemicals whose uses are mainly commercial must also try harder to make this a part of their economic equation. Often the balancing process may demand compromises in economics or function that were not previously necessary. But this change should be viewed as merely one part of the process of internalizing environmental costs. The desirability of integrating environmental considerations into business decisions and designing products to minimize their environmental impact are nothing new and have been acknowledged in CMA's Responsible Care program.

Resistance to biodegradation sees its ultimate expression in a broad class of compounds referred to as persistent organic pollutants or POPS. POPS are currently receiving considerable attention in a number of international fora in which the U.S. is engaged, including but not limited to the U.N. Economic Commission for Europe (ECE) Convention on Long-Range Transboundary Air Pollution (LRTAP); the U.S.-Canada International Joint Commission (IJC); the Global Program of Action on land-based marine pollution under the United Nations Environment Program (UNEP); and the Commission on Sustainable Development (CSD). The subset of POPS receiving most of the attention consists of halogenated compounds characterized by low water solubility, high fat solubility and a tendency to volatilize and migrate to colder latitudes, where they accumulate in fatty tissues of animals at the highest levels of the food chain.

Debate over these substances focuses on which chemicals should be targets for action and what actions to take, but **not** on the need to limit their release. Persistent chemicals are of concern because they may remain available to biota for long periods of time, and thus may build up in exposed organisms to levels that cause toxic effects even if releases to the environment are small or infrequent. Through the process of bioaccumulation levels may be achieved that appear safe on the basis of acute toxicity criteria, but which ultimately prove harmful to the organisms in chronic exposures, or even to other organisms that are consumers of the exposed organism. Human populations may be exposed to such chemicals through consumption of contaminated fish, and history contains numerous examples of cases in which accumulation of chemical substances from the ambient environment by fish has led to decreases in reproductive success in fish-eating birds.

Halogenated hydrocarbon insecticides like aldrin and DDT and their metabolites are now ubiquitous global contaminants. The obvious lesson from this is that we should not design, manufacture and release to the environment **new** substances that are so persistent. The less obvious but equally compelling message to be derived from the various examples given above is that **all** industrial chemicals that may be released in substantial amounts should reflect the principles of safe design to the extent practicable. There are four reasons why we must pursue this goal:

(i) **we cannot know in advance all of the possible toxic effects of released chemicals.** This is certainly true for new substances that have a minimum of test data, but it is also true for well-studied existing chemicals as witnessed by controversy over nonylphenol ethoxylates;

(ii) **consumption of a chemical, and therefore, potentially, its release to the environment, may increase significantly if the product is successful;**

(iii) **new uses may develop over time, and with them, a greater possibility of environmental release.** This is especially true for new substances because the original manufacturer typically envisions one or more specific uses, but the marketplace is the arena in which a chemical's utility is fully revealed. For example, a surfactant with a very limited market niche may find its way into broader uses in textile processing, papermaking or another industry that generates large volumes of wastewater; and

(iv) recognizing that we live in an era of global markets and of concern for the global environment, **chemical substances produced in the U.S. may be exported to other nations where environmental controls are less stringent.**

Table III shows the degree of connection of sewers to municipal wastewater treatment plants in Europe before 1989 (*30*) and is revealing in this regard. Coupled with this is the fact just noted that substances with certain properties may be transported regionally or globally from one nation to another even if banned from use in the latter. This is in fact happening with chemicals like DDT which are still used in developing nations. By incorporating into a molecule features that reduce toxicity and enhance biodegradation to nontoxic products, pollution is reduced at the source and the risk of environmental damage is thereby diminished.

Table III. Degree of Connection of Sewers to Wastewater Treatment Plants in Europe Before the German Reunification[a]

Nation	% Connected	Nation	% Connected
Portugal	13	United Kingdom	84
Spain	29	West Germany	87
Italy	30	Netherlands	90
France	50	Denmark	98
East Germany	58		

[a] From van Leeuwen (*30*).

Literature Cited

1. *Benign by Design. Alternative Synthetic Design for Pollution Prevention*; Anastas, P. T.; Farris, C. A., Eds.; ACS Symposium Series 577; American Chemical Society: Washington, DC, 1994.
2. Chapman, P. J. In *Workshop: Microbial Degradation of Pollutants in Marine Environments*; Bourquin, A. W.; Pritchard, P. H., Eds.; EPA report no. 600/9-79-012; U.S. Environmental Protection Agency: Gulf Breeze, FL, 1979; pp 28-66.
3. Gibson, D. T. In *The Handbook of Environmental Chemistry*; Hutzinger, O., Ed.; Springer-Verlag: Berlin, Germany, 1980, Vol. 2A; pp 161-192.
4. Dagley, S. *Ann. Rev. Microbiol.* **1987**, *41*, pp 1-23.
5. Scow, K. M. In *Handbook of Chemical Property Estimation Methods*; Lyman, W. J.; Reehl, W. F.; Rosenblatt, D. H., Eds.; McGraw-Hill: New York, NY, 1982; pp 9-1 to 9-85.
6. Howard, P. H.; Boethling, R. S.; Stiteler, W. M.; Meylan, W. M.; Hueber, A. E.; Beauman, J. A.; Larosche, M. E. *Environ. Toxicol. Chem.* **1992**, *11*, pp 593-603.
7. Boethling, R. S.; Howard, P. H.; Meylan, W. M.; Stiteler, W. M.; Beauman, J. A.; Tirado, N. *Environ. Sci. Technol.* **1994**, *28*, pp 459-465.
8. Howard, P.H.; Hueber, A.E.; Boethling, R.S. *Environ. Toxicol. Chem.* **1987**, *6*, pp 1-10.
9. Howard, P.H.; Sage, G.W.; LaMacchia, A.; Colb, A. *J. Chem. Inf. Comput. Sci.* **1982**, *22*, pp 38-44.
10. Niemi, G. J.; Veith, G. D.; Regal, R. R.; Vaishnav, D. D. *Environ. Toxicol. Chem.* **1987**, *6*, pp 515-527.
11. Gombar, V. K.; Enslein, K. In *Applied Multivariate Analysis in SAR and Environmental Studies*; Devillers, J; Karcher, W., Eds.; Kluwer: Boston, MA, 1991; pp 377-414.
12. Klopman, G.; Balthasar, D. M.; Rosenkranz, H. S. *Environ. Toxicol. Chem.* **1993**, *12*, pp 231-240.
13. Tabak, H. H.; Govind, R. *Environ. Toxicol. Chem.* **1993**, *12*, pp 251-260.
14. Boethling, R. S.; Gregg, B.; Frederick, R.; Gabel, N. W.; Campbell, S. E.; Sabljic, A. *Ecotoxicol. Environ. Saf.* **1989**, *18*, pp 252-267.
15. Weininger, D. *J. Chem. Inf. Comput. Sci.* **1988**, *28*, pp 31-36.
16. Swisher, R. D. *Surfactant Biodegradation*, 2nd ed.; Surfactant Science Series, Vol. 18; Marcel Dekker: New York, NY, 1987.
17. Painter, H. A. In *The Handbook of Environmental Chemistry*; Hutzinger, O., Ed.; Springer-Verlag: Berlin, Germany, 1992, Vol. 3F; pp 1-88.
18. Rapaport, R. A.; Eckhoff, W. S. *Environ. Toxicol. Chem.* **1990**, *9*, pp 1245-1257.
19. Fredell, D. L. In *Cationic Surfactants*; Cross, J.; Singer, E. J., Eds; Surfactant Science Series, Vol. 53; Marcel Dekker: New York, NY, 1994; pp 31-60.

20. Cross, J. In *Cationic Surfactants*; Cross, J.; Singer, E. J., Eds.; Surfactant Science Series, Vol. 53; Marcel Dekker: New York, NY, 1994; pp 3-28.
21. Boethling, R. S. In *Cationic Surfactants*; Cross, J.; Singer, E. J., Eds.; Surfactant Science Series, Vol. 53; Marcel Dekker: New York, NY, 1994; pp 95-135.
22. European Centre for Ecotoxicology and Toxicology of Chemicals (ECETOC). *DTDMAC:-Aquatic and Terrestrial Hazard Assessment. CAS No. 61789-80-8*; Technical Report No. 53; ECETOC: Brussels, Belgium, 1993.
23. Chemical Manufacturers Association (CMA). *Alkylphenol Ethoxylates. Human Health and Environmental Effects*; Interim Report; CMA: Washington, DC, 1991.
24. U.S. Environmental Protection Agency (EPA). *Cleaner Technologies Substitutes Assessment. Industry: Screen printing. Use Cluster: Screen Reclamation*; EPA report no. 744R-94-005; U.S. EPA: Washington, DC, 1994.
25. Naylor, C. G. *Environmental Fate of Alkylphenol Ethoxylates*; paper presented at the Annual Meeting of the U.S. Soap and Detergent Association, Boca Raton, FL, 1992.
26. Holt, M. S.; Mitchell, G. C.; Watkinson, R. J. In *The Handbook of Environmental Chemistry*; Hutzinger, O., Ed.; Springer-Verlag: Berlin, Germany, 1992, Vol. 3F; pp 89-144.
27. Cahn, A.; Lynn, Jr., J. L. In *Kirk-Othmer Encyclopedia of Chemical Technology*, 3rd ed.; Grayson, M.; Eckroth, D., Eds.; Wiley: New York, NY, 1983, Vol. 22; pp 332-432.
28. Jobling, S.; Sumpter, J. P. *Aquat. Toxicol.* **1993**, *27*, pp 361-372.
29. Gleisberg, D. In *The Handbook of Environmental Chemistry*; Hutzinger, O., Ed.; Springer-Verlag: Berlin, Germany, 1992, Vol. 3F; pp 179-203.
30. van Leeuwen, C.J. *Expert Meeting on Ultimate Biodegradability. Proceedings*; Consultative Expert Group Detergents-Environment: Bilthoven, the Netherlands, 1991; pp 37-46.

Chapter 9

Designing Aquatically Safer Chemicals

Larry D. Newsome, J. Vincent Nabholz, and Anne Kim

Health and Environmental Review Division, Office of Pollution Prevention and Toxics, U.S. Environmental Protection Agency, Mail Code 7403, 401 M Street, SW, Washington, DC 20460—0001

Aquatic species are essential components of aquatic ecosystems. Chemical substances that are toxic to aquatic organisms can disrupt aquatic food webs and threaten the survival of other parts of these systems, i.e., birds and mammals. Elucidation of structure-activity relationships (SARs) of the toxic effects caused by commercial chemical substances on aquatic species has led to a better understanding of the relationship between aquatic toxicity, chemical structure, and physicochemical properties, and provides a rational basis for designing environmentally safer chemicals. This chapter discusses the SARs of aquatic toxicity for a variety of industrial chemicals that include among others dyes, polymers, and surfactants, and provides insight into how substances can be redesigned or structurally modified such that aquatic toxicity is reduced.

Aquatic species have their own unique roles in ecosystems and are important for the subsistence of other species, including consumer and predator. Among other things, aquatic species comprise the components of food chains that lead to man. Green algae are the primary producers in aquatic ecosystems in that, through photosynthesis, they produce oxygen and synthesize carbohydrates and other foodstuffs. These substances are used by consumer species which in turn serve as the food source for predator species, including fish. Fish in turn, serve as a food source for wild mammals, birds, and man. Chemicals that are toxic to aquatic species may therefore put an ecosystem at unreasonable risk of harm, and can lead to disruption of some food chains. The survival of terrestrial species, including humans, is at least partially dependent upon aquatic organisms.

There are two general types of chemical-induced lethality in aquatic organisms: non-specific (i.e., narcosis), and specific. The majority of chemicals that are toxic to aquatic species are toxic by narcosis. Some have excess (or additional) toxicity due to a more specific mode of toxic action. Chemicals that are toxic from a specific mechanism are generally those that can react with cellular macromolecules, and may produce excess toxicity relative to the toxicity caused by narcosis, if a specific mechanism were not involved. Such chemicals may, for example, react covalently with enzymes. Chemicals with only a narcotic mode of toxic action are those that do

This chapter not subject to U.S. copyright
Published 1996 American Chemical Society

not react with cellular macromolecules, and represent a variety of chemical classes including chlorinated hydrocarbons, alcohols, ethers, ketones, weak organic acids and bases, and simple aromatic nitro-compounds, to name just a few *(1)*.

The mechanistic basis of narcosis toxicity is the ability of a chemical to diffuse across the biological membranes of an aquatic organism (e.g., gills in fish) and, once a high enough concentration is reached within the cells or in cellular membranes, cause non-specific perturbations in cellular function. These perturbations can lead to death if a sufficient concentration of a chemical has diffused into or across cellular membranes. Because cellular membranes have a higher lipid content, they are more readily penetrated by non-polar, lipid-soluble chemicals than by lipid-insoluble polar chemicals *(2, 3)*. Thus, the relative toxic potency of a non-polar substance that acts through a narcosis mechanism is a function of its lipophilicity. Narcosis, in principle, corresponds mechanistically to the same non-specific mode of action as induced by gaseous anesthetic drugs *(4)*. Narcosis toxicity represents baseline or minimum toxicity.

In contrast to chemicals that are only toxic by narcosis, some chemicals are toxic to aquatic organisms as a result of a chemical reaction or a specific interaction between the chemical (or its metabolite) with a critical cellular macromolecule. These chemicals exhibit excess toxicity to that of narcosis and is termed specific or reactive toxicity. For example, excess toxicity can be expected to result if a chemical (or its metabolite) can covalently bond to critical protein molecules (e.g., enzymes, DNA) or interact covalently with cellular receptor sites *(5-8)*. Examples of the types of chemicals that may exhibit excess (specific-type) toxicity to aquatic organisms are cyanogens, electrophiles (e.g., alkylating agents), and chemicals metabolized to electrophilic chemicals *(5-8)*.

Use of Structure-Activity Relationships to Predict Aquatic Toxicity.

During the 1970s it was realized that many industrial chemicals could cause significant toxic effects to humans and the environment. It was also realized that, unlike the requirement for premarket approval of drug chemicals by the Food and Drug Administration (FDA), no laws existed which required risk evaluation of industrial chemicals prior to their manufacture and use in commerce. As a result, the United States Congress passed the Toxic Substances Control Act (TSCA) in 1976. Unlike other federal statutes, which regulate a chemical risk after the chemical has been introduced into commerce, one of the major objectives of TSCA is to characterize and understand the risks a new chemical poses to humans and the environment *before* the chemical is introduced into commerce, so that such risk can be minimized or prevented. Consequently, the Act called for EPA to establish a list of chemicals already used in commerce, known as the TSCA Inventory. Section 5 of TSCA requires anyone who wishes to manufacture or import a chemical that is not on the TSCA Inventory to submit a Premanufacture Notice (PMN) of the "new" chemical to the Environmental Protection Agency (EPA), so that the Agency can evaluate the potential risks posed by the chemical. Under TSCA the Agency is allowed only 90 days to determine whether the new (PMN) chemical may or will present an unreasonable risk to human health or the environment. TSCA empowers the Agency to take appropriate regulatory action to protect the public and the environment against chemicals that in the Agency's opinion may or will present an unreasonable risk to human health and the environment. More than 30,000 PMN submissions have been reviewed by the EPA since the enactment of TSCA, with a current rate of more than 2,000 per year. More details of PMN review have been published elsewhere *(9-13)*.

TSCA requires as a minimum for Agency review that PMNs include chemical structure and identity, method of manufacture, production volume, use, disposal, and the identity and concentrations of any known byproducts or impurities *(7, 14, 15)*. TSCA does not require the submitter of a PMN to perform any chemical testing (including toxicity testing) prior to its submission to the Agency. Consequently, the majority of PMNs reviewed by the Agency contain little, if any, health or environmental toxicity test data. For example, only about 5% of PMN submissions contain aquatic toxicity data *(13)*. The sparsity of toxicological data available for new chemicals notwithstanding, the EPA is nonetheless charged with the responsibility of assessing the risks posed by all new chemicals.

In order to identify risks posed by new chemicals for which no data are available, and to do so under the strict time constraints prescribed by TSCA, the EPA bases many of its toxicity assessments on structure-activity relationships (SARs). SARs refer to the ability of a group of analogous chemicals to produce a particular biological effect, and the influence that the structural differences between the chemicals have on relative potency in producing the biological effect. (A more detailed definition and discussion of SARs are provided in Chapter 2.) SARs have been used for decades by medicinal chemists as an aid in the design of new drugs. The EPA has used SARs since 1979, as an aid in assessing the risks of new chemicals to human health and aquatic species, when the TSCA-mandated premanufacture review of new chemicals by the EPA began. With respect to the EPA's use of SARs, structural analogues of a new chemical for which toxicity test data are available are identified and used as surrogates for assessing the toxicity of the new chemical. New chemicals that are structurally analogous to highly toxic chemicals may be regarded by the EPA to be highly toxic as well. Conversely, new chemicals that are structurally analogous to chemicals that are known to be essentially non-toxic or to have mild toxicity may be regarded by the EPA as having low toxicity. Fish, daphnids, and green algae are the three aquatic groups routinely used by the EPA as surrogates for aquatic organisms during the SAR-based aquatic toxicity assessment of new chemicals.

Quantitative Structure-Activity Relationships. A more rigorous method for estimating the toxicity of chemicals is known as quantitative structure-activity relationships (QSARs). A QSAR is a regression model that correlates toxicity with one or more physicochemical properties for a series of analogous chemicals. The relationship between molecular structure and toxicity is represented in the form of a regression equation, which expresses the toxicity of the chemical. These physicochemical properties may include, for example, octanol-water partition coefficient (usually expressed as its logarithm, log P, and also commonly known as log K_{ow}), water solubility, dissociation constant (pKa), molecular weight, and percent amine nitrogen (excluding anilines, unless they are quaternized, and nitrogen-containing chemicals that are not basic at pH 7, i.e., amides, ureas). By quantifying structure-activity relationships, one can predict toxicity more precisely than is possible by visual comparison with analogs. (A detailed discussion of QSARs is given in Chapter 2).

The critical aspect in quantifying the structure-activity relationships of a given series of chemicals is the identification of the physicochemical property (or properties) that accurately correlates structure with biological activity. For QSAR equations developed to estimate aquatic toxicity, log P is often used as a descriptor because it governs to a large extent the transport and distribution of a chemical between biophase and water. The critical aspect in *using* a QSAR equation to predict the biological activity of an untested chemical is that one must be reasonably certain that the QSAR equation chosen for the prediction is appropriate. With respect to aquatic toxicology, prediction of the toxicity of a substance that is likely to be from a specific mechanism should not be performed with a QSAR equation derived from analogous chemicals whose aquatic toxicity is due to narcosis unless a minimum estimate of toxicity is desired. Such use of a QSAR equation may lead to an underestimate of toxicity *(16-20)*. The EPA's Environmental Effects Branch (EEB) has developed QSAR equations for a wide variety of chemical classes, and uses these equations to assess aquatic toxicity of chemicals *(4, 7, 8, 16-21)*.

Considerations in Designing Aquatically Safer Chemicals: Modification of Physicochemical Properties and Structure.

The preceding section discusses fundamental concepts of aquatic toxicity and approaches that may be used to assess aquatic toxicity. It is important to stress that there are a number of physical and chemical factors that can influence aquatic toxicity. These factors include water solubility, lipophilicity, color, formation of inner salts, acidity, basicity, molecular size, minimum cross-sectional diameter, and physical state, to name a few. Knowledge of how these factors relate to aquatic toxicity can aid the chemist in designing chemicals that are less toxic to aquatic organisms. For example, designing a chemical such that it has very poor solubility (less than 1 ppb) in water or by increasing the log P to greater than 8 will in most cases decrease its bioavailability and therefore its toxicity to aquatic organisms. In addition, the chemist can use these physical and chemical factors in combination with QSAR equations developed for predicting aquatic toxicity as a tool to design less toxic chemicals. Here a chemist can take chemicals that are planned for development and predict their aquatic toxicity using a QSAR equation previously developed for the chemical class to which the planned chemicals belong. If, for example, log P is the descriptor in the QSAR equation, the chemist can predict toxicity for each planned substance using the substance's log P value. The chemist can then choose the least toxic substance for further development or, if necessary, may wish to design a new analogous substance that will have a log P value that when used in the QSAR equation corresponds to a substance of lower toxicity.

Generic Modification of Physicochemical Properties and Chemical Structure. The remainder of this chapter will discuss some of the known physicochemical and structural factors common to most chemicals related to aquatic toxicity, and provide insight into how these factors may be modified through structural manipulation for purposes of lowering aquatic toxicity. Unless otherwise specified, the term "chemical(s)" refers to nonionic non-reactive organic chemical monomers, and the term "toxicity" assumes a narcosis (non-specific) mechanism.

Octanol-Water Partition Coefficient (Log P or Log Kow). The octanol-water partition coefficient (log P) is a term used to express a substance's lipophilicity. It is the physicochemical property most frequently used to estimate the aquatic toxicity of organic chemicals. This is because log P often correlates well with biological activity. It is helpful that many of the discrete chemicals from the TSCA Inventory are organic substances for which log P can be calculated (22). In general, for nonionic organic chemicals that exhibit only a narcotic mode of toxic action and are not dyestuffs, polymers, or surfactants, acute toxicity (lethality) and chronic toxicity will increase exponentially with increasing lipophilicity until about log P of 5. Above log P of 5, toxicity following acute exposure decreases exponentially because bioavailability begins to decrease. Only chronic toxicity following chronic exposure occurs between log P of 5 and 8 for nonionic organic chemicals. For substances with a log P of 8 or greater, toxicity seen during chronic exposures becomes essentially nonexistent. This is because the aqueous solubilities of such chemicals are very low and, consequently, these substances have poor bioavailability.

"Narcotic" chemicals such as aliphatic alcohols, chlorinated benzenes, ketones, disulfides, and "reactive" chemicals such as acrylates, and esters demonstrate acute aquatic toxicity up to log P of 6. However, some reactive chemicals, such as, aliphatic amines and surfactants that have log P values greater than 8 may still exhibit acute toxicity. Log P cut-offs and magnitude of toxicity following acute exposures have been identified for several chemical classes (25), and are shown in Figure 1. On the other hand, chemicals with low log P values are not sufficiently lipophilic to enter the cellular membranes of aquatic organisms, and therefore, are not bioavailable and are of low toxicity. For example, chemicals with molecular weights of less than 200 daltons and have a log P of 2 or lower have low toxicity to aquatic species, i.e., acute median lethal concentration (LC_{50}) values greater than 100 mg/L.

Thus, for chemicals whose aquatic toxicity is due to narcosis, chronic and acute aquatic toxicity is minimal when log P is greater than 8 or less than 2, respectively. Chemists can reduce the potential for aquatic toxicity by designing chemicals such that they will have molecular weights of less than 200 daltons and log P values less than 2, or log P greater than 8, regardless of molecular weight. A number of computer programs such as CLOGP (Pomona College) and LOGKOW (Syracuse Research Corporation) are available for the estimation of log P, and can be of great aid to chemists. To decrease log P, polar substituents such as carboxylic, alcoholic, or other water soluble groups can be added to chemicals, and the resultant log P of the modified substance can be calculated. On the other hand, log P can be increased by adding hydrophobic groups such as halogens, phenyl rings, and alkyl groups.

Water Solubility. Water solubility is another very important physicochemical property with respect to a chemical's bioavailablity and aquatic toxicity. As log P increases water solubility decreases, and vice versa. Narcotic chemicals which are either very poorly (e.g., less than 1 part per billion) or highly soluble (miscible) in water are generally not sufficiently bioavailable to be significantly toxic to aquatic organisms. As a general rule, chemicals having water solubility less than 1 ppb are essentially non-toxic to aquatic species due to low bioavailability. It is difficult to state quantitatively what the upper water solubility value is in which aquatic toxicity becomes insignificant. Thus, to reduce aquatic toxicity, chemists should try to incorporate structural modifications that makes a chemical either water insoluble or miscible.

A word of caution here is that even seemingly subtle changes in a molecule, such as the selective placement of a methyl group, can greatly influence water solubility (and, thus, lipophilicity and toxicity). This is an important point to consider if one must design a relatively small substance and is attempting to minimize aquatic toxicity by incorporating structural modifications that lower water solubility. For example, tertiary amyl alcohol **1** is 98 g/L more water soluble than its isomer *n*-pentanol **2** *(23),* and the larger molecular weight 2-aminobutyric acid **3** is 44 g/L more soluble than 2-aminopropionic acid **4** *(24).* These differences in water solubility that result from relatively subtle structural differences can be ascribed, in both cases, to hydrogen bonding caused by chain folding. These chemicals' higher water solubilities reduces their aquatic toxicities to fish, daphnids, and algae by about one half based on predicted values using a neutral organic QSAR *(25).* It is true that, because the methyl group is lipophilic, it usually lowers water solubility, and this example is an exception to the rule. In addition, linear alkyl groups are preferred over branched groups in the redesign of safer chemicals due to their potential to biodegrade more quickly than branched alkyls.

Property differences reported by the EPA *(26)* on safer substitutes based on QSAR using alcohols to reduce water solubility has led to the patented synthesis of new alcohol-substituted terpene cleaning agents *(27)* . The advantages of using an alcohol-substituted terpene rather than the terpene hydrocarbons are listed in the terpene industry's marketing advertisement, shown in Figure 2. The chemical's redesign by the addition of one alcohol group reduced aquatic toxicity by a factor of 40 and also improved several other properties.

Molecular Size and Weight. Another way to decrease the toxicity of a substance is to design a new analog that is larger in molecular weight. The new analog will be less toxic to aquatic organisms if the new chemical's molecular weight is increased while holding all other factors constant. Generally, as molecular weight increases toxicity decreases. At a molecular weight of 1000 daltons, uptake will be

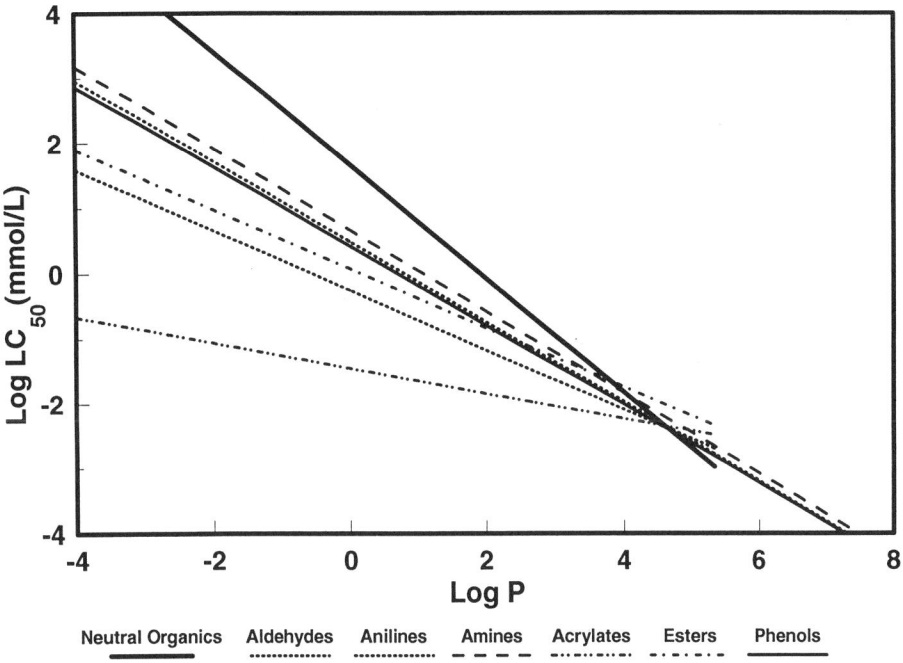

Figure 1. Octanol-Water Partition Coefficient (log P) Cut-Offs and Predicted Magnitude of Fish Acute Toxicity (expressed as median lethal concentration, LC_{50}) for Several Chemical Classes Using Equations from ref 25.

PROPERTY DIFFERENCES	
Tarksol® (terpene alcohol)	**d-Limonene** (terpene hydrocarbon)
1.) Flash point >200 degrees F. (TCC)*	1.) Flash point 128 degrees F. (TCC)*
2.) Controlled odor	2.) Odor increases with oxidation
3.) Aquatic toxicity- 41 mg/L	3.) Aquatic toxicity-1 mg/L
4.) Synthetic renewable supply	4.) Weather & demand
5.) High solvency kb>500	5.) Low solvency kb<100
6.) Emulsion Fl. Pt. >212 degrees F.	6.) Emulsion Fl. Pt. still 128 degrees F.
7.) Can be heated (OSHA regulation)	7.) Can't be heated (OSHA regulation)
8.) No residue	8.) Oxidized d-limonene leaves residue
9.) Less toxic to humans	9.) Moderate toxicity to humans

* Tag Closed Cup is the DOT regulated flash point method for test.

Figure 2. Property Differences Between d-Limonene (terpene hydrocarbon) and Tarksol (terpene alcohol). Note the large difference in fish toxicity. (Reproduced with permission from ref. 27).

1: (CH₃)₃C-OH (tert-butanol structure) **2**: CH₃-CH₂-CH₂-CH₂-OH (n-butanol structure)

3: CH₃-CH₂-CH(NH₂)-COOH **4**: (CH₃)₂C(NH₂)-COOH

negligible because the chemical will not diffuse across the respiratory membranes of aquatic organisms. The aquatic toxicity of chemicals also tend to decrease if molecular size (e.g., minimum cross-sectional diameter) is sufficiently increased. This is because as molecular size increases the molecules become too large to be taken up through respiratory membranes of aquatic organisms. Substances with large molecular sizes are polymers with molecular weights more than 1000 daltons, and large cross-sectional diameter monomeric chemicals.

Chemicals with minimum cross-section diameters greater than 10Å are too large for passive diffusion and uptake through the respiratory membranes of aquatic organisms *(28-31)*. Neutral phthalocyanine dyes, for example, are of low acute and chronic toxicity to aquatic organisms because their minimum cross-sectional diameters are greater than 10Å. The aquatic toxicity of chemicals with molecular weights less than 1000 daltons can be reduced by redesigning the chemical such that its cross-sectional diameter is greater than 10Å.

Ion Pairs. Chemical salts exist as strong ion pairs when an ion and its counterion are associated strongly with one another. These chemicals dissociate weakly or not at all in water at pH 7 and 25 °C, and, consequently, may have low water solubilities and low aquatic toxicity. However, some strong ion pairs, suich as that between a cationic surfactant and an anionic surfactant, may be self-dispersing in water. If a soluble and charged chemical can be converted to a strong ion pair and still retain its usefulness in a use application, then the resulting substance will be less toxic to aquatic organisms. For example, a soluble cationic surfactant can be highly toxic to fish, aquatic invertebrates, and algae. In addition, some pesticides are formulated with either cationic or anionic surfactants i.e., inert ingredients to increase their bioavailability. If the same pesticides could be formulated with the cationic surfactant: anionic surfactant strong ion pair (assuming a 1:1 ratio), then the toxicity of the surfactant(s), could be reduced more than 100-fold. Here the formation of a strong ion pair to disperse chemical pesticides result in a safer commercial product, as opposed to a surfactant/dispersant that is toxic to aquatic organisms.

Zwitterions. Zwitterions and are substances that contain positively and negatively charged groups. Acid Blue 1 **(5)** is an example of a zwitterionic substance. Zwitterions generally have low toxicity to aquatic organisms, provided that the substance has balanced charges and does not have surfactant properties (i.e., alkyl hydrophobic chains connected to positive and negative charged groups). (For further

[Structure 5: a triphenylmethane-type dye with $\overset{+}{N}(CH_2CH_3)_2$ group on one ring, SO_3^- and $Na^+ \, {}^-O_3S$ substituents on another ring, and $N(CH_2CH_3)_2$ on the third ring]

5

discussions see Amphoteric Surfactants.) In general, most zwitterions are not surfactants, and toxicity to fish and daphnids is low. However, algal toxicity has been observed at less than 10 mg/L for some chemicals *(32)*.

Chelation. Chelation is the process in which a metal cation becomes bound within an organic molecule. In chelation, two or more polar electron-donating atoms (e.g., oxygen, nitrogen, etc.) within an organic molecule form a complex with a multivalent metal (e.g., Co^{2+}, Ni^{2+}, Cu^{2+}, Zn^{2+}, Mg^{2+}, Ca^{2+}) to form a ring-like structure. In general, chelation through oxygen and nitrogen ligands takes place only when a five- or six-membered ring can be formed. Chelation can also occur through sulfur atoms with the formation of a stable four-membered ring. Substances that can chelate polyvalent metals are often toxic to algae in soft water situations because they deprive the algae of essential nutrient elements, e.g., Ca^{2+}, Mg^{2+}, or Fe^{2+}. Therefore, algal toxicity can be reduced if a chelator is bound with a polyvalent metal before exposure to algae takes place or if the chemical is released to surface waters which have moderate to high hardness.

Modification of Chemical Structure via Chemical Class

Narcosis versus Excess Toxicity. Organic chemicals that are aquatically toxic due to a specific mode(s) of toxic action are more toxic on a molar basis than related chemicals with similar lipophilicities but whose general mode of toxic action is via narcosis. Examples of chemicals that are toxic by specific mechanisms include: electrophiles such as epoxides, alkyl halides, acrylates, and aldehydes; certain esters; dinitrobenzenes; and thiols, to name a few. The electrophilic chemicals may combine covalently with nucleophiles located in receptors of intracellular macromolecules. This bonding results in a cellular change that is difficult to reverse, and results in irreversible toxicity. (Narcotic toxicity can be reversible, depending upon the extent of exposure of the aquatic organism to the substance). Some chemicals that are not electrophilic are metabolized to chemicals that are electrophilic following their absorption into the biological systems of aquatic species *(7)*. These chemicals are termed "proelectrophiles".

The aquatic toxicity of a chemical that is toxic from a specific mechanism is usually considerably greater than that predicted for the substance using a QSAR equation developed for chemicals that are toxic by narcosis *(4, 7)*. The chemicals that are toxic from a specific mechanism are those that show "excess toxicity," because the actual toxicity is in "excess" of predicted toxicity if a narcosis mechanism is assumed. Thus, one must be careful not to use a QSAR equation that assumes narcosis to predict the toxicity of a chemical that is electrophilic or is likely to be metabolized to electrophilic substances. Specific mechanisms resulting in excess toxicity to aquatic organisms have been discussed in great detail, and guidance as to when and when not to use narcosis-based QSARs is available *(4-8, 18-20)*.

Excess toxicity has been observed for a variety of chemicals examined in a study which assessed acute toxicity by measuring the 24-hour (24-h) median lethal concentration (LC_{50}) in goldfish *(7)*. In this study acute toxicity was predicted using QSARs which assumed only narcosis as the mechanism of toxicity. Large differences were observed between predicted values and actual values for some of the chemicals. Examples of these chemicals are shown in Table I, and include chemicals that contain either a double bond alpha to a carbonyl group, epoxide moieties, or an alkyl group with a terminal chloro substituent. These chemicals are electrophilic, and the excess toxicity observed for these chemicals was ascribed to their ability to react with endogenous nucleophiles. Separate QSARs have been developed for these chemical classes *(25)*.

Excess toxicity has also been demonstrated for some reactive chemicals within a series of ionizable chemicals. A QSAR study investigating the toxicity of aliphatic amines to fish *(20)* showed that tripropargylamine **6** was 84 times more toxic than was predicted by the equation for an aliphatic amine with comparable log P and molecular weight. Excess toxicity was also observed with chemicals that contain allylic and propargylic alcohol moieties *(4)*. The excess toxicity demonstrated for these chemicals was attributed to their metabolism to electrophilic substances.

6

Several QSARs derived from the examination of a phenol data set revealed that dinitro phenol substituents showed excess toxicity relative to other phenols in the data set *(33)*. In this set 2,5-dinitrophenol and 2-*sec*-butyl-4,6-dinitrophenol were 4-fold and 76-fold more toxic, respectively, than predicted. Support of the excess toxicity of certain phenols can be seen when using a QSAR that assumes only narcosis type toxicity, their observed toxicity exhibit excess than that predicted *(25)*.

Structural Modifications That Decrease Excess Toxicity. The toxicity of chemicals that are toxic by a specific mechanism can be decreased by making structural modifications that sterically hinder them such that they can no longer reach or react with their target sites (i.e., reducing their toxicity to only narcosis). For example, the sterically hindered cyclic aliphatic amine **7** is 25-fold less toxic to fish than predicted by the QSAR equation developed for aliphatic amines *(32)*. Another example of how steric hindrance decreases aquatic toxicity is apparent with 2-hydroxyethyl methacrylate (**8**) which is sterically hindered compared to the 2-hydroxyethyl acrylate (**9**). The 96-h fish LC_{50} values are 227 and 4.8 mg/L respectively. Both substances, however, have comparable commercial usefulness.

A decrease in toxicity due to the addition of a carboxylic acid group is demonstrated by comparing the toxicity of phenol (**10**) to that of salicylic acid (**11**).

Table I. Examples of Substances Whose Aquatic Toxicities are Greater than that Predicted by Narcosis-Based (Baseline Toxicity) QSAR [a]

chemical name	log P [c]	24-hour median lethal concentration (LC_{50}), mg/L: [b]		T_e [f]
		observed [d]	predicted [e]	
3-Chloro-2-methylpropene	1.85	14	156	11
1,3-Dichloro-2-propanol	0.20	680	11800	17
Phenyl glycol monoethyl ether acetate	1.13	69	1480	22
Allyl chloride	0.65	160	4090	26
3-Chloro-1-propanol	0.007	170	13700	81
Ethylene glycol monomethyl ether acetate	0.12	190	13000	69
Propylene oxide	-0.27	170	16600	97
Acrylamide	0.86	460	83000	180
Allyl glycidyl ether	-0.33	78	37600	480
Ethylene oxide	-0.79	90	43800	490
Allyl bromide	1.59	< 0.8	390	> 490
1-Chloro-2,3-epoxypropane	-0.21	23	22700	990
Allyl alcohol	-0.25	1	15700	15700
Pentaerythritol triallyl ether	-1.6	100	1840000	18400
Acrolein	0.10	< 0.08	6500	> 81000

[a] For more details see ref. 7. [b] These toxicity data pertain to goldfish. [c] Logarithm of the octanol-water partition coefficient. [d] Determined experimentally.
[e] Predicted using QSAR that presume a narcosis mechanism of toxicity. These predicted data represent baseline toxicity values. [f] Extent of excess toxicity, obtained by dividing predicted LC_{50} by observed LC_{50}.

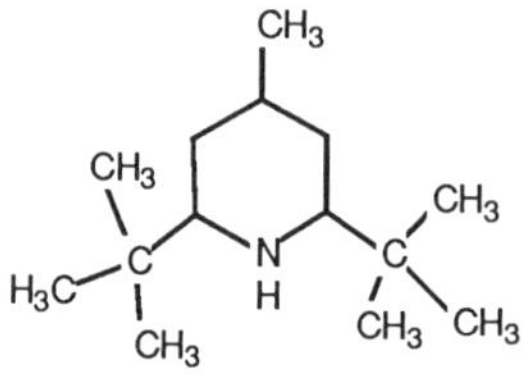

7

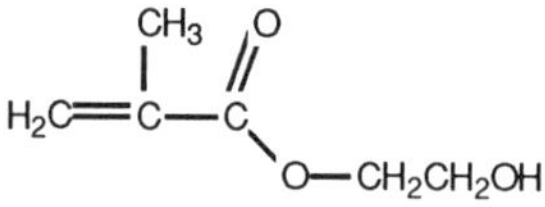

8 H$_2$C=C(CH$_3$)—C(=O)—O—CH$_2$CH$_2$OH

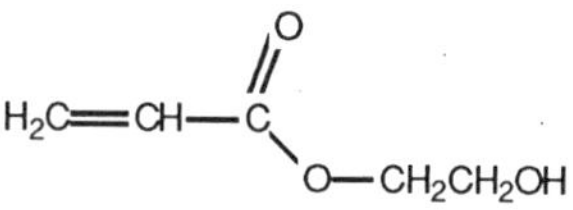

9 H$_2$C=CH—C(=O)—O—CH$_2$CH$_2$OH

The guppy 96-h LC$_{50}$ for phenol is 47 mg/L *(33)*. The fathead minnow 96-h LC$_{50}$ is greater than 1000 mg/L for salicylic acid at pH 7 *(32)*. Toxicity is decreased more than 21-fold by the addition of the carboxylic acid group. This decrease in toxicity with the addition of a carboxylic acid group has also been observed with aliphatic amines *(32)*.

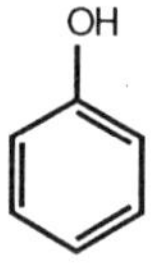

10 phenol

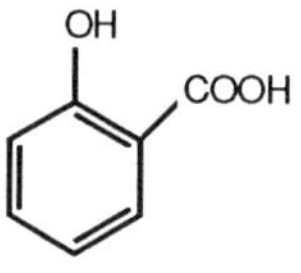

11 salicylic acid

Dyes. Organic dyes may be divided into four classes based on the type of groups present: nonionic (neutral dyes); anionic (negatively charged dyes or acid dyes); amphoteric (mixture of positive and negative charges on the same molecule); and cationic (positive charged). Analysis of more than 200 dyes *(34)* indicated that for dyes which are charged (anionic, cationic, and amphoteric) no correlation (QSAR) has been found between aquatic toxicity and physicochemical properties. Therefore, instead of using QSARs, scientists predict aquatic toxicity using the nearest analog SAR method for which data are available. For neutral dyes, however, QSARs for other chemical classes e.g., phenols and anilines can be used to predict aquatic toxicity. The toxicity of dyes and insight into structural modifications that reduce their toxicity are discussed below.

Nonionic (Neutral) Dyes. Nonionic dyes are dyes that contain no ionic groups. The theory of narcosis and excess toxicity discussed in the preceding paragraphs also applies to neutral dyes. Disperse dyes, a particular type of neutral dye, are poorly soluble in water (and typically have high melting points) and consequently have low aquatic toxicity. Disperse Blue 79 (**12**) is an example of a disperse dye whose physical and chemical properties include a calculated log P of 3.6, and a melting point range of 138 - 157 ºC.

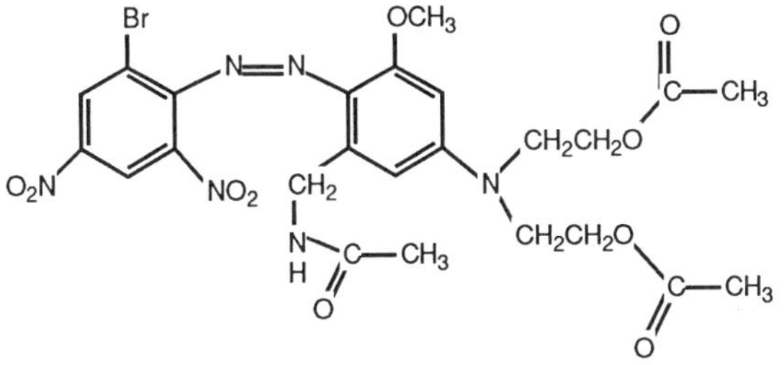

12

To design safer neutral dyes, chemists should design dyes with a water solubility below 1 ppb (i.e., high log K_{ow} and/or high melting point). At this point bioavailability becomes essentially nil, resulting in no toxic effects at saturation even for chronic exposures, (i.e., 20-d to 90-d), and generally no significant bioconcentration in aquatic organisms.

If a neutral dye is required to be soluble rather than insoluble, then the dye can be more safely re-designed by adding polar substituents such as four or more repeated ethoxylated groups, alcohols, or ketones. A reminder to avoid adding groups that will exhibit excess toxicity is noted here and discussed earlier in the section on modification of physicochemical properties. Neutral dyes with molecular weights more than 1000 daltons or a minimum cross-sectional diameter of greater than 10Å will have reduced toxicity to fish and daphnids regardless of their aqueous water solubilities. Also dyes with molecular weights more than 1000 daltons can have attached groups (i.e., anthraquinones, dinitro aromatics, naphthols) that otherwise would bestow excess toxicity, as long as the toxicity of these groups is associated with systemic rather than surface-active toxicity (such as localized cationic dyes) and still show low toxicity. Generally, chemicals that show surface-active toxicity irritate and damage the gills of fish and respiratory membranes of aquatic organisms.

Anionic Dyes. Acid dyes described here are those dyes with one or more acid groups present (e.g., **13**). Acid dyes are chemicals that also include chelated metal atoms and are presented in the next section (i.e., under Metalized Acid Dyes).

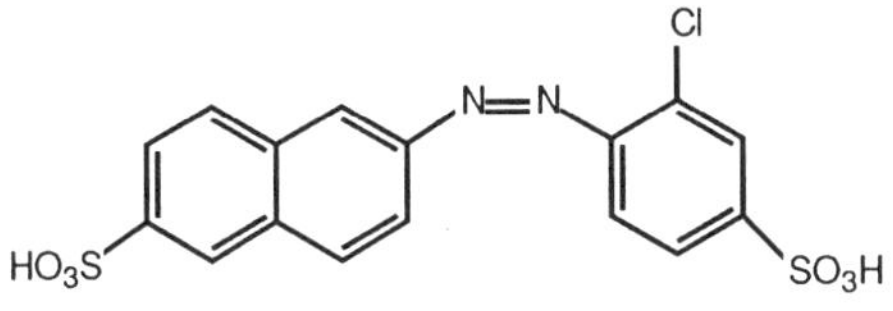

13

Common structures for most acid and amphoteric dyes (discussed in a later category) include but are not limited to anthraquinones, naphthols, and dinitrobenzenes. For systemic toxicity to be observed in aquatic organisms, molecular weights must be less than 1000 daltons. Mono-acid and di-acid dyes with molecular weights less than 1000 daltons and without a chelated metal can show moderate toxicity toward fish, invertebrates, and algae. Dyes with azo bonds, as an example, attached to groups that normally would bestow or exhibit excess toxicity (described above) have been associated with moderate aquatic toxicity. Acid dyes with ≥ 3 acid groups generally show low toxicity to fish and daphnids.

The best way to design mono- and di- acid dyes of lower toxicity is to increase the molecular weight to more than 1000 daltons. This results in reduced toxicity to fish and invertebrates, regardless of toxic functional groups (i.e., dinitro, phenols, anthraquinones) as long as they are not surface-active. However, if dyes are required to have molecular weights less than 1000 daltons, then add three or more acid groups. In general, acid dyes with three or more acid groups show low toxicity toward fish and daphnids, (i.e., acute LC_{50} values more than 100 mg/L), and only indirect effects (shading) toward green algae.

Most acid dyes, regardless of the number of carboxylic acid groups present, show moderate toxicity to algae due to the indirect effect of shading. Analysis of available data *(12)* has suggested the effect of acid dyes on algae growth is not the result of direct toxicity, but results as an indirect effect of "algal shading" caused by the dye's color. Algal shading is a term used to indicate the process where a dye, because of its color, inhibits the utilization of sunlight by algae for photosynthesis, thus diminishing algae growth. The EPA's Office of Pollution Prevention and Toxics (OPPT) has decided that indirect (i.e., shading) toxicity is not indicative of a chemical's potential to cause an unreasonable risk for two reasons: (1) growth of algae resumes quickly as soon as the dye is diluted, and (2) the release of colored effluents in the U.S. generally results in immediate public outcry and quick regulatory action by local officials *(12)*.

Metalized Acid Dyes. A metalized dye is a dye that is chelated to a metal, such as that represented by **14**.

Most metalized dyes are acid dyes. Frequently, iron, copper, cobalt, aluminum, nickel, chromium, and zinc are the metals associated with metalized dyes. These dye molecules are complexed to metals but sometimes contain residual uncomplexed free metal, that is toxic to aquatic organisms. Among metalized acid dyes free of uncomplexed metal ions, only aluminum, chromium, and cobalt are toxic to aquatic organisms *(32)*. The mechanism of their toxicity is unknown.

The discussion for designing safer acid dyes (preceding section) also applies to metalized acid and amphoteric dyes. Thus, increasing the molecular weight above 1000 daltons will usually decrease aquatic toxicity. In addition, chemists should use metals such as iron, copper, or zinc, and avoid the use of aluminum, chromium and cobalt whenever possible. Dyes should be made to eliminate residual unchelated metal in the overall dye product.

Cationic Dyes. Cationic dyes are a chemically diverse group of positively charged dyes ranging in molecular weight from 200 to greater than 1000 daltons. The positive charge may appear on a carbon, nitrogen, phosphorus, oxygen, or sulfur atom. Some cationic dyes are associated with metal complexes, and water solubilities vary depending on their counterion. Delocalized cationic dyes are those whose positive charge is not located on a particular atom, often as a result of conjugation between heteroatoms such as nitrogen (also sulfur, oxygen, and phosphorus). An example of a delocalized cationic dye is represented by structure **15**.

15

Localized cationic dyes contain one or more positive charges permanently located on particular atoms, such as quaternary nitrogens or triply bonded sulphur. An example of a localized cationic dye is represented by structure **16**.

16

Many cationic dyes are extremely toxic to aquatic organisms. The toxicity of localized cationic dyes appears to be primarily due to their ability to bind to membrane surfaces of aquatic organisms and disrupt membrane function *(35)*. However, once absorbed they can cause systemic toxicity. Delocalized cationic dyes are not as positive as localized cationic dyes and consequently have weaker surface active properties. The toxicity of delocalized cationic dyes is believed to be due more to systemic effects and less to disruption in surface membrane function. Evidence of this is suggested by the ability of some localized cationic dyes to bind to humic acid more tightly than dyes with delocalized charge. For example, fish acute toxicity in the presence of 10 mg/L of humic acid was reduced less than 10-fold for delocalized cationic dyes, but more than 20-fold for dyes with localized charge. Based on the testing of cationic dyes under Section 5 of TSCA, localized cationic dyes appear to bind more strongly to humic acid relative to delocalized cationic dyes. These observations suggests that chemicals with localized positive charges function more as a surface-active agent on biological membranes than chemicals with delocalize positive charge.

Delocalized cationic dyes with molecular weights of about 1000 daltons or greater are generally less toxic than ones with molecular weights below 1000 daltons, due to their inability to be taken up by aquatic organisms. For delocalized cationic dyes that contain nitrogens which participate in the delocalization of the positive charge, the extent of substitution on the nitrogens appears to affect toxicity: delocalized cationic dyes that contain primary nitrogens are generally less toxic than those that contain secondary nitrogens, which are less toxic than those that contain tertiary nitrogens.

The best approach to designing delocalized cationic dyes that have reduced toxicity is to increase molecular weights to more than 1000 daltons. Another approach is to limit the extent of substitution on the nitrogens if molecular weight cannot be held to greater than 1000 daltons.

For localized cationic dyes, decreases in toxicity occur as the number of positive charges of the dye decreases and all other factors are held constant. Localized cationic dyes that have one positive charge per molecule are less toxic than localized cationic dyes that have two positive charges, and so forth. One can reduce the toxicity of a localized cationic dye by limiting the number of localized positive charges or redesigning the dye to be amphoteric. A localized cationic dye that contains a single positive charge should be less toxic than one that contains two positive charges, and so forth. In principle, designing localized or delocalized cationic dyes such that log P is greater than 8 will ensure limited water solubility and bioavailability. This however may not be feasible since the cationic nature of these dyes adds considerable polarity and solubility which may be difficult to reduce through structural modifications.

Amphoteric Dyes. Amphoteric dyes (e.g.,17) are dyes that can behave as either an acid or a base, and are often characterized by the presence of positive and negative charges within the same molecule. The toxicity of amphoteric dyes depends on the ratio of cations to anions in the dye. If the ratio favors the anion, then the dye will have lower toxicity (i.e., nearer to the acid dyes). For example, the addition of a carboxylic acid group to an amphoteric dye will reduce the toxicity to fish and daphnids about 10-fold *(32)*. Adding a sulfonic acid group rather than a carboxylic acid group, will further reduce the toxicity *(32)*. Guidelines for designing an aquatically safer amphoteric dye are: (1) the anion to cation ratio must favor the anion; (2) sulfonic acid groups are preferred over carboxylic acid groups; (3) maximize molecular weight to greater than 1000 daltons; and (4) maximize minimum cross-sectional diameter. Toxicologically, amphoteric dyes behave similarly to acid dyes if charge is balanced or anion to cation ratio is > 1.0.

Surfactants. Surfactants are characterized here as substances that contain both hydrophobic and hydrophilic moieties, which enable the substances to form micelles in water. There are four general types of surfactants, based on charge: cationic (positive charge); anionic (negative charge); nonionic (neutral or no charge); and amphoteric (negative and positive charges). These surfactants are represented by structures **18-21**, respectively.

Surfactants are toxic to aquatic species because they can disrupt the interface between biological membranes (which are chiefly lipid in content) and the aquatic environment (which is water). In simple terms, the aquatic toxicity of surfactants is largely due to their ability to disintegrate biological membranes to the extent that their structure-specific components are loosened and are no longer functional. Surfactants with molecular weights less than 1000 daltons exhibit surface active toxicity and can be systemically toxic to aquatic organisms. Surfactants that have molecular weights greater than 1000 daltons are too large to be absorbed, and consequently exhibit surface active toxicity only. In addition to acute toxicity, surfactants with molecular weights below 1000 daltons can also be absorbed into aquatic organisms and cause chronic toxicity as well.

The relative toxicological potency of a surfactant is dependent on balance between the hydrophobic component and the hydrophilic component. In general, toxicity is higher when the hydrophobic component is relatively large compared to the hydrophilic component. As the hydrophobic component becomes smaller relative to the hydrophilic component, toxicity generally decreases *(10)*. Surfactants with melting points of 35 °C will be self-dispersing in warm water only (e.g., 50 °C) and will not function as surfactants at environmental temperatures (20 °C). Such surfactants will generally not be acutely toxic to aquatic organisms, and only chronic toxicity is possible.

17

In the late 1960s and early 1970s surfactants were voluntarily designed by the detergent industry to be less persistent (i.e., easily biodegradable). Apparently many of the surfactants used up until that time persisted in the environment, which resulted in uncontrollable foaming in sewage treatment facilities and rivers, and also led to the dispersion of pathogenic bacteria. The design of less persistent surfactants was accomplished by using linear rather than branched alkyl chains as hydrophobes (see chapter by Boethling). Although these new surfactants were less persistent, they were more toxic to aquatic organisms when compared to the more persistent surfactants.

The challenge in designing safer surfactants is to design them such that they neither persist in the environment or are toxic to aquatic species. It may be possible to reduce or even eliminate the persistence and toxicity of surfactants if they are designed such that they degrade (biotically or abiotically) readily to substances that neither persist or have surfactant properties, or otherwise are not aquatically toxic. It may be possible to accomplish this by incorporating a double bond in a surfactant's alkyl chain. This modification is expected to make the surfactant more susceptible to biotic (bacterial) degradation. To enhance abiotic degradation, the surfactant should be designed such that it contains labile functional groups that do not require bacteria for decomposition. Ideally, the half-life of such a surfactant should be similar to the duration of time required for the surfactant to carryout its intended use. This will ensure that the surfactant will never reach the natural environment intact or, if it does, will not last very long. This could be accomplished, perhaps, by incorporating a peroxy (-O-O-) group between the hydrophobic and hydrophilic portions. The half-life of a peroxy moiety under environmental conditions is not expected to exceed an hour. Peroxy groups between the hydrophobic and hydrophilic portion of the surfactant will ensure that the

18; R = cationic surfactant: -CH(CH₂OH)(NH⁺X⁻)(CH₃)

19; R = anionic surfactant: -C₆H₄-SO₃⁻ X⁺

R = nonionic surfactant: -(CH₂CH₂O)ₓ-H

21; R = amphoteric surfactant: -CH(CH₂COO⁻)(NH⁺)(CH₃)

substance is rapidly degraded to non-surfactant substances. Similarly, degradation to non-surfactant substances can also be enhanced with the use of an ester group between the hydrophobic and hydrophilic moieties. Other approaches to limiting the aquatic toxicity of surfactants include designing them such that their molecular weights are above 1000 daltons.

Alkylphenols (branched and unbranched) are commonly used as surfactants or as feedstocks in the manufacture of alkylphenol ethoxylate-type surfactants. The latter substances are biodegraded to the alkylphenol feedstock. Alkylphenols are highly toxic to aquatic species. Therefore, it is advisable that alkylphenols not be used as surfactants or as starting materials for surfactant manufacture, unless absolutely necessary.

Polyanionic Monomers. Polyanionic monomers are substances below 1000 daltons that contain one or more acid groups at each end of a short alkyl chain and which behave like chemical chelators, e.g., ethylenediaminetetraacetic acid (EDTA). Polyanionic monomers may contain polar groups such as carboxylic acid, silicic acid, phosphoric acid, sulfonic acid and thiols. The acid groups on a polyanionic monomer may be the same or mixed. The monomeric nucleus may include carbon, silicon, oxygen, sulfur, and nitrogen, or mixtures of these elements.

The salts of polyanionic monomers (e.g., sodium) are very water soluble and generally exhibit low toxicity to fish and daphnids but are often moderately toxic to green algae in the standard algal bioassay which is done with growth/test medium which is soft (hardness = 15 to 24 mg/L as $CaCO_3$). It has been shown that the algal toxicity is due to over-chelation of nutrient elements (i.e., calcium, magnesium, iron) needed for algal growth. Manipulation of chelation potency can be done by changing the type of acid group as well as the distance between the acid groups. For example, diethylenetriaminepentaacetic acid (DTPA) **22**, used as a chelator has five acid groups present and each connected by three atoms in any direction. DTPA inhibits algal growth at 1 mg/L.

The design of safer polyanionic monomers can be accomplished by: 1) varying the type of acid (i.e., carboxylic, sulfonic, or phosphoric); and 2) regulating the distance between them to determine which chemical designs are less toxic to aquatic

$$^-OOC-CH_2-N\begin{matrix}CH_2CH_2-N\diagup^{CH_2COO^-}_{CH_2COO^-}\\ \diagdown CH_2CH_2-N\diagup^{CH_2COO^-}_{CH_2COO^-}\end{matrix} \quad 5\ Na^+$$

22

organisms and still act as a chelator. The design objective for polyanionic monomers is to weaken the chelation potency so that the algal 96-h EC_{50} is greater than 100 mg/L (based on 100% active ingredient). Polyanionic monomers with a weaken chelation potential will protect algae and still be useful for industrial applications.

Organometallics. Many organometallics such as those used as antifouling agents, pesticides, and fungicides are toxic to aquatic organisms. Of the alkyl metals, the most toxic members include alkyl mercurials (e.g., methyl mercury), organotin (e.g., triethyltin), and organolead (e.g., tetramethyl lead) substances. To prevent toxicity of the metal, an organometallic chemical should be designed by increasing: 1) the log P to greater than 8; 2) the melting point; or 3) the molecular weight to greater than 1000 daltons.

Inorganic Metallics. A variety of inorganic metallic substances are essential for all forms of life. These substances are taken up by living cells and their uptake is strictly regulated because most (or all) of them are toxic in when consumed in excess. For example, inorganic phosphate is essential for energy production and utilization, but has the potential to over stimulate the growth of green algae and cause algal blooms and eutrophication in freshwater environments. Excessive algae growth in turn can block sunlight from reaching aquatic species below the surface. When the algae die they are decomposed by bacteria which use up available dissolved oxygen, producing anaerobic conditions. Oxygen depletion of the water is detrimental to fish and aquatic invertebrates, and these species die of suffocation. As a result, use of inorganic phosphates in consumer detergents have been severely limited or banned in 13 states, the District of Columbia, and several countries and municipalities *(36)*. Unlike the other chemical categories discussed thus far, there are no general ways of designing safer inorganic metallic substances. However, those inorganic chemicals known to be toxic or cause environmental damage (e.g., phosphates) can be avoided or not used.

Conclusion

The development and use of SARs to predict the toxicity of chemicals to aquatic species provides a rational tool for designing safer chemicals. The examples presented in this chapter exemplify what range of physicochemical properties and structural requirements are necessary to reduce aquatic toxicity of industrial chemicals. Although the relationships of aquatic toxicity with chemical structure and physicochemical properties have been used by the EPA's Office of Pollution Prevention and Toxics (OPPT) to predict toxicity for purposes of assessing environmental risk, chemists may also find these relationships and ideas useful to design safer chemicals.

Acknowledgments

The authors wish to thank Drs. Maurice Zeeman, and Richard Clements for their encouragement during the preparation of this paper; Dr. Gordon Cash for his help in chemistry; and Dr. Carol A. Farris for editorial comments.

Dedication

This paper is dedicated to the memory of David W. Johnson, an outstanding chemist and always hard working contributor in the Environmental Effects Branch, Health and Environmental Review Division, Office of Pollution Prevention and Toxics, whose expert work and generous spirit have allowed for the success of the SAR/QSAR activities in OPPT.

Disclaimer

This document has been reviewed by the Office of Pollution Prevention and Toxics, U. S. EPA and approved for publication. Approval does not signify that the contents necessarily reflect the views and polices of the Agency nor does mention of trade names or commercial products constitute endorsement or recommendation for use.

Literature Cited

1. Blum, D.J.W.; Speece, R.E. *Environ. Sci. Technol.*, **1990**, *24*, pp 284-293.
2. Seydel, J.; Schaper, K. In *Chemische Struktur und Biologische Aktivitat von Wirkstoffen: Methoden der Quantitativen Struktur-Wirkung Analyse.* Verlag Chemie, Weinhiem, F. R. G., **1979**.
3. Hanch, C.; Leo, A. In *Substituent Constants for Correlation Analysis in Chemistry and Biology,* Wiley-Interscience, New York, NY. **1979**.
4. Lipnick, R. L.; Johnson, D. E.; Gilford, J. H.; Bickings, C. K.; Newsome, L. D. *Environ. Tox. and Chem.* **1985**, *4*, pp 281-296.
5. Newsome, L. D.; Lipnick, R. L.; Johnson, D. E. In *QSAR in Environmental Toxicology;* Kaiser, K. L. E., Ed.; D. Reidel: Dordrecht, Holland, **1984**; pp 279-299.
6. Lipnick, R. L. In *QSAR in Toxicology and Xenobiochemistry;* Tichy, M., Ed.; Elsevier: Amsterdam, the Netherlands, **1985**; Vol. 8; pp 39-52.
7. Lipnick, R. L.; Watson, K. R.; Strausz, A. K. *Xenobiotica.* **1987**, *17*, pp 1011- 1025.
8. Lipnick, R. L. *Environ. Tox. Chem.* **1989**, *8*, pp 1-12.
9. Zeeman, M. G.; Gilford, J. In *Environmental Toxicology and Risk Assessment;* Landis, W. G.; Hughes, J. S.; Lewis, M. A. Eds.; Special Technical Publication 1179; American Society for Testing and Materials: Philadelphia, PA, **1993**, pp 7-21.
10. Nabholz, J. V.; Miller, P.; Zeeman, M. In *Environmental Toxicology and Risk Assessment;* Landis, W. G.; Hughes, J. S.; Lewis, M. A. Eds.; Special Technical Publication 1179; American Society for Testing and Materials: Philadelphia, PA, **1993**, pp 40-55.
11. Clements, R. G.; Nabholz, J.V.; Johnson, D. E.; Zeeman, M.G. In *Environmental Toxicology and Risk Assessment;* Gorsuch, J. W.; Dwyer, F. J.; Ingersoll, C. G.; La Point, T. W. Eds. ; Special Technical Publication 1216; American Society for Testing and Materials: Philadelphia, PA, **1993**, Vol. 2; pp 555-570.
12. Nabholz, J. V.; Clements, R. G.; Zeeman, M. G.; Osborn, K. C.; Wedge, R. In *Environmental Toxicology and Risk Assessment;* Gorsuch, J. W.; Dwyer, F. J.; Ingersoll, C. G.; La Point, T. W. Eds.; Special Technical Publication 1216; American Society for Testing and Materials: Philadelphia, PA, **1993**, Vol. 2; pp 571-590.
13. Zeeman, M. G. In *Fundamentals of Aquatic Toxicology: Effects, Environmental Fate, and Risk Assessment;* Rand, G., Ed.; Taylor & Francis: Washington, DC, **1995**; pp 703-715.
14. Nabholz, J. V. *The Sci. of the Total Environ.* **1991**, *109/110*, pp 31-46.

15. Zeeman, M. G.; Nabholz, J. V.; Clements, R. G. In *Environmental Toxicology and Risk Assessment;* Gorsuch, J. W.; Dwyer, F. J.; Ingersoll, C. G.; La Point, T. W. Eds.; Special Technical Publication 1216; American Society for Testing and Materials: Philadelphia, PA, **1993**, Vol. 2; pp 523-539.
16. Lipnick, R. L.; Dunn, W. J. III. In *Quantitative Approaches to Drug Design;* Dearden, J. C., Ed.; Elsevier: Amsterdam, the Netherlands, **1983**; pp 265-266.
17. Newsome, L. D.; Johnson, D. E.; Cannon, D. J.; Lipnick, R. L. In *QSAR in Environmental Toxicology-II;* Kaiser, K. L. E., Ed.; D. Reidel: Dordrecht, Holland, **1987**; pp 231-250.
18. Lipnick, R. L.; Pritzker, C. S.; Bentley, D. L. In *QSAR in Drug Design and Toxicology;* Hadzi, D.; Jerman-Blazic, B., Eds.; Elsevier: Amsterdam, the Netherlands, **1987**; pp 301-306.
19. Clements, R. G.; Johnson, D. E.; Lipnick, R. L.; Nabholz, J. V.; Newsome, L. D. In *Estimating Toxicity of Industrial Chemicals to Aquatic Organisms Using Structure Activity Relationships;* Clements, R., Ed.; EPA-560-6-88-001, U. S. Environmental Protection Agency, Office of Toxic Substances, Health and Environmental Review Division: Washington, DC., **1988**.
20. Newsome, L. D.; Johnson, D. E.; Lipnick, R. L.; Broderius, S. J.; Russom, C. L. *The Sci. of the Total Environ.* **1991**, *109/110,* pp 537-551.
21. Newsome, L.; Johnson, D.; Nabholz, J. V. In *Environmental Toxicology and Risk Assessment;* Landis, W. G.; Hughes, J. S.; Lewis, M. A. Eds.; Special Technical Publication 1179; American Society for Testing and Materials: Philadelphia, PA, **1993**, pp 413-426.
22. Clements, R. G.; Nabholz, J. V.; Zeeman, M.; Auer, C. M. In *SAR and QSAR in Environmental Research,* **1995**, Vol. 3, pp 203-215.
23. Ginnings, P. M.; Baum, R. *J. Am. Chem. Soc.* **1937**, *59,* pp 1111-1113.
24. Cohn, E. J.; McMeekin, T.L.; Edsall, J. T.; Weare, J. H. *J. Am. Chem. Soc.* **1934**, *56,* pp 2270-2282.
25. Clements, R.G.; Nabholz, J. V. *ECOSAR:* A Computer Program for Estimating the Ecotoxicity of Industrial Chemicals Based on Structure Activity Relationships, U. S. EPA, OPPT (7403), Technical Publication, 748 R 93 002, **1994**.
26. United States Environmental Protection Agency (USEPA), Aqueous and Terpene Cleaning: Interim Report, Available from TSCA Assistance Information Service, tel. 202-554-1404, fax 202-554-5603, Washington, DC.
27. *Precision Cleaning;* Witter Publishing Corporation, Vol. 4, No. 2, 1996, p 20.
28. Anliker, R.; Clarke, E. A.; Moser, P. *Chemosphere* **1981**, *10,* pp 263-274.
29. Opperhuizen, A. *Aquatic Tox. and Environ. Fate;* Poston, T. M.; Purdy, R., Eds.; Special Technical Publication 921; American Society for Testing and Materials: Philadelphia, PA, **1986**, Vol. 9; pp 304-315.
30. Gobas, F. A. P. C.; Lahittete, J. M.; Garofalo, G.; Shiu, W. Y.; Mackay, D. *J. Pharm. Sci.* **1988**, *77,* pp 265-272.
31. Opperhuizen, A.; van der Velde, E. W.; Gobas, F. A. P. C.; Liem, A. K.; van der Steen, J. M. D. *Chemosphere* **1985**, *14, 11/12,* pp 1871-1896.
32. Nabholz, J. V. *The OTS PMN ECOTOX Data Base;* Environmental Effects Branch, Health and Environmental Review Division (7403), U. S. EPA, Washington, DC., **1990**.
33. Saarikoski, J.; Viluksela, M. *Ecotox. and Eviron. Safety.* **1982**, *6,* pp 501-512.
34. Auer, C. M.; Nabholz, J. V.; Baetcke, K. P. *Environ. Health Perspect.,* **1990**, *87,* pp 183-197.
35. Albert, A. *Selective Toxicity The Physico-Chemical Basis of Therapy,* Chapman and Hall Ltd: New York, NY., **1981**.

36. U.S. E.P.A. Phosphates Detergents: An Evaluation of the Benefits and Costs of Eliminating Their Use in the United States, *Environmental Results Branch,* Office of Policy, Planning, & Evaluation, U.S. Environmental Protection Agency, Washington, DC., **1992.**

APPLICATIONS AND EXAMPLES

Chapter 10

Designing Safer Nitriles

Stephen C. DeVito

Office of Pollution Prevention and Toxics, U.S. Environmental Protection Agency, Mail Code 7406, 401 M Street, SW, Washington, DC 20460

Nitriles represent an important class of chemical substances that have broad commercial utility. A comprehensive search and review of the literature to identify studies pertaining to toxic effects caused by nitriles revealed that certain nitriles are acutely toxic (lethal) or may produce osteolathyrism. A retrospective analysis of these studies, particularly those that describe toxic mechanisms and provide structure-activity relationship data, enabled an understanding of why certain nitriles are highly toxic while others are not. From this understanding, structural modifications that reduce toxicity became apparent. This paper describes how safer (less toxic) nitriles can be designed. Chemists who design nitriles will find this paper useful for designing safer, commercially efficacious nitriles. The general approach and strategy described herein for the design of safer nitriles can be used for the design of safer substances belonging to other chemical classes as well.

The design of a safer substance begins with the concomitant consideration of structural features that are needed for commercial usefulness along with any toxicological effects that are caused by such features. In addition, the toxic effects that can be caused by other structural features intended to be present in a planned substance needs to be considered as well. Chemists developing a particular type or class of substance for a specific purpose (e.g., dye, surfactant) are aware of the structural features that are necessary for use function, and will design substances such that they contain these structural features. In many instances, however, chemists are probably not aware of the toxic effects associated with the structural features or the chemical class to which a planned new chemical belongs. Consequently, the toxic properties are likely to be retained in the new substance. If a structural feature that is necessary for commercial use is likely to bestow toxicity, the chemist should consider making molecular modifications that will reduce the toxicity but will not affect the substance's commercial usefulness.

But how does one know when a planned substance is likely to be toxic? The potential for a planned substance to be toxic may be inferred from what is known about other chemicals that contain similar structural features. Whether one is striving to design a new, less toxic analog of an existing substance, or develop a series of substances for a particular commercial purpose, it is imperative that the chemist be aware of the toxicity-related information available for the existing substance, or the class to which the series of substances belongs. To design a safer chemical, the need to identify and familiarize oneself with any toxic properties and toxicity data related to the chemical class is just as important as a familiarity with the chemistry of use. How does one identify such information? Moreover, how can one use such information to design safer chemicals? Chapter 2 provides a list of reference sources that are extremely useful for identifying toxicity-related information. Chapter 2 also describes several ways in which safer chemicals can be designed and, using many examples, demonstrates and provides insight into how one can infer, from available information, structural modifications that can reduce toxicity without affecting commercial efficacy. Data that pertain to mechanism of toxicity, metabolism, and structure-activity relationships for a series of analogous compounds are particularly useful to assess the toxicity of a new or planned substance, and for inferring structural modifications that will reduce toxicity.

Using nitriles as an example of an important commercial class of substances, the present chapter answers the above questions more fully, and demonstrates how safer chemicals can be designed from an analysis of the toxicity-related information available in the literature. More specifically, this paper will demonstrate the importance of the following principles of designing safer chemicals:

- the need to first familiarize oneself with any toxicological or health related data available for existing analogous substances (i.e. substances belonging to the same class as the intended safer chemical);

- when toxicological effects are known for the chemical class to which a planned new substance belongs, how these toxic effects can be minimized by ultilizing available toxicological data (e.g., toxic mechanisms, structure-activity data, metabolism data, etc.) to infer structural modifications that will reduce or prevent the same toxicity in the new substance.

The approach described herein for the design of safer nitriles can be applied to the design of safer chemicals of any chemical class.

Nitriles

Nitriles are organic substances that contain the cyano ($C\equiv N$) group. Nitriles have wide commercial application, that includes their use as solvents, synthetic intermediates, pharmaceuticals, and monomers, to name a few. To establish if any toxic effects are associated with nitriles, a literature search was conducted to identify information pertaining to toxic properties of nitriles. This began with a search of the Chemical Abstracts Database, starting with the first Decennial Subject Index (1907), and searching each subsequent decennial subject index. Using the subject heading of "nitriles", the decennial indices were searched for literature citations pertaining to toxicity or health-related studies of nitriles. In addition, toxicological textbooks *(1,2)* were searched similarly. These reference sources were also searched for toxicity-related data pertaining to specific, commercially important nitriles (e.g., acetonitrile, propionitrile) because it was felt that studies on these nitriles may also contain data on other nitriles as well. The literature search revealed that much has been published on the toxicity of nitriles. Literature references containing toxicity information on nitriles were obtained and analyzed. It was found that, as a class, there are two toxic properties associated with nitriles. These are acute lethality and osteolathyrism. Some nitriles possess both of these toxic properties. Each of these properties are discussed separately below.

Acute Lethality of Nitriles

Certain nitriles are quite potent in causing acute lethality in humans and animals *(3,4)*. Consequently, a considerable number of studies have been conducted to determine the acute lethality of a variety of nitriles used commercially *(3-18)*. Table I summarizes some of the available acute lethality data measured during these studies, and pertains specifically to acute lethality measurements made in mice following oral administration of each nitrile listed. Acute lethality is expressed as the acute median lethal dose (LD_{50}). The lower the LD_{50} value, the more toxic the substance.

It can be seen from Table I that as a class nitriles vary broadly in their ability to cause acute lethality, and that subtle differences in structure can dramatically affect toxic potency. For example acetonitrile (**1**) is not very acutely lethal (LD_{50} = 6.55 mmol/kg) whereas its homolog propionitrile (**2**) is ten-fold greater in toxicity (LD_{50} = 0.65 mmol/kg). Increasing the carbon chain length by one more methylene ($-CH_2-$) group (i.e., butyronitrile, **3**) however, only slightly increases toxicity. In fact, the toxicity decreases sharply as the number of $-CH_2-$ groups increases above two, as can be seen in comparing the LD_{50} values of **3-7**. It is also

Table I. Mouse Acute Toxicity Data, Log P, and Theoretical Metabolic Rate Constants of Some Nitriles

compound no.	name	structure	log P [a]	k_α [b] (× 10^{-24})	$k_{\alpha corr}$ [c] (× 10^{-24})	LD$_{50}$ [d] (log (1/LD$_{50}$)) exp	calcd [e]
1	acetonitrile	CH$_3$CN	-0.39	4.0	4.0	6.55[f] (-0.816)	3.42 (-0.534)
2	propionitrile	CH$_3$CH$_2$CN	0.14	287.9	251.0	0.65[f] (0.187)	0.89 (0.051)
3	butyronitrile	CH$_3$(CH$_2$)$_2$CN	0.66	310.0	58.4	0.57[f] (0.244)	1.15 (-0.059)
4	valeronitrile	CH$_3$(CH$_2$)$_3$CN	1.19	324.9	24.7	2.30[f] (-0.362)	1.54 (-0.188)
5	capronitrile	CH$_3$(CH$_2$)$_4$CN	1.72	324.9	16.9	4.77[f] (-0.679)	2.25 (-0.352)
6	caprylonitrile	CH$_3$(CH$_2$)$_6$CN	2.78	348.8	8.1	14.09[f] (-1.149)	8.71 (-0.940)
7	pelargonitrile	CH$_3$(CH$_2$)$_7$CN	3.31	348.8	5.5	14.79[f] (-1.170)	23.28 (-1.367)
8	isobutyronitrile	(CH$_3$)$_2$CHCN	0.44	8422.2	8288.5	0.37[f] (0.432)	0.34 (0.470)
9	3-methylbutyronitrile	(CH$_3$)$_2$CHCH$_2$CN	1.06	97.4	0.98	2.80[f] (-0.447)	3.31 (-0.520)
10	4-methylvaleronitrile	(CH$_3$)$_2$CH(CH$_2$)$_2$CN	1.59	349.9	6.6	5.02[f] (-0.701)	2.61 (-0.416)
11	2-methylbutyronitrile	CH$_3$CH$_2$CH(CH$_3$)CN	0.97	5811.7	4288.2	0.29[f] (0.538)	0.39 (0.398)
12	malononitrile	NCCH$_2$CN	-1.20	5.4	5.4	1.80[g] (-0.255)	7.59 (-0.880)
13	succinonitrile	NC(CH$_2$)$_2$CN	-0.82	28.9	28.9	1.62[g] (-0.210)	3.13 (-0.495)
14	glutaronitrile	NC(CH$_2$)$_3$CN	-0.96	70.6	41.7	2.83[g] (-0.452)	3.34 (-0.524)
15	adiponitrile	NC(CH$_2$)$_4$CN	-0.43	226.9	113.5	1.59[g] (-0.201)	1.53 (-0.185)
16	pimelonitrile	NC(CH$_2$)$_5$CN	0.11	495.0	202.1	1.03[g] (-0.013)	0.95 (0.023)
17	chloroacetonitrile	ClCH$_2$CN	0.22	107.0	107.0	1.84[f] (-0.265)	1.07 (-0.029)
18	3-chloropropionitrile	Cl(CH$_2$)$_2$CN	0.20	45.8	4.9	0.57[f] (0.244)	2.32 (-0.366)
19	4-chlorobutyronitrile	Cl(CH$_2$)$_3$CN	0.38	131.1	16.2	0.52[f] (0.284)	1.64 (-0.213)
20	3-(dimethylamino)-propionitrile	(CH$_3$)$_2$N(CH$_2$)$_2$CN	-0.24	788.9	500.2	15.30[h] (-1.185)	9.19 (-0.963)
21	3-(isopropylamino)-propionitrile	(CH$_3$)$_2$CHNH(CH$_2$)$_2$CN	0.12	515.2	0.005	19.40[h] (-1.288)	13.37 (-1.126)
22	3,3'-iminodi-propionitrile	HN(CH$_2$CH$_2$CN)$_2$	-0.92	300.0	0.041	27.50[i] (-1.439)	17.99 (-1.255)
23	3-butenenitrile	CH$_2$=CHCH$_2$CN	0.12	12703.4	12703.4	1.00[f] (0.000)	0.34 (0.474)
24	phenylacetonitrile	C$_6$H$_5$CH$_2$CN	1.56	34808.9	34808.9	0.39[f] (0.409)	0.25 (0.598)
25	3-phenylpropionitrile	C$_6$H$_5$CH$_2$CH$_2$CN	1.55	703.9	3.3	0.89[f] (0.051)	3.03 (-0.481)
26	3-hydroxypropionitrile	HO(CH$_2$)$_2$CN	-1.10	70.2	0.82	48.72[f] (-1.688)	10.62 (-1.026)

[a] Octanol/water partition coefficient. [b] Theoretical reaction rate constants for cytochrome P450-mediated α-hydrogen atom abstraction (ref. 22). [c] Theoretical reaction rate constants statistically corrected for metabolism at other positions (ref. 22). [d] Acute oral median lethal dose in mice (mmol/kg). [e] Calculated by eq. 1. [f] Obtained from ref. 11. [g] Obtained from ref. 13. [h] Obtained from ref. 3. [i] Obtained from ref. 29.

interesting to note the influence of methyl substitution on toxicity. Methyl substitution of butyronitrile (3) in the 3-position greatly reduces toxicity, as can be seen when comparing the toxicity of 3 to that 9. However, methyl substitution of butyronitrile in the 2-position nearly doubles toxicity, as evident in comparing the toxicity of 3 to that of 11.

Mechanism of Acute Lethality. It has been observed that exposure of humans and experimental animals to the more acutely toxic nitriles results in signs and symptoms similar to that of cyanide poisoning, implicating free cyanide as the cause of lethality (3,4). Lang observed in 1894 that cyanide is in fact released from nitriles following their administration to mammals: large amounts of thiocyanate (a metabolite of cyanide) appeared in the urine of dogs and rabbits following oral administration of several nitriles (5). Subsequent studies by other investigators have shown that cyanide is released from a wide variety of nitriles in other species and is associated with the acute lethality of these substances (6-13).

Ohkawa and co-workers proposed that cyanide release from nitriles results from the metabolism of these substances in the liver. They proposed that cyanide liberation from nitriles results from cytochrome P450-mediated hydroxylation of the carbon atom alpha (α) to the cyano (C≡N) group to form a cyanohydrin intermediate, which rapidly decomposes to liberate cyanide and the corresponding carbonyl compound (7). Subsequent studies have confirmed the role of hepatic cytochrome P450 enzymes in the release of cyanide from various nitriles following the administration of these substances to laboratory animals (7-17).

Generally speaking, nitriles that are metabolized most readily or most quickly at the carbon atom alpha to the cyano group (i.e. the α-carbon) are more toxic than nitriles that are metabolized more slowly at this position (11, 12). Thus, the mechanism for the acute lethality of nitriles is directly related to their propensity to release cyanide as result of metabolic bioactivation by cytochrome P450 enzymes. As discussed elsewhere in this book (see chapter by Jones), radical formation is a requisite step in cytochrome P450 hydroxylations, and it is now known that cytochrome P450-mediated hydroxylation proceeds by a stepwise process involving radical substrate intermediates (18-20). With respect to the toxicity of nitriles, this would involve radical formation on the α-carbon, followed by subsequent hydroxylation to the cyanohydrin intermediate. Cytochrome P450-mediated release of cyanide from nitriles is depicted in Figure 1.

Not surprisingly, nitriles which liberate cyanide more quickly are more toxic. The less toxic nitriles (the ones which have mouse oral LD_{50} values larger than about 2 mmol/kg) are less toxic because they tend to release cyanide more slowly and to a much lesser extent than nitriles that have LD_{50} values below 2 mmol/kg (11,12). It is important to emphasize that cytochrome P450-mediated hydroxylation can occur at other carbon positions (i.e. postions that are not alpha to the cyano group) as well. Hydroxylation at positions other than the α–carbon does not result in cyanide release and represents a detoxication pathway. In fact, it is likely that most nitriles, even the toxic ones, are hydroxylated at multiple carbon positions. The toxic nitriles, however, are those in which metabolism at the α–carbon predominates, and the less toxic nitriles are those in which metabolism at non-α–carbon positions predominate (21,22). This point is discussed in greater detail in the next section and elsewhere in this chapter.

Structure-Activity Relationships of Acute Lethality. The structure-activity relationships of the acute toxicity of nitriles have been reviewed(21). Because the acute toxicity of nitriles is a function of their ability to undergo cytochrome P450-mediated hydroxylation on the carbon atom alpha to the cyano group, and that the hydroxylation is a radical-based reaction (Figure 1), the acute toxicity of nitriles is expected to relate to structural features that influence α–carbon radical stability. Logically, structural features which are expected to increase α–carbon radical stability are likely to favor α–hydrogen atom abstraction. The more quickly α–hydrogen atom abstraction occurs, the more quickly cyanohydrin formation occurs and the more quickly cyanide is released. The more quickly cyanide is released, the more toxic the nitrile is expected to be.

In examining the toxicity data in Table I one can observe a direct relationship between acute toxicity and the relative ease of radical formation (hydrogen atom abstraction) on the α–carbon. It appears that the differences in the toxicities between the nitriles in Table I can be ascribed to the type of radical that can be formed at the α–carbon. For example, the α–carbons of isobutyronitrile (**8**) and 2-methylbutyronitrile (**11**) are tertiary (3º) and, of course, will form 3º radicals at the α–carbon. These nitriles are considerably more toxic than propionitrile (**2**) and butyronitrile (**3**), which form less stable secondary (2º) radicals on their α–carbon atoms. Propionitrile, on the other hand, is considerably more toxic than acetonitrile (**1**) because the latter substance forms a less stable primary (1º) radical at the α–carbon. The large difference in toxicity between chloroacetonitrile (**17**) and **1** is likely to be due to the fact that the chloro substituent of **17** is expected to promote cyanohydrin formation because it will stabilize the radical formed at the α–carbon (23), and thereby favor its formation, when compared to **1**.

Phenylacetonitrile (**24**) contains benzylic hydrogens at the α–carbon. These hydrogens are particularly easy for cytochrome P450 enzymes to remove because the resultant carbon radical is benzylic, and well stabilized through resonance of the phenyl ring. In fact, this substance is one of the most toxic of all the nitriles listed in Table I.

Thus, the differences in the acute toxicities of the nitriles in Table I are attributable to the most likely position of cytochrome P450-mediated metabolism. The carbon atoms most likely to be hydroxylated by cytochrome P450 are those which will form the more stable carbon radical. While structural features that favor radical formation at the α–carbon tend to increase toxicity, it also appears that structural features that favor radical formation at other carbon atoms decrease acute toxicity. For example, the α–carbons of nitriles **3, 4, 9** and **10** are secondary, but **9** and **10** are considerably less toxic than **3** and **4**, respectively. This is because nitriles **9** and **10** contain tertiary carbon atoms elsewhere in their structure, and cytochrome P450-mediated hydroxylation is expected to occur preferentially at these tertiary carbon atoms rather than the secondary carbons. Hydroxylation at the tertiary carbon of **9** or **10** will not result in cyanide release.

For nitriles with multiple CH_2 (i.e. secondary) carbons, the large decrease in toxicity observed as the number of CH_2 groups increases (e.g., nitriles **3-7**) is likely to be due to the fact that these extra methylene units offer additional sites of metabolism that compete for metabolism at the α–carbon. It has been shown that release of cyanide from nitriles **4-7**, and **9-10** is very slow *(11,12)* and is likely to be the reason why these nitriles are considerably less acutely toxic than nitriles such as **2-3, 8, 11**, and **24**, which release cyanide much more quickly (i.e. are metabolized predominately at the α–carbon) *(11,12)*.

A similar point can be made in comparing the extreme differences in toxicities of propionitrile **(2)** to 3-hydroxypropionitrile **(26)**. Placement of the hydroxy group onto the terminal carbon of propionitrile (to give **26**) results in a very large decrease in toxicity (Table I). Although the products of metabolism of **26** have not been fully studied, one investigation did not detect any cyanide in the blood of rats administered **26** at a dose of 3 g/kg (route not specified) *(24)*. It is known that alcoholic (C-OH) carbons are easily metabolized *(25)* and it seems likely, therefore, that metabolism of **26** occurs predominately (if not entirely) at the C-OH carbon. Thus, the 3-hydroxy group of **26** directs cytochrome P450 mediated metabolism away from the α–carbon, thereby reducing the acute toxicity of this substance relative to **2**.

It is noteworthy that there is a very large difference in toxicity between **26** and its isomer 2-hydroxypropionitrile **(27)**, shown on the next page. The rat oral LD_{50} of **27** is 1.23 mmol/kg *(28)*. (Mouse oral LD_{50} data for this substance are not available). On the other hand, **26** has very low toxicity: the rat oral LD_{50} of **26** is 45 mmol/kg *(2)*. The reason for the high toxicity of **27** is because this substance is a cyanohydrin, the bioactivated form (Figure 1), and as such decomposes readily to release cyanide.

Similar to hydroxy-substituted nitriles, the toxicity of amino-substituted nitriles can vary widely, depending upon the location of the amino group. Acute toxicity data for some α–nitriles and β–amino

CH₂—CH₂—C≡N
|
OH

26

rat oral LD_{50} = 45 mmol/kg

CH₃—CH—C≡N
|
OH

27

rat oral LD_{50} = 1.23 mmol/kg

nitriles are presented in Table I and Table II. The β–amino nitriles shown in Tables I and II are derived from propionitrile. The toxicity data in Table I were measured in mice, whereas the data in Table II were measured in rats. It is readily apparent that the β–aminopropionitriles (**20-22, 32, 33**) are considerably less acutely toxic than propionitrile (**2**). The low toxicity of **20-22, 32** and **33** compared to **2** may be explained from the influence of the amino substituents directing metabolism away from the carbon atom alpha to the cyano group. Mumtaz et al. reported that in rats approximately 44% of orally administered 3-(dimethylamino)-propionitrile (**20**) is excreted unchanged, and 3-aminopropionitrile and cyanoacetic acid are the only urinary metabolites (*26*). Keiser et al. reported that 3-aminopropionitrile is almost completely excreted in the urine as cyanoacetic acid when administered to mice or rats (*27*). These studies clearly indicate that beta (β)–amino nitriles are metabolized at the carbon atom adjacent to the amino group, and not the carbon alpha to the cyano group. Thus, it appears that cyanide is not released to any large extent from β–amino nitriles.

However, nitriles substituted with an amino group at the carbon alpha to the cyano group are quite toxic (Table II). The acute toxicity of both aminoacetonitrile (**28**) and 2-(dimethylamino)acetonitrile (**29**), for example, are greater than the acute toxicity of acetonitrile (**1**) by approximately 100-fold (Table II). Similarly, the acute toxicities of the butyronitrile-derived α–amino nitriles **30** and **31** are substantially greater than that of butyronitrile (**3**). Thus, the presence of an α–amino group in a nitrile greatly increases acute toxicity. Unlike the β–amino nitriles described above, metabolism data for α–amino nitriles **28-31** are unavailable and consequently the reason for their extreme toxicity is not as readily apparent. What is particularly interesting, however, is that **30** and **31** contain α–methyl groups as well: no α–hydrogen atoms are present. In addition, the 3-position of **31** contains a tertiary hydrogen atom. These structural characteristics of **30** and **31** are expected to *decrease* acute toxicity if cytochrome P450-mediated α–hydroxylation were the basis of toxicity.

The reason for the high toxicity of α–amino nitriles may be due to their chemical similarity to cyanohydrins: the amino group of an α–amino nitrile may be behaving similarly to the α–hydroxy group of a cyanohydrin, and as such is expected to release cyanide rapidly without

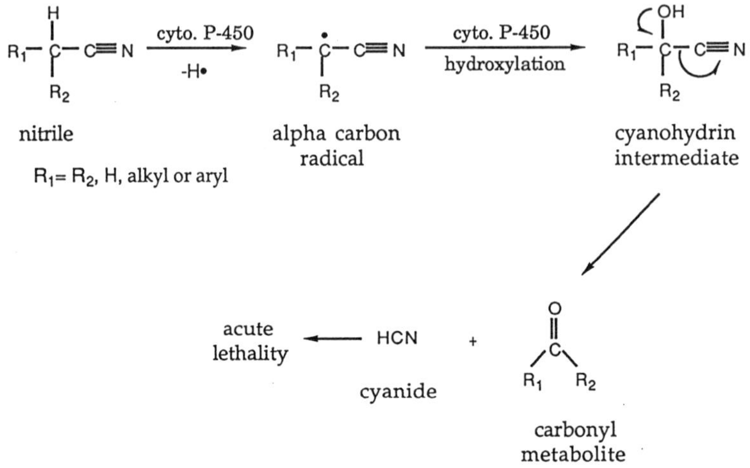

```
         H
         |                cyto. P-450        •                 cyto. P-450            OH
    R₁ — C — C≡N      ———————→      R₁ — C — C≡N      ———————————→      R₁ — C — C≡N
         |                 -H•              |                hydroxylation            |
         R₂                                  R₂                                       R₂
```

nitrile   alpha carbon   cyanohydrin
                           radical   intermediate

$R_1 = R_2$, H, alkyl or aryl

```
                                                              O
                                                              ‖
     acute    ←——— HCN    +                                   C
     lethality                                              /   \
                          cyanide                         R₁     R₂
```

carbonyl
metabolite

Figure 1. Mechanism of Acute Toxicity of Nitriles: Cytochrome P450-Mediated Cyanide Release.

Table II. Acute Toxicity (LD_{50}) Data of Some Alpha (α)-, and Beta (β)-Amino Nitriles.

compound			Oral LD_{50} (mmol/kg)	
name	no.	structure	mice	rats
acetonitrile[a]	1	CH_3CN	6.55[b]	78[c]
propionitrile[a]	2	CH_3CH_2CN	0.65[b]	3.26[c]
butyronitrile[a]	3	$CH_3CH_2CH_2CN$	0.56[b]	3.18[c]
α-Aminonitriles				
aminoacetonitrile	28	NH_2CH_2CN		0.47[d]
2-(dimethylamino)acetonitrile	29	$(CH_3)_2NCH_2CN$		0.6[e]
2-amino-2-methylbutyronitrile	30	$CH_3CH_2C(CH_3)(NH_2)CN$		0.76[f]
2-amino-2,3-dimethylbutyronitrile	31	$CH_3CH(CH_3)C(CH_3)(NH_2)CN$		0.74[g]
β–Aminonitriles				
3-aminopropionitrile	32	$NH_2CH_2CH_2CN$		4.3[h]
3-methylaminopropionitrile	33	$CH_3NHCH_2CH_2CN$		41.6[c]
3-(dimethylamino)propionitrile	20	$(CH_3)_2NCH_2CH_2CN$	15.3[c]	26.5[c]
3,3'-iminodipropionitrile	22	$HN(CH_2CH_2CN)_2$	27.5[i]	24.7[i]

[a] Included for comparison purposes. [b] Obtained from ref. 11. [c] Obtained from ref. 3. [d] As sulfate salt (LD_{50} = 100 mg/kg), ref. 30. [e] Obtained from ref. 31. [f] Obtained from ref. 32. [g] Obtained from ref 29. [h] As fumarate salt (LD_{50} = 800 mg/kg), obtained from ref. 33. [i] Obtained from ref. 33.

the need for cytochrome P450 metabolism. The possibility of α–amino nitriles behaving similarly to cyanohydrins is supported by the high toxicity of these substances, including the rapid onset of toxic symptoms, and that the presence of structural features expected to reduce toxicity (if cytochrome P450-mediated bioactivation were required) do not appear to do so.

Summary of Structure-Acute Lethality Relationships. The structure-acute lethality relationships of nitriles become apparent from an understanding of the mechanism of acute toxicity of these substances. The acute lethality of a nitrile can be ascribed to structural factors that influence the ease in which cytochrome P450 enzymes can catalyze radical formation at the α–carbon atom relative to other carbon atoms within the substance. The general toxicity pattern, in decreasing order, with respect to the type of α–carbon radical formed following α–hydrogen atom abstraction is benzylic ≈ 3° > 2° > 1°. The presence of a hydroxy or a substituted or unsubstituted amino group on the α–carbon atom greatly increases toxicity, whereas the presence of these substituents at other carbon positions greatly reduces acute toxicity.

Quantification of Structure-Acute Lethality Relationships. As discussed in Chapter 2, quantification of structure-activity relationships enables precise prediction of biological activity (toxicological or pharmacological) directly from structure. Quantitative structure-activity relationships are particularly useful for the design of safer chemicals because one can quantitatively predict the toxicity of untested or planned substances, and observe the influence of structural differences between analogous substances on toxic potency. Lipophilicity is an important factor of biological activity of a substance because it is a descriptor of bioavailability (this is discussed in Chapter 2). For this reason log P (octanol/water partition coefficient; a physicochemical expression of lipophilicity) is widely used, either alone or in combination with other physicochemical properties, as a descriptor in the quantification of structure-activity relationships.

Several investigators have attempted to quantify the structure-acute lethality relationships of nitriles by correlating into a regression equation LD_{50} data with certain physicochemical descriptors *(11-13, 21, 22)*. Tanii and Hashimoto *(11-13)* had limited success in correlating the LD_{50} data of many of the nitriles in Table I with log P. It was found that log P was not a good overall descriptor of acute lethality of nitriles: they observed that the acute lethality of certain nitriles correlate well with log P, whereas others do not. Those nitriles in which acute lethality correlates with log P only correlate when divided into subgroups; each subgroup having its own regression equation. These regression equations are very limited for predicting the acute toxicity of untested nitriles because how is one to know if the nitrile whose toxicity is to be predicted is one in which acute lethality correlates with lipophilicity. Even if this where known, it is not clear which of the regression equations is the most appropriate to use. What is more useful is a single regression equation that accurately quantifies the structure-acute lethality relationships of a large group of structurally diverse nitriles.

A possible reason why the acute lethality data of nitriles does not correlate well in a single regression equation in which log P is the only term may be because the log P term is not a descriptor of the *mechanism* of acute lethality of nitriles. From Table I one can readily see that there are nitriles that have similar lipophilicity but differ greatly in acute lethality. Phenylacetonitrile (24), for example, has essentially the same lipophilicity as 10 but is over 10-fold more acutely toxic. The same is true in comparing 12 with 26, 11 with 9, or 11 with 4. The differences in the toxicities of these substances is related to their differences in the type of carbon radical formed at the carbon atom alpha to the cyano group following cytochrome P450-mediated hydrogen atom abstraction (metabolism).

From the LD_{50} data of the nitriles in Table I, Grogan and co-workers (22) developed a single quantitative structure-activity relationship equation (equation 1) that uses log P and a descriptor ($k_{\alpha corr}$) of cytochrome P450-mediated hydrogen atom abstraction at the α–carbon, relative to hydrogen atom abstraction at other carbons.

$$\log (1/LD_{50}) = -0.16(\log P)^2 + 0.22(\log P) + 0.11(\ln k_{\alpha corr}) + 5.67 \quad (1)$$

$$n = 26 \qquad r = 0.85$$

In equation 1 n is the number of nitriles used and r is the correlation coefficient. The $k_{\alpha corr}$ term is derived from k_α, which is a descriptor of cytochrome P450-mediated hydrogen atom abstraction at the α–carbon only and does not consider cytochrome P450-mediated hydrogen atom abstraction at other carbon atoms. The k_α term is calculated from ionization potential, ground-state and radical heats of formation (22). The $k_{\alpha corr}$ term takes into account the likelihood of cytochrome P450-mediated hydrogen atom abstraction occurring at all non α–carbon atoms relative to the likelihood of hydrogen atom abstraction at the α–carbon, and is a much better descriptor of overall cytochrome P450 metabolism than is k_α. The acute toxicity data of the nitriles in Table I correlated well with equation 1 (22).

Table I contains the k_α and $k_{\alpha corr}$ values for each nitrile used in the derivation of equation 1. The meaningfulness of these descriptors with respect to acute toxicity of nitriles becomes apparent upon examination of the data provided in Table I. Nitriles whose k_α values are considerably higher than their corresponding $k_{\alpha corr}$ values (e.g. nitriles 4-7, 10, 20-22) are those that are metabolized predominantly at non-alpha carbon positions and, consequently, are less toxic. Nitriles with high $k_{\alpha corr}$ values (e.g.

nitriles **8, 11, 23, 24**) are metabolized predominately at the α–carbon, and are quite toxic. Thus, these descriptors give an indication of the extent of cytochrome P450 hydrogen atom abstraction that is expected to occur at the α–carbon relative to other carbons in the molecule. For an untested or planned nitrile, these descriptors (as well as log P) can be derived directly from structure *(22)* and, using equation 1, one can estimate the LD_{50} value (acute toxicity) of the nitrile.

The Design of Less Acutely Toxic Nitriles. How can the information discussed in the preceding paragraphs be used to design nitriles that are less acutely toxic? The above analysis of the literature sources pertaining to the acute toxicity of nitriles provides insight into how less acutely toxic nitriles can be designed. More specifically, the studies that have explored and elucidated the mechanism of acute lethality enable a clearer understanding of the relationship between structure and acute lethality. Guidelines for the design of less acutely toxic nitriles are provided below.

When designing a nitrile, add structural features that will prevent or minimize cytochrome P450-mediated hydroxylation at the α–carbon. Remember that the acute toxicity of a nitrile is directly related to structural factors that influence the ease in which cytochrome P450 enzymes catalyze radical formation at the α–carbon atom relative to the other carbon atoms within the substance. Acute toxicity is enhanced when the α–carbon can form a stable radical, such as when the α–carbon is tertiary or benzylic. Therefore, one should avoid having tertiary or benzylic α–carbons in a nitrile. If the α–carbon must be tertiary or benzylic for purposes of intended use, add other substituents that compete with, or direct cytochrome P450 metabolism away from the α–carbon. This could be accomplished by making another (i.e. non-α) carbon in the nitrile benzylic, tertiary, or alcoholic (i.e., add an OH group). Or, if allowable from a use standpoint, design a nitrile that has a quarternary α–carbon (i.e., an α–carbon that is bonded to three other carbon atoms, in addition to the cyano group). Such a nitrile should have low acute toxicity because it does not contain an α–hydrogen and thus cannot form a cyanohydrin intermediate upon cytochrome P450 metabolism.

Unless absolutely necessary for intended use, do not add hydroxy or amino groups (substituted or unsubstituted) to the α–carbon, because these nitriles will be highly acutely toxic. As discussed earlier, nitriles that are cyanohydrins or contain an amino group (substituted or unsubstituted) on the α-carbon are highly acutely toxic because they release cyanide readily, without the need for bioactivation from cytochrome P450. Because such nitriles do not require cytochrome P450-bioactivation to release cyanide, substituents added to redirect metabolism away from the α–carbon will have little, if any, effect in reducing acute toxicity. There is very little that can be done to reduce the toxicity of a cyanohydrin or α–amino nitrile except, perhaps, to incorporate structural modifications intended to minimize absorption (see Chapter 2).

Calculate the log P, k_α and $k_{\alpha corr}$ values for a planned nitrile and, using equation 1, estimate its LD_{50} value. Details for calculating log P, k_α and $k_{\alpha corr}$ and using equation 1 are provided in reference 22. Equation 1 will be very helpful in assessing the acute toxicity of a planned nitrile or a series of nitriles, and for assessing the effects of structural modifications on acute toxicity.

When considering structural modifications that are intended to reduce acute toxicity (lethality), be careful not to choose and incorporate structural modifications that will reduce acute toxicity, but will bestow other toxicity, such as osteolathyrism or neurotoxicity. In addition to acute lethality, osteolathyrism is another other health risk posed by certain nitriles. At least two nitriles are known to be neurotoxic. One needs to be careful not to make molecular modifications that will reduce acute lethality, but will make a nitrile an osteolathyrogen or a neurotoxicant. For example, addition of a dimethylamino group to the 3-position of propionitrile (**2**), to give **20**, results in a substantial decrease in acute toxicity; **20** is substantially less acute toxic than **2** (Tables I and II). However, **20** is highly neurotoxic whereas **2** is not. Thus, although addition of the dimethylamino group to the terminal carbon of **2** greatly reduces the acute lethality of **2**, this structural modification bestows neurotoxicity. A discussion of the nitriles that cause osteolathyrism and neurotoxicity is provided below.

Nitrile-Induced Osteolathyrism

Osteolathyrism is a medical condition characterized by hernias, dissecting aortic aneurysms, lameness, skeletal deformities such as exostoses and kyphoscoliosis, and slowing or cessation of body growth. These medical complications are, of course, quite serious and often lead to premature death. Osteolathyrism occurs when there is interference with the ability of collagen and elastin to undergo the crosslinking that normally occurs with aging. Thus, the condition is more likely to occur (and is more serious) during the early stages of growth and development (e.g., childhood), where collagen and elastin have not yet fully crosslinked, than it is during later stages of life, when the majority of crosslinking of collagen and elastin has already occurred.

Striking characteristics of osteolathyrism in laboratory animals include: weakening of the arteries, tendinous and ligamentous attachments, epiphyseal plates, skin, cartilage; delayed wound healing; and marked deformations of the skeleton *(35-37)*. Osteolathyrism is caused by exposure to certain chemical substances during early stages of life, particularly during gestation, infancy and early childhood. Certain hydrazines, hydrazides, semicarbazides and nitriles, for example, are known to cause osteolathyrism. Several comprehensive reviews on osteolathyrism and the chemical substances that cause it are available *(38-40)*.

The first indication that nitriles can induce osteolathyrism came when the condition was observed in animals that ingested seeds of the leguminous plants of the genus *Lathyrus*, which includes many kinds of peas *(38)*. A crystalline substance producing the signs and symptoms of osteolathyrism was isolated from *Lathyrus Odoratus* almost simultaneously by Schilling and Strong *(41)* and Dasler *(42)*. Using degradation and unambiguous synthesis, Schilling and Strong*(43)* later identified this substance to be b-(N-g-L-glutamyl)-aminopropionitrile, of which the osteolathyritic principle is 3-aminopropionitrile (**32**). Diets containing the sweet pea *Lathyrus Odoratus* or **32** produce the signs and symptoms of osteolathyrism in several animal species *(35, 44-52)*. Although 3-aminopropionitrile-induced osteolathyrism has not, to date, been reported in humans, the ability of this substance to induce this condition in many species strongly suggests that humans exposed to this substance may be at risk of developing osteolathyrism.

The literature reveals that many other nitriles have been tested for inducing osteolathyrism in a variety of mammalian and non-mammalian species *(35, 38, 47, 53-58)*. These nitriles are listed in Table III. Also shown in Table III is a relative ranking of the potency of these nitriles to induce osteolathyrism. This relative ranking was based on a retrospective analysis of the experimental results of the above studies. As can be seen from Table III, of the many nitriles tested, only a few are known to produce osteolathyrism. Of these, **32** is by far the most studied with respect to causing osteolathyrism *(35, 38, 47-73)*.

Mechanism of Nitrile-Induced Osteolathyrism. The majority of studies conducted to elucidate the mechanism of nitrile-induced osteolathyrism have utilized **32** and, to a lesser extent, aminoacetonitrile (**28**) *(54, 74-77)*. The early studies that utilized **32** found that the osteolathyrism induced by this substance resulted from an interference with intermolecular crosslinking within collagen and elastin fibrils *(59-63)*. To enable a better understanding of the mechanism of nitrile-induced osteolathyrism, a brief discussion of collagen and elastin physiology is provided below.

Collagen is the principal protein of connective tissue. Its major function is to provide strength and maintain the structural integrity of tissues and organs. Collagen fibrils are made up of tropocollagen subunits that are composed of three polypeptide chains. The tropocollagen subunits are connected to each other by covalently crosslinked lysine or 5-hydroxylysine residues resulting from the action of lysine oxidase, a monoamine oxidase requiring copper and pyridoxal phosphate as cofactors. The crosslinking increases with age. Elastin, like collagen, is

Table III. Summary of the Results of Studies that have Tested Nitriles for Osteolathyrism [a]

compound name	no.	structure	relative potency in causing osteolathyrism[b]
acetonitrile	1	CH_3CN	0
propionitrile	2	CH_3CH_2CN	0
butyronitrile	3	$CH_3CH_2CH_2CN$	0
malononitrile	12	$NCCH_2CN$	0
succinonitrile	13	$NCCH_2CH_2CN$	0
phenylacetonitrile	24	$C_6H_5CH_2CN$	0
3-hydroxypropionitrile	26	$HOCH_2CH_2CN$	0
aminoacetonitrile	28	NH_2CH_2CN	+++
2-aminopropionitrile	34	$CH_3CH(NH_2)CN$	0
3-aminopropionitrile	32	$NH_2CH_2CH_2CN$	+++
3-(methylamino)propionitrile	33	$CH_3NHCH_2CH_2CN$	0/+ [c]
3-(dimethylamino)propionitrile	20	$(CH_3)_2NCH_2CH_2CN$	0
3-amino-2-methylpropionitrile	35	$NH_2CH_2CH(CH_3)CN$	+++
3,3'-iminodipropionitrile	22	$HN(CH_2CH_2CN)_2$	0/+ [d]
3,3',3''-iminotripropionitrile	36	$N(CH_2CH_2CN)_3$	0
4-aminobutyronitrile	37	$NH_2CH_2CH_2CH_2CN$	0
ethylaminodipropionitrile	38	$CH_3CH_2N(CH_2CH_2CN)_2$	0
benzylaminopropionitrile	39	$C_6H_5CH_2NHCH_2CH_2CN$	0
phenylethylaminopropionitrile	40	$C_6H_5(CH_2)_2NHCH_2CH_2CN$	0
3,3'-oxydipropionitrile	41	$O(CH_2CH_2CN)_2$	0
3,3'-thiodipropionitrile	42	$S(CH_2CH_2CN)_2$	0

[a] Results were obtained from refs. 35, 38, 47, 53-58, and reflect effects observed in at least one species. [b] Relative potency is based on an analysis of the data presented in these studies, and are strictly qualitative. The scale used here with respect to causing osteolathyrism is as follows: 0 = not found to cause osteolathyrism; + = weak; ++ = moderately potent; +++ = highly potent. [c] Experimental results are ambiguous. Most studies show that this substance is non-osteolathyrogenic. [d] Experimental results are ambigous. Some studies show this substance to be negative for osteolathyrism, while other studies show that it is positive.

rich in glycine and alanine but, unlike collagen, contains few prolines and consequently has elastic properties. It is present in ligaments and the walls of major arteries. The basic subunit of elastin fibrils is tropoelastin. Tropoelastin subunits are connected by the crosslinking of lysine residues, resulting from the action of lysine oxidase.

Covalent crosslinking in collagen and elastin takes place almost entirely through amino and/or aldehyde groups. The aldehyde groups are formed by lysine oxidase-catalyzed oxidative deamination of the epsilon (ε) amino group of lysine or 5-hydroxylysine. The crosslinks occur by either imine formation between the aldehyde and amino group or by aldol condensation.

The metabolic turnover of collagen and elastin is greatest during the developmental and growth periods of mammals, and gradually becomes very low in adult life. As an animal matures the imine moieties are modified, presumably by reduction or condensation with other tropocollagen or tropoelastin subunits. The crosslinks become chemically more stable and the collagen or elastin fibers become increasingly rigid with age *(78-92)*. A decrease in crosslinks, as in osteolathyrism, leads to an increase in collagen and elastin fragility and, eventually, the signs and symptoms of osteolathyrism (discussed earlier).

In mice and rats, the osteolathyrogenic effects of **32** occur after chronic administration rather than acute administration, and predominantly in the weanling rather than the adult mouse or rat *(27, 50, 64)*. Several studies found that **28** produced a very pronounced osteolathyrogenic effect in the fetuses of pregnant rats *(76, 35, 47)*. It had the strongest effect on the fetal skeleton and cardiovascular system. Collectively, these studies strongly indicate that the osteolathyrogenic actions of **28** and **32** result primarily from the inhibition of new covalent crosslink formation rather than by rupturing previously formed crosslinks.

Inhibition of Lysine Oxidase. Page and Benditt concluded that 3-aminopropionitrile-induced osteolathyrism results from the inhibition of oxidative deamination of lysyl residues, and not from direct binding of **32** to collagen *(68, 70)*. Other investigators subsequently reported that **32** irreversibly inhibits lysine oxidase *(69,71,76,77)*. It is now well established that the osteolathyritic properties of **28** and **32** are due to the ability of these substances to irreversibly inhibit lysine oxidase *(56, 58, 77, 93)*. It is believed that the osteolathrytic properties of **22, 33,** and **35** are also due to lysine oxidase inhibition *(40, 58)*. As discussed above, lysine oxidase is a highly important enzyme because it functions during the initial phase of collagen and elastin formation. Its inhibition effectively results in the complete blockade of collagen and elastin biosynthesis, and accounts for the extreme toxicity of substances that cause osteolathyrism from its inhibition.

Several studies have been undertaken to elucidate the specific mechanism by which **28** and **32** irreversibly inhibit lysine oxidase *(73, 76, 93)*. Maycock and co-workers *(76)* reported that the mechanism of action of enzymes that oxidize amines to aldehydes (e.g., lysine oxidase) involves enzyme-mediated hydrogen atom abstraction from the carbon atom that is oxidized (i.e. attached to the amine). Thus, the presence of at least one hydrogen atom on the carbon that is attached to the amine is required for the oxidation. They found that aminoacetonitrile inactivates rabbit plasma amine oxidase *in vivo*. The compound showed no substrate activity but was a potent irreversible inhibitor *(76.* Experiments with [1-^{14}C]aminoacetonitrile showed that the enzyme became covalently labeled *(76).* The irreversible inhibition is thought to result from proton abstraction of the α-carbon which leads to the formation of a reactive ketenimine (Figure 2). The ketenimine intermediate, which is highly electrophilic, most likely undergoes attack from a nucleophile located within the enzyme active site resulting in covalent bond formation and inactivation of the enzyme.

Rando *(73)* suggested a similar mechanism for the inhibition of lysine oxidase by **32**. Rando proposed that the amino group of **32** reacts with the aldehyde moiety of pyridoxal phosphate (a cofactor for lysine oxidase) to form an imine. Enzyme-mediated hydrogen abstraction from the -CH$_2$- group adjacent to the imino group liberates a cyanide ion and yields a reactive imino-enamine product. It was postulated that the imino-enamine product undergoes a 1,4-addition reaction with some nucleophile located within the enzyme active site. In a later study, Tang and co-workers *(93)* found that cyanide is not liberated during the inhibition of lysine oxidase by **32**, and proposed a slightly different mechanism of inhibition that is shown in Figure 3. They proposed that following imine formation of **32** with pyridoxal phosphate a hydrogen atom from the carbon atom attached to the cyano group is removed by the enzyme, yielding a ketenimine intermediate (Figure 3). This ketenimine intermediate reacts as described in the preceding paragraph, resulting in irreversible inhibition of lysine oxidase. It seems plausible that irreversible inhibition of lysine oxidase by **28** may also involve initial imine formation with pyridoxal phosphate.

Although it is generally accepted that imine formation between the amino group of **32** and pyridoxal phosphate occurs during the lysine oxidase inhibition cascade, it is not clear as to whether it is a requisite step for ketenimine formation *(56, 58,93)* . The amino groups of nitriles such as 3-(methylamino)propionitrile **(33)** and 3,3'-iminodipropionitrile **(22)** are substituted and cannot form imine adducts with pyridoxal phosphate, yet these substances have been reported to cause osteolathyrism (Table III). The osteolathyritic effects of **22** and **33**, however, are somewhat ambiguous. Some studies have found these substances to be inactive as osteolathyrogens. In addition, of these two substances, only **22** inhibits lysine oxidase *in vivo* (i.e., in the body) and the inhibition is largely reversible *(58)*. Clearly, the more potent amino nitriles are those which contain an unsubstituted amino group and, hence, can form imines. It is also noteworthy that both **22** and **33** are known to be metabolized to 3-aminopropionitrile **(32)** *(58)*, and this may explain or at least contribute to their osteolathyrogenic properties observed in some studies. Thus, although it is not entirely clear as to whether initial imine formation is essential for lysine

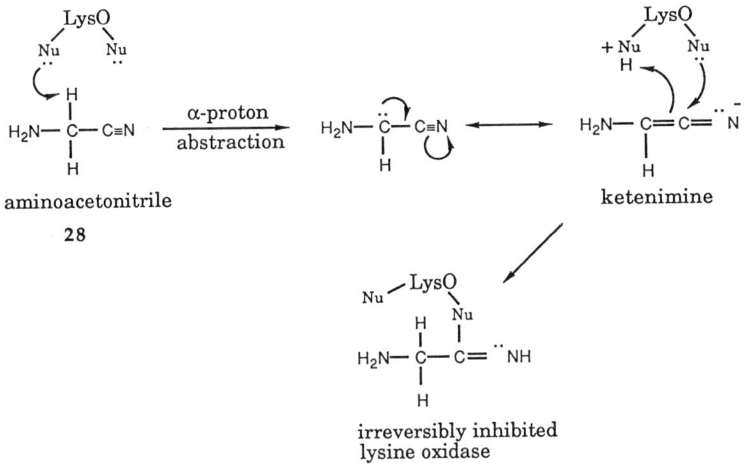

Figure 2. Mechanism of Irreversible Inhibition of Lysine Oxidase (LysO) by Aminoacetonitrile, **28** (adapted from Maycock, et. al., ref. 76).

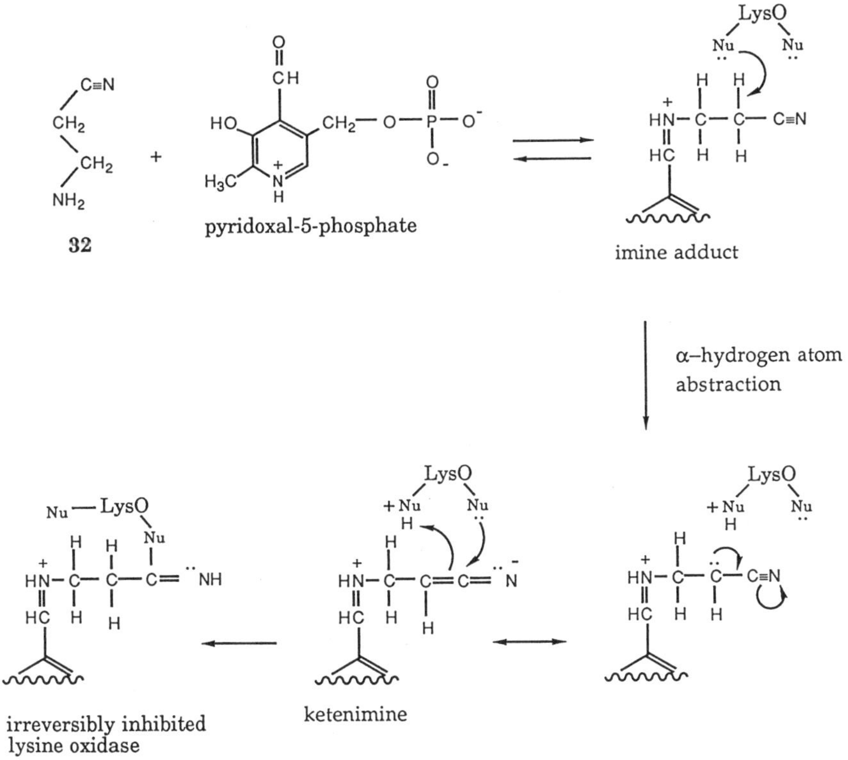

Figure 3. Mechanism of Irreversible Inhibition of Lysine Oxidase (LysO) by 3-Aminopropionitrile, **32** (adapted from refs. 73 and 93).

oxidase inhibition, it at least appears to have considerable significance with regard to osteolathyrogenic potency.

Structure-Activity (Osteolathyricity) Relationships of Nitriles. As can be seen from Table III, only a few nitriles are known to cause osteolathyrism. Unlike the relatively large amount of single specie acute toxicity data available for nitriles (Table I), the osteolathyritic data compiled in Table III is comparatively limited for discerning structure-osteolathyricity relationships. This is because only a few nitriles are known to cause osteolathyrism and because the data used to derive the relative potencies in Table III are largely qualitiative and were collected in a variety of species. Nonetheless, a retrospective analysis of the osteolathyricity data available for nitriles and other substances, in conjunction with the mechanism of nitrile-induced osteolathyrism, provides some insight into the relationships between nitrile structure and osteolathyricity. This insight can be used for assessing the likelihood of a planned or untested nitrile having osteolathyritic properties, and for the design of nitriles that will not cause osteolathyrism.

The general requirements for a nitrile to cause osteolathyrism are a cyano group and an amino group. Nitriles that do not contain an amino group are not osteolathyritic. Acetonitrile (**1**) and propionitrile (**2**) for example do not produce osteolathyrism, whereas aminoacetonitrile (**28**) and 3-aminopropionitrile (**32**) are potent osteolathyrogens (Table III). Amines that do not contain a cyano group, such as methylamine (CH_3NH_2), also do not cause osteolathyrism *(47,54)*. Thus, both an amino group and a cyano group are essential requirements for a nitrile to cause osteolathyrism, and are consistent with the proposed mechanisms for lysine oxidase inhibition. It is also interesting to point out that 3-hydroxypropionitrile (**26**), the hydroxy isostere of **32**, is not osteolathyritic. This also is consistent with the mechanism proposed for irreversible inhibition of lysine oxidase by **32**, in that **26** cannot form an imine.

Another important general requirement for a nitrile to cause osteolathyrism is the presence of at least one hydrogen atom on the carbon atom adjacent to the cyano group. This requirement is inferred from the mechanisms of nitrile-induced inhibition of lysine oxidase (Figures 2 and 3). Nitriles that do not contain hydrogen atoms at the carbon atom adjacent to the cyano group cannot form lysine oxidase-catalyzed ketenimine intermediates and, logically, should not cause osteolathyrism.

The distance between the amino and cyano groups appears to be important for osteolathyritic activity. While **28** and **32** are potent osteolathyrogens, 4-aminobutyronitrile (**37**) does not produce osteolathyrism. Thus, osteolathyritic activity is lost when the number of carbon atoms separating the cyano group from the amino group is three. Presumably, amino nitriles in which the number of carbon atoms separating the cyano group from the amino group is greater than three will also not cause osteolathyrism. Experiments need to be conducted to confirm this presumption.

The position of the amino group also appears to be important with respect to osteolathyritic activity. With the exception of 37, nitriles that have an amino group on their terminal carbon (e.g., 28 or 32) are potent osetolathyrogens, whereas osteolathyrogenic activity is lost when the amino group is located on a non-terminal carbon. This can be seen, for example, in comparing the osteolathyritic potency of 32 to that of 34; 32 is highly potent whereas 34 is inactive. The reason for these differences this is unclear. It is also not clear from the available data whether this is a function of the amino group being non-terminal or its distance from the cyano group. The amino groups of 28, 32 and 34 are primary and, thus, capable of forming imine intermediates with pyridoxal phosphate. In addition, each of these nitriles contain at least one hydrogen atom of the carbon atom adjacent to the cyano group, and are expected to be removable by lysine oxidase to form a ketenimine intermediate. One would expect, therefore, that 34 and 37 would be capable of causing osteolathyrism. The apparent inability of 34 and 37 to cause osteolathyrism is not explainable from present knowledge, but could be related to steric factors.

Substitution of the amino group greatly influences osteolathyritic potency. The most potent osteolathyritic nitriles are those that contain a primary amino (i.e., NH_2) group on their terminal carbon. Nitriles that have either a tertiary amino group (e.g., 20, 36, and 38), or a secondary amino group that contains a large subtituent (e.g., 39, 40) do not cause osteolathyrism. Nitriles that have a secondary amino group that contains a relatively small substituent, such as seen in 33 or 22, may still be osteolathyritic, but are less potent. These observations are generally consistent with the mechanisms proposed for nitrile-induced inhibition of osteolathyrism shown in Figures 2 and 3. The nitriles that contain primary amino group can undergo imine formation with pyridoxal phosphate, whereas tertiary and and secondary nitriles cannot.

As discussed earlier, it is not entirely clear if initial imine formation is required for lysine oxidase inhibition but it certainly appears that nitriles that can form imine intermeditates are potent osteolathyrogens (with the exception of nitriles 34 and 37, which are not osteolathyrogens for other reasons) compared to nitriles such as 22 and 33, which cannot form imine intermediates. As stated previously, in laboratory animals nitriles 22 and 33 and known to be metabolized to the potent osteolathyrogen 32, and this may explain or at least contribute to their observed osteolathyritic properties. It is interesting to note that nitriles 41 and 42, the oxygen and thio isosteres (respectively) of 22 cannot be metabolized to 32 and are not osteolathyrogenic (Table III).

The Design of Non-Osteolathyritic (Safer) Nitriles. Just as an understanding of the structure-activity and mechanistic data pertaining to the acute toxicity of nitriles enables one to infer structural modifications that can be used for the design of less acutely toxic nitriles, so can an understanding of the structure-activity and mechanistic data pertaining to the osteolathyricity of nitriles enable one to infer structural modifications for the design of nitriles that will not cause osteolathyrism.

When designing a nitrile, the only time one needs to be concerned about the possibility of osteolathyrism is when the nitrile will contain an amino group. Nitriles that do not contain an amino group are not osteolathyritic. Therefore, unless absolutely necessary, it is best not to incorporate an amino group onto a nitrile. In cases where a nitrile must contain an amino group, particularly a primary amino group, it is advisable that the amino group not be on a terminal carbon atom because at this position the amino group appears to bestow substantial osteolathyritic potency. If an amino group must be on a terminal carbon, there should be at least three carbon atoms that separate the amino group from the cyano group. Alternatively, one may also want to consider making the amino group a secondary amine that contains a large substituent (e.g., phenyl, benzyl), or a tertiary amine. Any of these structural characteristics are expected to minimize the potential for osteolathyrism. Placement of an amino group on a non-terminal carbon is not expected to bestow osteolathyritic properties. However, an amino group should never be placed on the carbon atom that is adjacent (alpha) to the cyano group because, as discussed earlier, such amino nitriles are highly acutely toxic.

Another structural characteristic that should mitigate the risk of osteolathyrism of an amino nitrile is the lack of hydrogen atoms on the carbon atom alpha to the cyano group. Because the mechanism by which amino nitriles induce osteolathyrism is believed to involve hydrogen atom abstraction at the carbon atom alpha to the cyano group (Figures 2 and 3), it is logical to assume that nitriles that do not contain hydrogen atoms at this position will not cause osteolathyrism. Replacement of the alpha hydrogens of osteolathyritic amino nitriles with, for example, methyl or ethyl groups should eliminate osteolathyritic potency.

Neurotoxic Nitriles

The literature search for identifying toxic properties of nitriles revealed that two nitriles, 3-(dimethylamino)propionitrile (**20**) and 3,3'-iminodipropionitrile (**22**), are neurotoxic. The neurotoxic signs and symptoms exhibited by **20** and **22** are distinct and different from one another, suggesting that their neurotoxic mechanisms are different. There are some nitriles that exhibit neurotoxicity in laboratory animals, but only when administered in high doses or under conditions that could only exist in a laboratory *(94,95)*. Except under conditions of extremely high exposure, it would seem unlikely that such nitriles would produce neurotoxicity in humans. Because only two nitriles exhibit neurotoxicity following exposure concentrations that are known or likely to be encountered in such places as the workplace, neurotoxicity is not a health concern for nitriles in general. This section summarizes the neurotoxic properties of **20** and **22**, and suggests structural modifications that may attenuate their neurotoxicity.

Neurotoxicity of 3-(Dimethylamino)propionitrile (20). The neurotoxic effects of 3-dimethylaminopropionitrile were first realized when an epidemic of urinary retention among workers exposed to the substance occurred in the spring of 1978 *(96)*. The outbreak occurred in a plant that manufactured automobile seat cushions from polyurethane foam that contained a catalyst which consisted of **20** in 95% concentration. Workers experienced difficulty in urinating and lost any sensation of bladder distention. Exposure to the catalyst was through direct skin contact with foam cushions and by inhalation of catalyst vapor released from the foam following curing. Initially, worker complaints were limited to urinary retention. However, the affected workers later complained of loss of sensory acuity in the lower half of the body, parasthesia, muscle weakness, and sexual dysfunction *(97)*. The urinary retention in these patients was attributed to effects on the peripheral nervous system. The inner lining (mucosa) of their bladders were normal *(97)*. Of eight patients who underwent neurological testing during recovery, seven had a subclinical sensory abnormality and lacked either detrusor reflex or normal sensation of bladder filling; three had prolonged sacral-evoked responses; and two of these three had limb motor neuropathies *(98)*. With cessation of use of the catalyst improvement occurred rapidly in nearly all cases, but ceased after one year. Repeat examinations two years later showed that those people with the most severe symptomatology had persistant urological, sexual and neurological abnormalities *(99)*. Based on these observations and a subsequent animal study, it was concluded that **20** was responsible for the neurotoxic effects in these workers *(100)*.

Jaeger and Plugge evaluated the acute urinary bladder toxicity of **20** in male rats *(101)*. It was observed that large doses rapidly produced central nervous system excitation that was followed by depression and death. Single oral doses of 0.31 or 0.62 mL/kg induced acute urinary bladder lesions in three days. Bladder changes consisted of massive transmural edema, acute ulcers and inflammation in essentially all animals, and hemorrhagic necrosis of the bladder wall in some of the animals tested. Jaeger and Plugge suggested that the bladder lesions may have been caused by the metabolite cyanoacetic acid. The ability of **20** to induce bladder damage in rats is probably not related to the urinary retention observed during occupational exposure, since workers exhibiting this effect had normal bladder mucosa as previously stated.

The neurotoxic mechanism of **20** has not yet been elucidated. Although **20** has been found to inhibit acetylcholinesterase *(102)* and the glycolytic enzymes glyceraldehyde-3-phosphate dehydrogenase and phosphofructokinase *(103)*, inhibition of these enzymes would not explain the urinary retention and the other neurotoxic effects caused by **20**. This substance does not inhibit monoamine oxidase *(104)*. Several authors have postulated that the neurotoxic effects of **20** are related to its metabolism. Ahmed and Farooqui have reported the presence of cyanoacetaldehyde and cyanoacetic acid in the urine of animals administered **20**, and suggested these metabolites are responsible for the neurotoxicity and bladder irritation, respectively *(105, 106)*. It seems unlikely, however, that these metabolites play a role in the neurotoxicity or bladder irritation of **20**. Both 3-aminopropionitrile (**32**) and 3-(methylamino)propionitrile (**33**) are metabolized to cyanoacetaldehyde and cyanoacetic acid, yet neither **32** or **33** are neurotoxic or produce bladder irritation *(104)*. Thus, the neurotoxic mechanism of **20** remains an interesting enigma.

Because the mechanism of neurotoxicity of **20** is unknown, and structure-activity (neurotoxicity) data are limited, it is difficult to infer a safer analog of **20**. A possibly safer analog of **20** is its homolog 4-(dimethylamino)butyronitrile. This substance does not produce urinary retention in rats *(102)*. Further studies need to be undertaken to assess more fully any neurotoxic properties of this substance. No neurotoxic effects have been reported for 2-(dimethylamino)acetonitrile (**29**), the other homolog of **20**. However this substance is not a safer substitute for **20** because it is considerably more acutely toxic (Table II).

Neurotoxicity of 3,3'-Iminodipropionitrile (22). The neurotoxic properties of 3,3'-iminodipropionitrile (**22**) were originally reported by Bachhuber and co-workers *(47)*. These investigators observed that rats fed diets containing 0.3% **22** developed muscle spasticity, weakness of the extremities, unsteady gait, and turned in circles. Similar neurotoxic effects have been reported in mice and birds *(107)*. Over the years, other neurotoxic that include hearing loss, and memory and learning deficits have been reported in laboratory animals *(108-115)*. No similarities to 3-(dimethylamino)propionitrile (**20**)-induced neurotoxicity have been reported, and the neurotoxic properties of **20** and **22** appear to be unrelated.

The mechanism of neurotoxicity of **22** has been explored by Sayre and co-workers *(116, 117)*. The neurotoxicity of **22** is believed to involve covalent bond formation of a metabolite of **22** with neurofilament fibers, resulting in the rapid dissociation of these fibers from their neuromicrotubules and subsequent neurotoxicity. These investigators postulate that **22** undergoes N-hydroxylation via flavin monooxygenase to N-hydroxy-3,3'-iminodipropionitrile (**43**) which, through a series of steps, is converted to the imine **44** (Figure 4). This latter substance undergoes tautomerization to dehydroiminodipropionitrile (**45**). It is believed that **45** undergoes direct attack by nucleophiles (e.g., an NH_2 group) located within the neurofilament fibers (Figure 4a). Alternatively, **45** may react with water to yield cyanoacetaldehyde (**46**) and **32** (Figure 4b). The former substance can react covalently with neurofilament fibers. The end result, in either case, are covalently modified neurofilament fibers and neurotoxicity. Of the two possible means of covalent modification neurofilament fibers (Figure 4a or 4b), the reaction of **45** with neurofilaments seems more plausible because substances such as **32** and **33** are also metabolized to cyanoacetaldehyde but these substances are not neurotoxic. If metabolism to

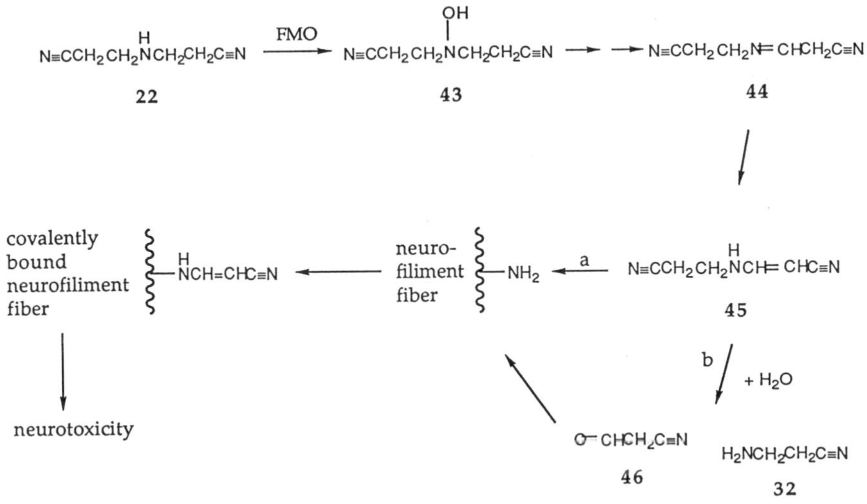

N≡CCH₂CH₂NHCH₂CH₂C≡N —FMO→ N≡CCH₂CH₂N(OH)CH₂CH₂C≡N → N≡CCH₂CH₂N=CHCH₂C≡N
         22   43   44

covalently
bound   —NHCH=CHC≡N ← neuro-filiment fiber—NH₂ ←ᵃ— N≡CCH₂CH₂NHCH=CHC≡N
neurofiliment   45
fiber

↓   ᵇ↓ + H₂O

neurotoxicity   O=CHCH₂C≡N   H₂NCH₂CH₂C≡N
                                            46   32

Figure 4. Proposed Mechanism of Neurotoxicity of 3,3′-Iminodipropionitrile, **22** (adapted from refs. *116* and *117*). FMO is flavin monooxygenase enzymes.

cyanoacetaldehyde were the basis of neurotoxicity caused by **22**, then **32** and **33** would be expected to be neurotoxic as well.

Assuming that the neurotoxic mechanism of **22** proposed by Sayre, *et al.* is correct, an approach that could be taken to design of a safer analog of **22** could be to incorporate structural modifications that prevent formation of **45**. Since formation of **45** involves loss of a hydrogen atom at either the 3- or 3'-carbon atom, replacement of the hydrogen atoms at these positions with, for example, methyl groups will prevent formation of **45** and, hence, the resultant neurotoxicity. An analog such as 3,3'-tetramethyl-3-3'-iminodipropionitrile (**47**), for example, is not expected to be neurotoxic.

$$N{\equiv}C-CH_2-\underset{\underset{CH_3}{|}}{\overset{\overset{CH_3}{|}}{C}}-\overset{H}{N}-\underset{\underset{CH_3}{|}}{\overset{\overset{CH_3}{|}}{C}}-CH_2-C{\equiv}N$$

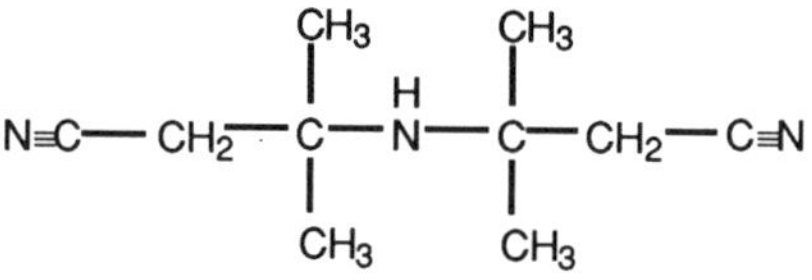

47

This substance is also not expected to cause osteolathyrism for reasons discussed previously. It is possible however, that **47** may be significantly acutely lethal because in this substance the carbon atoms attached to the cyano groups are secondary and more likely to undergo cytochrome P450-mediated hydroxylation (which would result in cyanohydrin formation and subsequent release of cyanide) than are the CH_3 carbon atoms. Equation 1 could be used to estimate the acute lethality **47**.

It is noteworthy to mention that the oxygen and sulfur isosteres of **22** (**41** and **42**, respectively) are not neurotoxic *(38)*. The inability of **41** or **42** to cause neurotoxicity may be ascribed to their inability to undergo flavin monooxygenase-mediated hydroxylation to products analogous to **43**, and supports the mechanism of neurotoxicity of **22** proposed by Sayre and co-workers (Figure 4). Neither **41** or **42** cause osteolathyrism (Table III), and are probably not acutely toxic (their oral LD_{50} values are likely to be similar to that of **22**). Thus both **41** and **42** are safer analogs of **22**. The use of isosteric substitution in the design of safer chemicals is described in Chapter 2.

$$N{\equiv}CCH_2CH_2\overset{H}{N}CH_2CH_2C{\equiv}N \quad \textbf{22}$$

$$N{\equiv}CCH_2CH_2OCH_2CH_2C{\equiv}N \qquad N{\equiv}CCH_2CH_2SCH_2CH_2C{\equiv}N$$

41 **42**

Conclusion

The literature search for identifying toxic effects caused by nitriles revealed that there are two general health concerns for this class of substances: acute toxicity (lethality) and osteolathyrism. Neurotoxicity is known to result from two nitriles. Chemists who design nitriles need to be aware of the possibility that these toxic effects may be caused by new nitriles, unless structural modifications that are intended to reduce or prevent toxicity are included during the design of new nitriles. A retrospective analysis of the literature studies has enabled the deduction of general structural modifications that prevent or reduce toxicity. Chemists who design nitriles can use these general structural modifications for the design new nitriles that are commercially efficacious and of low toxicity. One important caveat is that great care must be taken not to incorporate structural modifications that reduce or prevent one particular toxic effect but bestow another. This is why it is important to identify all toxic endpoints of concern, and understand, to the extent possible, any mechanistic and structure-activity data. An understanding of such information will greatly minimize the likelihood of incorporating structural modifications that reduce one particular type of toxic effect but cause another. The general strategy described herein for determining structural modifications for the design of safer nitriles can be applied to other chemical classes as well.

The important lesson to be learned from this chapter is that before designing chemical substances, chemists need to first familiarize themselves with all toxic properties that are associated with the chemical class to which the substance(s) under development belongs. Studies which describe or propose a mechanistic basis of toxicity are particularly useful for clarifying structure-activity relationships, and inferring structural modifications that mitigate toxicity. Once armed with such knowledge, the environmentally conscious chemist is in a much better position to design less toxic substances.

Disclaimer

This chapter was prepared by Dr. Stephen C. DeVito in his private capacity. The contents of this chapter do not necessarily reflect the views, rules or policies of the U.S. Environmental Protection Agency, nor does mention of any chemical substance necessarily constitute endorsement or recommendation for use.

Literature Cited

1. *Cassarett and Doull's Toxicology, the Basic Science of Poisons,* Fifth Edition. Klaassen, C.D., Ed; McGraw-Hill: New York, 1996.
2. *Patty's Industrial Hygiene and Toxicology,* 3rd Revised Edition, Wiley-Interscience, New York, 1982.

3. Zeller, Von H.; Hofmann, H. Th.; Theiss, A.M.; Hey, W. *Zentralbl. Arbeitsschutz* **1969**, *19*, pp 225-238.
4. Hartung, R. Cyanide and Nitriles. In *Patty's Industrial Hygiene and Toxicology*, Clayton, G.D.; Clayton, F.E., Eds.; 3rd Revised Ed., Wiley-Interscience, New York, 1982; Vol. 2C (Toxicology), pp 4845-4900.
5. Lang, S. *Arch. Exp. Path. Pharmak* **1894**, *34*, pp 247-258.
6. Stern, J.; Weil-Malherbe, H.; Green, R.H. *Biochem. J.* **1952**, *52*, pp 114-125.
7. Ohkawa, H.; Ohkawa, R.; Yamamoto, I.; Casida, J.E. *Pest. Biochem. Physiol.* **1972**, *2*, pp 95-112.
8. Willhite, C.C.; Smith, R.P. *Toxicol Appl. Pharmacol.* **1981**, *59*, pp 589-602.
9. Silver, E.H.; Kuttab, S.H.; Hasan, T.; Hassan, M. *Drug Metab. Disp.* **1982**, *10*, pp 495-498.
10. Ahmed, A.A.; Farooqui, M.Y.H. *Toxicol. Lett.* **1982**, *12*, pp 157-163.
11. Tanii, H.; Hashimoto, K. *Arch. Toxicol.* **1984**, *55*, pp 47-54.
12. Tanii, H.; Hashimoto, K. *Toxicol. Lett.* **1984**, *22*, pp 267-272.
13. Tanii, H.; Hashimoto, K. *Arch. Toxicol.* **1985**, *57*, pp 88-93.
14. Johannsen, F.R.; Levinskas, G.J. *Fundam. Appl. Toxicol.* **1986**, *7*, pp 690-697.
15. Yoshikawa, H. *Med. Biol.* **1968**, *77*, pp 1-4.
16. Davis, R.H. Cyanide detoxification in the domestic fowl. In *Cyanide in Biology*; Vennesland, B.; Conn, E.E.; Knowles, C.J.; Westley, J.; Wissing, F. Eds.; Academic Press, London, 1981, pp 51-60.
17. Froines, J.R.; Postlethwait, E.M.; LaFuente, E.J.; Liu, W.C.V. *J. Toxicol. Environ. Health* **1985**, *16*, pp 449-460.
18. Groves, J.T.; McClusky, G.A.; White, R.E.; Coon, M.J. *Biochem. Biophys. Res. Commun.* **1978**, *81*, pp 154-160.
19. White, R.E.; Groves, J.T.; McClusky, G.A. *Acta Biol. Med. Ger.* **1979**, *38*, pp 475-482.
20. Guengerich, P.F.; MacDonald, T.L. *Acc. Chem. Res.* **1984**, *17*, pp 9-16.
21. DeVito, S.C.; Pearlman, R.S. *Med. Chem. Res.* **1992**, *1*, pp 461-465.
22. Grogan, J.; DeVito, S.C.; Pearlman, R.S.; Korzekwa, K.R. *Chem. Res. Toxicol.* **1992**, *5*, pp 548-552.
23. Estabrook, R.W.; Werringloer, J. In *ACS Symposium Series*, 44th Ed., D.M. Jerina, Ed. American Chemical Society: Washington DC, 1977, pp 1.
24. Korshunov, Y.N. *Gig Primen., Toksikol. Pestits. Klin. Otravlenii*, **1970**, No. 8, pp 398-403 (see Chemical Abstracts 77:122773).
25. Parkinson, A. In *Cassarett and Doull's Toxicology, the Basic Science of Poisons*, Fifth Edition. Klaassen, C.D.; McGraw-Hill: New York, 1996, pp 113-186.
26. Mumtaz, M.M.; Farooqui, M.Y.H.; Ghanayem, B.I.; Ahmed, A.E. *Toxicol. Appl. Pharmacol.* **1991**, *110*, pp 61-69.
27. Keiser, H.R.; Harris, E.D.; Sjoerdsma, A. *Clin. Pharm. Ther.* **1967**, *8*, pp 587-592.
28. *Registry of Toxic Effects of Chemical Substances* (on-line version).

29. EPA document 8(e) HQ-0988-0754.
30. *Dangerous Properties of Industrial Materials*, 7th Edition, Vol. II, Sax, N.I.; Lewis, R.J., Eds., Van Nostrand Reinhold: New York, N.Y., 1989.
31. Smyth, H.F.; Carpenter, C.P.; Weil, C.S.; Urbano, P.C.; Striegel, J.A. *Am. Ind. Hyg. Assoc. J.* **1962**, *23*, pp 95-107.
32. EPA document OTS-50-8001950 (Microfiche No. OTS 205790; P-80-311)
33. EPA document 8(e)HQ-0988-0754
34. *Registry of Toxic Effects of Chemical Substances*, 1985-86, Sweet, D.V., Ed.; Government Printing Office, Washington, D.C., 1988.
35. Stamler, F.W. *Exp. Biol. and Med.* **1955**, *90*, pp 294-298.
36. Kalliomaki, L.; Yli-Pohja, M.; Kulonen, E. *Experientia* **1957**, *13*, p 495.
37. Dasler, W. *J. Nutr.* **1954**, *53*, pp 105-113.
38. Selye, H. *Revue Canadienne De Biologie* **1957**, *16*, pp 1-82.
39. Barrows, M.V.; Simpson, C.F.; Miller, E.J. *Quart. Rev. Biol.* **1974**, *49*, pp 101-128.
40. Roy, D.N. *Nutr. Abstr. Rev. (Ser. A).* **1981**, *51*, pp 691-706.
41. Schilling, E.D.; Strong, F.M. *J. Am. Chem. Soc.* **1954**, *76*, p 2848.
42. Dasler, W. *Science* **1954**, *120*, pp 307-308.
43. Schilling, E.D.; Strong, F.M. *J. Am. Chem. Soc.* **1955**, *77*, pp 2843-2845.
44. Geiger, B.J.; Steenbock, H.; Parsons, H.T. *J. Nutr.* **1933**, *6*, pp 427-442.
45. Ponseti, I.V.; Baird, W.A. *Amer. J. Pathol.* **1952**, *28*, pp 1059-1078.
46. Ponseti, I.V.; Shepard, R.S. *J. Bone and Joint Surg.* **1954**, *36a*, pp 1031-1058.
47. Bachhuber, T.E.; Lalich, J.J.; Angevine, D.M.; Schilling, E.D.; Strong, , F.M. *Proc. Soc. Exp. Biol. Med.* **1955**, *89*, pp 294-297.
48. Lalich, J.J. *Arch. Path.* **1956**, *61*, pp 520-524.
49. Clemmons, J.J.; Angevine, D.M. *Am. J. Path.* **1957**, *33*, pp 175-187.
50. Julian, M.; Pieraggi, M. TH.; Bouissou, H. *Pharmacol. Res. Commun.* **1979**, *11*, pp 501-508.
51. Simpson, C.F. *Exp. Molec. Pathol.* **1982**, *37*, pp 382-392.
52. Simpson, C.F.; Cardeilhac, P.T. *Proc. Soc. Exp. Biol. Med.* .**1983**, *172*, pp 168-172.
53. Wawzonek, S.; Ponseti, I.V.; Shepard, R.S.; Wiedenmann, L.G. *Science.* **1955**, *121*, pp 63-65.
54. Levene, C.I. *J. Exp. Med.* **1961**, *114*, pp 295-310.
55. Roy, D.N.; Lipton, S.H.; Bird, H.R.; Strong, F.M. *Proc. Soc. Exp. Biol. Med.* **1960**, *103*, pp 286-289.
56. Levene, C.I.; Carrington, M.J. *Biochem. J.* **1985**, *232*, pp 293-296.
57. Ronchetti, I.P.; Contri, M.B.; Fornieri, C.; Quaglino Jr., D.; Mori, G. *Connect. Tissue Res.* **1985**, *14*, pp 159-167.
58. Wilmarth, K.R.; Froines, J.R. *J. Toxicol. Environ. Health* **1992**, *37*, pp 411-423.
59. Levene, C.I. *Fed. Proc.* **1958**, *17*, p 445.

60. Levene, C.I.; Gross, J. *Fed. Proc.* **1959**, *18*, p 90.
61. Gross, J.; Levene, C.I. *Am. J. Path.* **1959**, *35*, pp 687-688.
62. Levene, C.I.; Gross, J. *J. Exp. Med.* **1959**, *110*, pp 771-790.
63. Martin, G.R.; Piez, K.A.; Lewis, M.S. *Biochim. Biophys. Acta* **1963**, *69*, pp 472-479.
64. Davies, I.; Schofield, D. *Exp. Geront.* **1980**, *15*, pp 487-494.
65. O'Dell, B.L.; Elsden, D.F.; Thomas, J.; Partridge, S.M. *Nature.* **1966**, *209*, pp 401-402.
66. Page, R.C.; Benditt, E.P. *Proc. Soc. Exp. Biol. Med.* **1967**, *124*, pp 454-459.
67. Lalich, J.J. *Science.* **1958**, *128*, pp 206-207.
68. Page, R.C.; Benditt, E.P. *Proc. Soc. Exp. Biol. Med.* **1967**, *124*, pp 459-465.
69. Narayanan, A.S.; Siegel, R.C.; Martin, G.R. *Biochim. Biophys. Res. Commun.* **1972**, *46*, pp 745-751.
70. Page, R.C.; Benditt, E.P. *Lab. Inv.* **1972**, *26*, pp 22-26.
71. Chvapil, M.; Misiorowski, R.; Eskelson, C. *J. Surg. Res.* **1981**, *31*, pp 151-155.
72. Fleisher, J.H.; Speer, D.; Brendel, K.; Chvapil, M.*Toxicol. Appl. Pharmacol.* **1979**, *47*, pp 61-69.
73. Rando, R.R. *Acc. Chem. Res.* **1975**, *8*, pp 281-288.
74. Clemmons, J.J. *Fed. Proc.* **1958**, *17*, p 432.
75. Schmidt, W.; Wendler, D. *Folia Morphol.* **1982**, *30*, pp 146-148.
76. Maycock, A.L.; Suva, R.H.; Abeles, R.H. *J. Am. Chem. Soc.* **1975**, *97*, pp 5613-5614.
77. Pinnell, S.R.; Martin, G.R. *Proc. Nat. Acad. Sci. U.S.,* **1968**, *61*, pp 708-716.
78. Piez, K.A. *Ann. Rev. Biochem.* **1968**, *37*, pp 547-570.
79. Gallop, P.M.; Blumenfield, O.O.; Seifter, S. *Ann. Rev. Biochem.* **1972**, *41*, pp 617-672.
80. Bailey, A.J.; Robins, S.P. *Sci. Prog.* **1976**, *63*, pp 419-444.
81. Bailey, A.J.; Robins, S.P.; Balian, G. *Nature.* **1974**, *251*, pp 105-110.
82. Sandberg, L.B. *Int. Rev. Connect. Tissue Res.* **1976**, *7*, pp 159-210.
83. Bailey, A.J. *J. Clin. Path.* (Suppl.) **1978**, *12*, pp 49-58.
84. Bornstein, P.; Piez, K.A. *Biochemistry* **1966**, *5*, pp 3460-3473.
85. Bornstein, P.; Kang, A.H.; Piez, K.A. *Biochemistry.* **1966**, *5*, pp 3803-3812.
86. Bornstein, P.; Kang, A.H.; Piez, K.A. *Proc. Nat. Acad. Sci. U.S.* **1966**, *55*, pp 417-424.
87. Piez, K.A.; Martin, G.R.; Kang, A.H.; Bornstein, P.*Biochemistry.* **1966**, *5*, pp 3813-3820.
88. Kang, A.H.; Bornstein, P.; Piez, K.A.*Biochemistry.* **1967**, *6*, pp 788-795.
89. Lent, R.; Franzblau, C.*Biochem. Biophys. Res. Commun.* **1967**, *26*, pp 43-50.
90. Miller, E.J.; Martin, G.R.; Piez, K.A. *Biochem. Biophys. Res. Commun.* **1964**, *17*, pp 248-253.

91. Anwar, R.A.; Oda, G. *Biochim. Biophys. Acta.* **1967**, *133*, pp 151-156.
92. Pinnell, S.R.; Martin, G.R. *Proc. Nat. Acad. Sci. U.S.* **1968**, *61*, pp 708-716.
93. Tang, S-S.; Trackman, P.C.; Kagan, H.M. *J. Biol. Chem.* **1983**, *258*, pp 4331-4338.
94. Tanii, H.; Hayashi, M.; Hashimoto, K. *Neurotoxicology* **1989**, *10*, pp 157-166.
95. ibid. *Arch. Toxicol.* **1990**, *64*, pp 231-236.
96. Pestronk, A.; Keogh, J.P.; Griffin, J.W.*Neurology* **1979**, *29*, p 540.
97. Kreiss, K.; Wegman, D.H.; Niles, C.A.; Siroky, M.B.; Krane, R.J.; Feldman, R.G. *JAMA* **1980**, *243*, pp 741-745.
98. Keogh, J.P.; Pestronk, A.; Wertheimer, D.; Moreland, R. *JAMA* **1980**, *243*, pp 746-749.
99. Baker, E.L.; Christiani, D.C.; Wegman, D.H.; Siroky, M.; Niles, C.A.; Feldman, R.G. *Scand. J. Work Environ. Health.* **1981**, *70 (Suppl. 4)*, pp 54-59.
100. Gad, S.C.; McKelvey, J.A.; Turney, R.A. *Drug and Chem. Toxicol.* **1979**, *2*, pp 223-236.
101. Jaeger, R.J.; Plugge, H. *J. Environ. Pathol. Toxicol.* **1980**, *4*, pp 555-562.
102. Froines, J.R. *Biochem. Pharmacol.* **1985**, *34*, pp 3185-3187.
103. Froines, J.R.; Watson, A.J.*Toxicol. Environ. Health* **1985**, *16*, pp 469-479.
104. Wilmarth, K.R.; Froines, J.R. *J. Toxicol. Environ. Health* **1991**, *32*, pp 415-427.
105. Ahmed, A.E.; Farooqui, M.Y.H. *Pharmacologist* **1984**, *26*, p 199.
106. Mumtaz, M.M.; Farooqui, M.Y.H.; Ghanayem, B.I.; Ahmed, A.E. *Toxicol. Appl. Pharmacol.* **1991**, *110*, pp 61-69.
107. Hartmann, H.A.; Stich, H.F. *Fed. Proc.* **1957**, *16*, pp 358 359.
108. Peele, D.B.; Allison, S.D.; Crofton, K.M. *Toxicol. Appl. Pharmacol.* **1990**, *105*, pp 321-332.
109. Schulze, G.E.; Boysen, B.G. *Fundam. Appl. Toxicol.* **1991**, *16*, pp 602-615.
110. Crofton, K.M.; Knight, T. *Neurotoxicol. and Teratology* **1991**, *13*, pp 575-581.
111. Genter, M.B.; Llorens, J.; O'Callaghan, J.P.; Peele, D.B.; Morgan, K.T.; Crofton, K.M. *J. Pharmacol. and Experim. Therap.* **1992**, *263*, pp 1432-1439.
112. Goldey, E.S.; Kehn, L.S.; Crofton, K.M. *Hearing Research* **1993**, *69*, pp 221-228.
113. Llorens, J.; Dememes, D.; Sans, A. *Toxicol. Appl. Pharmacol.* **1993**, *123*, pp 199-210.
114. Llorens, J.; Crofton, K.M.; O'Callaghan, J.P. *J. Pharmacol. and Experim. Therap.* **1993**, *265*, pp 1492-1498.
115. Genter, M.B.; Deamer, N.J.; Cao, Y.; Levi, P.E. *J. Biochem. Toxicol.* **1994**, *9*, pp 31-39.
116. Morandi, A.; Gambetti, P.; Arora, P.K.; Sayre, L.M.; *Brain Research* **1987**, *437*, pp 69-76.
117. Jacobson, A.R.; Coffin, S.H.; Shearson, C.M.; Sayre, L.M. *Molecular Toxicol.* **1987**, *1*, pp 17-34.

Chapter 11

Designing an Environmentally Safe Marine Antifoulant

G. L. Willingham and A. H. Jacobson

Rohm and Haas Company, 727 Norristown Road, Spring House, PA 19477

Tin-based antifoulants came under regulatory advisement in the 1980's and we began to examine isothiazolones as an environmentally safe alternative to tin compounds. Our search generally followed the paradigm that environmental risk is a function of toxicity and exposure. We realized that it would be extremely difficult to design a marine antifoulant that was toxic to fouling organisms (e.g. tube worms and barnacles) but non-toxic to their related non-target organisms (e.g. mussels and oysters). Thus we looked for a compound that would have reduced exposure - a short environmental half-life and/or partition rapidly into a matrix of limited bioavailability. The results of this examination yielded 4,5-dichloro-2-n-octyl-4-isothiazolin-3-one (DCOI), a very efficacious compound with a half-life in a marine environment of less than one hour and limited bioavailability of metabolites due to their strong association with sediment.

Fouling is the result of the growth of a variety of marine plants and animals on submerged structures. The fouling of a ship's hull causes increased hydrodynamic drag, resulting in increased fuel consumption, decreased ship speed, increased costs to clean and service the vessel, and increased time out of service. The total annual cost due to fouling is estimated to significantly exceed a billion dollars with a major expense being the increased consumption of fuel (estimated at several hundreds of thousands of gallons) used to overcome the hydrodynamic drag.

Organotin biocides are effective at preventing fouling, but environmental concerns have been raised regarding their use (*1,2*). Since the discovery of their ecotoxicological problems in the mid to late 1970's, many countries have banned the use of tin in antifoulants on ships less than 25 meters in length. Japan has effectively banned the use of tin as an antifoulant in all marine applications. In the United States, tin usage on vessels longer than 25 meters is restricted to formulations with a certifiable leach rate of less than 4 $\mu g/cm^2/day$. Worldwide, more stringent restrictions and/or registration cancellations are certain to follow. New antifoulants are needed which are efficacious at preventing fouling, but which meet stringent ecotoxicological requirements.

The ideal antifoulant must prevent fouling from a wide variety of marine organisms while causing no harm to non-target organisms. The environmental risk from the introduction of a new marine antifoulant active ingredient can be described by the following function:

$$Environmental\ Risk = f\ (Toxicity\ /\ Exposure)$$

It would be extremely difficult to design an effective marine antifoulant that was efficacious against a wide range of organisms but non-toxic to the systematically related non-target organisms. Thus we sought to minimize environmental risk by developing an active ingredient with reduced environmental exposure. This reduced exposure could be expressed by rapid degradation/metabolism of the active ingredient and/or partition into a matrix such as the sediment which would limit the bioavailability.

Experimental

Biological Testing Against Fouling Organisms. Compounds were evaluated for their control of fouling by algae and barnacles and water solubility was estimated by membrane screening as described in Miller and Lovegrove (3). Biological testing against marine algae, bacteria, diatoms, and barnacle larvae was carried out as previously described (4).

Aquatic Microcosms. The procedures used in these studies have been described in detail elsewhere (5). Briefly, sediment and seawater from the Chesapeake Bay (York River) were placed into Erlenmeyer flasks. In the anaerobic microcosm, anaerobic conditions were assured by pretreating the sediment and seawater with glucose in a stoppered flask for 30 days prior to dosing. Flasks were dosed with $^{13/14}C$ DCOI at a nominal dose of 0.05 ppm. The seawater and sediment were separated by centrifugation or filtration. The sediment was Soxhlet extracted with dichloromethane:methanol (9:1/v:v) for 48 hours and then methanol for an additional 24 hours. The Soxhlet insoluble residue was characterized by separation into humin, humic acid and fulvic acid (6).

Radiocarbon compounds detected in the Soxhlet solutions were chromatographically analyzed by HPLC as described previously (5). Structures of isolated metabolites were confirmed by GC-MS (Hewlett-Packard, Model 5985) or DCI-MS (Finnigan, Model TSQ 46).

Photolysis and Hydrolysis. Hydrolysis and photolysis studies were performed following the EPA prescribed methods (7).

Fish Bioconcentration. The procedure used in determining the bioconcentration factor and quantitation of parent compound in bluegill sunfish has been presented previously (4).

Dissipation Model. A model was employed in order to assess the fate and maximum concentration of DCOI that might theoretically occur in an aquatic marine environment following implementation of the compound as a marine antifoulant. Three harbors were chosen as example environments: New York harbor, San Diego bay, and Norfolk harbor. Modeling was accomplished using the Exposure Analysis Modeling System, EXAMS (Version 2.92) (8).

Toxicology Testing. Detailed procedures of the toxicological testing appeared previously (9).

Results and Discussion

Biological Testing Against Fouling Organisms. The isothiazolin-3-one class of compounds has been shown to have high activity against bacteria, fungi, and algae, and has demonstrated biological activity in some aquatic organisms. Additionally, biological degradation of these compounds was rapid and they showed a strong affinity for soil and sediment (10). Based on these results, we began to search for an isothiazolone that would provide broad spectrum control of marine fouling organisms, have sufficiently low water solubility to provide long term control, bind tightly to the sediment to limit bioavailability, and degrade rapidly in the environment.

A screening program was initiated to identify appropriate candidates which had excellent biological efficacy. In order to provide broad spectrum control, the antifoulant must have activity against both *soft* fouling organisms (e.g., algae, diatoms) and *hard* fouling organisms (e.g., barnacles). The screening model we used is shown in Figure 1.

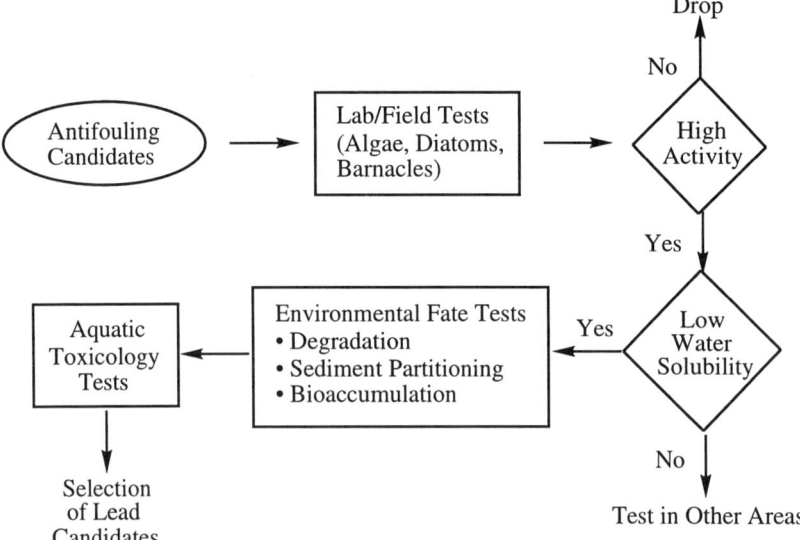

Figure 1. Screening model for new antifoulants

Membrane screening was carried out at Newton Ferrers, Devon, England. This technique evaluated control of algae and barnacle fouling and estimated the water solubility of the compounds.

From this screening, the basic potential of a test compound as an antifoulant was assessed. Activity was noted as:

1) Effective - prevents major fouling organisms from settling and/or persisting;
2) Some - allows organisms to settle but in considerably reduced quantity compared to the non-toxic control; and
3) None - settlements are very similar to those of the non-toxic control.

Over 140 isothiazolones were screened in field exposure tests. Test data was used to determine the structure-activity relationships for substitution at the 4 and 5 positions in the isothiazolone ring. The influence of halogenation at these positions on activity, particularly at the 5 position, was apparent (*3*). In addition, halogenation lowers the water solubility, which is essential to prevent rapid leaching from the antifouling paints.

Further testing revealed the 2-(*n*-alkyl)-substituted isothiazolones to be particularly promising candidates. Results in Table I show the effect of the *n*-alkyl chain length on biological activity and water solubility of 4,5-dichloro-2-*n*-alkyl-4-isothiazolin-3-ones. Based on the superior biological activity and low water solubility, 4,5-dichloro-2-*n*-octyl-4-isothiazolin-3-one (DCOI) was selected for further studies.

Table I. Effect of *n*-alkyl Chain Length on the Biological Activity and Water Solubility of 4,5-dichloro-2-*n*-alkyl-4-isothiazolin-3-ones[a]

Alkyl -$(CH_2)_x$-H	Algae[b]	Barnacle[b]	Estimated Aqueous Solubility[c] (ppm)
x=4	++	0	1000
x=6	++	++	120
x=8[d]	++	++	2
x=10	++	0	1
x=12	0	+	1
x=14	0	0	1

[a] Reproduced with permission from ref. 3.
[b] Activity; (++) Effective; (+) Some; (0) None.
[c] Solubility measured at 25°C by ultraviolet absorption analysis.
[d] DCOI.

DCOI was tested against several major fouling organisms. Data has been presented elsewhere (4). Data against three of the major marine fouling organisms is presented in Table II. The marine diatom, *Amphora coffeaeformis,* is a major fouling organism. Many diatoms contribute to the formation of a slime layer and are major sources of ship fouling. DCOI was shown to be highly efficacious against *Amphora coffeaeformis.*

The most ubiquitous marine fouling algae is the species *Enteromorpha intestinalis.* DCOI showed excellent activity against *Enteromorpha intestinalis.* The major hard fouling species are barnacles. Data shown in Table II demonstrate that DCOI activity against barnacle larvae was considerable, although it was significantly less than the activity against slime formers and marine algae. Based on the promising activity shown in these initial results, studies on the environmental fate and toxicology of DCOI were conducted.

Table II. Efficacy of DCOI Against Marine Fouling Organisms

Organism	Fouling class	LD_{50} (ppm)
Amphora coffeaeformis	Marine diatoms	3.4×10^{-3}
Enteromorpha intestinalis	Marine algae	2.0×10^{-3}
Balanus amphitrite	Barnacles (larvae)	3.4×10^{-1}

Half-life in the Environment. The half-lives of DCOI in various environmental matrices are listed in Table III.

Table III. Summary of Half-lives in Various Environmental Matrices[a]

Matrix	Half-life (hrs)
Photolysis	322
Hydrolysis	
pH 5	216
pH 7	>720
pH 9	288
Aerobic aquatic microcosm	1
Anaerobic aquatic microcosm	1
Water	
Seawater	<24
Synthetic seawater	>720
Synthetic seawater + algae	<3
Reagent water	>720

[a] Reproduced with permission from ref.4.

Biological degradation, as demonstrated in the aquatic microcosms, is over 200 times faster than hydrolysis or photolysis. The predominance of biological degradation is further demonstrated in the water samples. DCOI is essentially stable in synthetic seawater, whereas it degrades rapidly in seawater samples obtained from the Atlantic and Pacific Oceans, or synthetic seawater spiked with algae. DCOI was also stable in laboratory reagent water (from a Millipore Milli-Q-Water System) which is sterile. In addition, the half-life of DCOI was determined in natural seawater samples collected from coastal sites in New Jersey and California. The half-lives were all less than 24 hrs and were directly proportional to the microbial concentration (reference 9 and unpublished data from A. Jacobson). The faster half-life observed in microcosm compared to water is due to the sediment being naturally enriched in microbial activity (5).

Environmental Partitioning. The distribution of applied ^{14}C-activity in the microcosms between the seawater and sediment (both Soxhlet soluble and insoluble fractions) has been documented elsewhere (5). After separation of the treated sediment and seawater phases over 92% of the ^{14}C-activity, for both aerobic and anaerobic microcosms, was in the sediment (Soxhlet soluble and insoluble). A majority of the activity was associated with the Soxhlet insoluble residue. In the aerobic and anaerobic studies, $^{14}CO_2$ increased throughout the study and by day 30 comprised 24% and 6% of the applied dose, respectively. The recovery of ^{14}C label averaged 103.6 ± 10.8% and 88.1 ± 9.6% for the aerobic and anaerobic studies, respectively.

Since the Soxhlet insoluble residue contained a majority of the activity, it was further analyzed by exhaustive extraction with 0.25 N HCl, followed by 1 N NaOH. The results appear in Table IV. Treatment with acid had virtually no effect on the bound residues. Overnight extraction with NaOH released a portion of the bound residues with practically all the base soluble residues being associated with the humic acid fraction rather than the fulvic acid fraction. However, even with the severe base treatment, over 50% of the bound residues were associated with the insoluble humin. Thus a large percentage of the applied ^{14}C-activity is tightly bound to the sediment and probably not bioavailable. The activity is not parent compound since the application of DCOI to sterilized sediment demonstrated that over 90% of the applied activity could be removed by Soxhlet extraction (5).

Table IV. Exhaustive Extraction of Sediment Bound Residues

	Percent of Applied ^{14}C Detected				
Day	Soxhlet Insoluble	HCl Soluble	Humin	Humic Acid	Fulvic Acid
Anaerobic					
0	21.0	0.1	14.5	13.4	0.2
Aerobic					
0	53.4	0.1	40.7	22.1	0.3
30	60.3	0.1	29.3	25.0	0.5

Metabolic Pathway of DCOI in the Aquatic Sediment. Metabolite identification has been presented elsewhere (4,5). A metabolic pathway is presented in Figure 2. The initial reaction has been shown to be cleavage of the N-S bond by either nucleophiles (11,12) or photoisomerization (13). The N-methyl isothiazolones were shown to have a similar metabolic pathway in the environment (10).

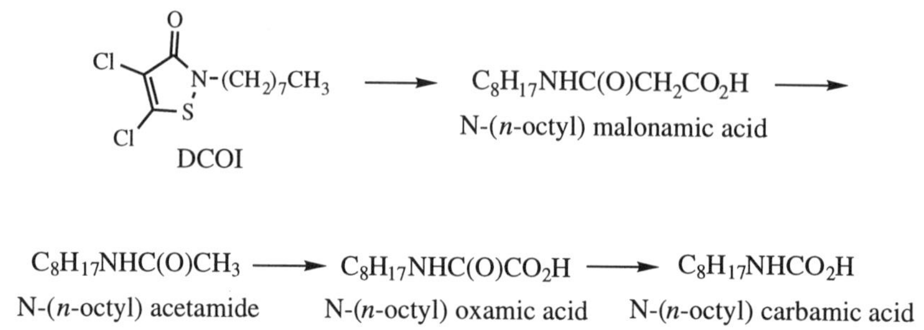

```
        O
    Cl  ‖
      ╲ ╱ N-(CH₂)₇CH₃
       ╲│
       ╱ S
    Cl
       DCOI
```
→ $C_8H_{17}NHC(O)CH_2CO_2H$ →
N-(n-octyl) malonamic acid

$C_8H_{17}NHC(O)CH_3$ → $C_8H_{17}NHC(O)CO_2H$ → $C_8H_{17}NHCO_2H$
N-(n-octyl) acetamide   N-(n-octyl) oxamic acid   N-(n-octyl) carbamic acid

Figure 2. Metabolic pathway for DCOI in aquatic sediment.[a]
[a]Reproduced with permission from ref.4.

Concentration of DCOI in Fish. Since the logarithm of the octanol/water partition coefficient (log P) for DCOI is 2.85, very little bioaccumulation would be expected. In bluegill sunfish, less than 1% of the bioaccumulated ^{14}C-activity in viscera and fillets was parent compound(4).

Characterization of Fish Metabolites. As described previously (4), the metabolites are ring-cleaved compounds with a significant quantity being associated with proteins, especially cysteine residues.

Toxicology in Nontarget Aquatic Organisms. A detailed discussion of the aquatic toxicology of DCOI has been presented previously (9). Summaries of the acute and subchronic results appear in Tables V and VI, respectively. From these two tables it can be concluded that DCOI is an acute toxin to marine organisms but not a chronic or reproductive toxin. The acute toxicity of DCOI is not surprising since the organisms tested are closely related to those which would foul the bottom of a ship's hull. It would be expected that an efficacious marine antifoulant would show such acute toxicities.

Table V. Acute Toxicology of DCOI in Aquatic Organisms

Species	Test	ppb
Bluegill sunfish	96 hr LC50	14
Daphnia magna	48 hr LC50	9
Bay mussel - adult	96 hr LC50	850
American oyster - embryo	48 hr LC50	24
Fiddler crab	96 hr LC50	1312
Mysid shrimp	96 hr LC50	5
S. costatum	96 hr LC50	20

Table VI. Subchronic Toxicology of DCOI in Aquatic Organisms

Species	Test[a]	ppb
Sheepshead minnow	NOEL	6
(Early life stage)	LOEL	14
	MATC	9
Daphnia magna	EC_{50}	1
(21 day chronic)	NOEL	0.63
	LOEL	1
	MATC	0.83

[a]NOEL, No Observable Effect Level; LOEL, Lowest Observable Effect Level; and MATC, Maximum Allowable Toxic Concentration.

Modeling Study. EXAMS, a computerized mathematical modeling program (8), was employed to study the environmental fate and dissipation of DCOI contained in a coating painted on ship bottoms. The three harbors chosen, New York harbor, San Diego Bay and Norfolk harbor, represent some of the busiest harbors in the world. The environmental parameters appearing in Table VII were taken from previous publications (4,5,9).

Table VII. Environmental Fate Inputs for EXAMS Model

Parameter	End Point
Water Solubility	4.7 ppm
Degradation	
Seawater	24 hrs
Sediment	1 hr
Photolytic	322 hrs
Bioconcentration Factor	13.6 (DCOI)
Adsorption Coefficient	1666
Vapor Pressure	7.4×10^{-6} torr

The results of the modeling study are presented in Table VIII. The leach rate of DCOI from a painted surface was determined from submersion of painted panels. A representative high end value was 1.0 µg DCOI/cm² painted surface/day. The values obtained in the water column were at least 1.5 orders of magnitude less than the MAEC (Maximum Acceptable Environmental Concentration) of 0.63 ppb based on the *Daphnia magna* subchronic study. The major contributing factor to the variation in DCOI concentration predicted in the harbor is due to hydraulic flushing. San Diego Bay is an enclosed harbor with very little fresh water influx, and flushing is accomplished primarily by tidal action. New York harbor, on the other hand, has a tremendous amount of fresh water influx coming from the Hudson and East rivers.

Table VIII. Results from the Environmental Dissipation Model for DCOI

Harbor	Leach Rate μg/cm²/day	DCOI Detected Water (ppt)	Sediment (ppb)
San Diego	0.1	0.9	
	1.0	9.4	1.8
	5.0	47.0	
New York	0.1	0.07	
	1.0	0.7	0.2
	5.0	3.6	
Norfolk	0.1	0.5	
	1.0	5.0	4.6
	5.0	25.0	

Comparison of DCOI and Tributyltin oxide (TBTO). TBTO has been used as an antifouling agent for decades. It has recently come under regulatory scrutiny due to toxicological concerns (*14*). The results in Table IX present a comparison of the environmental fate and toxicology of DCOI and TBTO. The values for TBTO are representative averages from a World Health Organization publication (*14*). The results in Table IX demonstrated that unlike DCOI, TBTO is persistent in the environment due to its long degradation time and will bioaccumulate. While both compounds are toxic to nontarget marine organisms, TBTO has been shown to be a reproductive toxin. The predicted concentration of TBTO in the environment was determined by the EXAMS model described above. The predicted concentration of TBTO in the environment, 0.01 to 1 ppb, could be in excess of the MAEC of 20 ppt based on a United Kingdom water quality standard (*14*).

Table IX. Comparison of the Environmental Fate of DCOI and TBTO

Property	DCOI	TBTO
Half-life		
Seawater	<24 hours	up to 5 months
Sediment	1 hour	6 - 9 months
Bioaccumulation Factor	Nil	up to 10,000X
Chronic Toxicity	None	effects at low ppt levels
MAEC[a]	0.63 ppb	0.020 ppb
Predicted Environmental Concentration	0.01 ppb	0.01 - 1 ppb

[a] Maximum Acceptable Environmental Concentration

Conclusions

In order to be a reasonable alternative to organotins, a new antifoulant must be efficacious against a wide range of fouling organisms as well as pose no significant risk to the environment. DCOI was shown in laboratory tests to have excellent activity against a broad spectrum of fouling organisms. This activity was confirmed in exposure tests. In seawater DCOI and its metabolites bind strongly and essentially irreversibly to sediment. It is rapidly degraded in both seawater and sediment by microorganisms to ring-opened structures with greatly reduced toxicity. Bioconcentration studies in bluegill sunfish demonstrate that bioaccumulation of DCOI is essentially nil. These results demonstrate that DCOI can effectively prevent fouling while causing no adverse effects to the environment.

Literature Cited

1. Beaumont, A.R. and M.D. Budd, *Mar. Pollut. Bull.* **1984**, *15*, pp. 402-405.
2. Huggett, R.J., M.A. Unger, P.F. Seligman and A.O. Valkirs, *Environ. Sci. Technol.* **1992**, *26*, pp. 232-237.
3. Miller, G.A. and T. Lovegrove, *J. Coatings Technol.* **1980**, *52*, pp 69-72.
4. Willingham, G.L. and Jacobson, A., In *The Proceedings of the Third Asia-Pacific Conference of the Paint Research Association,* International Centre for Coatings Technology, **1993**, Paper 14, pp. 1-13.
5. Jacobson, A., Mazza, L.S., Lawrence, L.J., Lawrence, B., Jackson, S and Kesterson, A., In *Pesticides in the Urban Environments, Fate and Significance,* Racke, K.D. and Leslie, A.R., Eds; A.C.S. Symp. Series 552, **1993**.
6. U.S. Environmental Protection Agency, Federal Register *40*, 26803 (**June 1975**).
7. U.S. Environmental Protection Agency, Pesticide Assessment Guidelines, Subdivision N-Chemistry: Environmental Fate, **1982**.
8. Burns, L.A., Exposure Analysis Modeling System, U.S. Environmental Protection Agency, Athens, GA, **1981**.
9. Shade, W.D., Hurt, S.S., Jacobson, A.H., and Reinert, K.H., In *Environmental Toxicology and Risk Assessment; 2nd Volume ASTM STP 1173*; Gorsuch, J.W., Dwyer, F.W., Ingersoll, C.M., and LaPoint, T.W., Eds; American Society for Testing and Materials, Philadelphia, PA, **1994**.
10. Krzeminski, S.F., Brackett, C.K. and Fisher, J.D., *J. Agric. Food Chem.* **1975**, *23*, pp. 1060-1068.
11. Crow, W.D. and Leonard, N.J., J. Org. Chem. **1965**, *30*, pp. 2060-2065.
12. Crow, W.D. and Gosney, I., *Aust. J. Chem.* **1967**, *20*, pp. 2729-2736.
13. Rokach, J. and Hamel, P., *J.C.S. Chem. Comm.* **1979**, *1979*, p. 786.
14. World Health Organization, International Programme on Chemical Safety, Environmental Health Criteria, **1990**, *116*.

Chapter 12

Imine–Isocyanate Chemistry: New Technology for Environmentally Friendly, High-Solids Coatings

Douglas A. Wicks and Philip E. Yeske

Industrial Chemicals Division, Coatings Research, Bayer Corporation, 100 Bayer Road, Pittsburgh, PA 15205-9741

> The chemical industry faces a major challenge in the design and manufacture of new environmentally-friendly chemicals. Within the coatings industry, these new chemicals must play a role in pollution prevention as well. The need for development of new raw materials which reduce solvent demand and yet maintain coating performance is of primary importance. The inherent high reactivity and toxicity of primary amines has limited their applicability in low volatile organic compound content (low VOC), high solids coatings. This paper details the design and use of ketimines and aldimines as blocked primary amines in such formulations. Characterized by low toxicity, viscosity and reactivity relative to the parent amine, imines are shown to be excellent reactive resins for high solids coatings. Aspects of imine-isocyanate chemistry presented include the direct reaction of imines with polyisocyanates as well as the relative hydrolytic stability of aldimines versus ketimines. The impact of each of these aspects upon pollution prevention is also discussed. This paper demonstrates how safer chemicals can be designed without affecting efficacy.

As the 1990's unveil a new era of increasingly stringent environmental and safety regulations in conjunction with increased consumer demands for ecologically friendly materials, many traditional coatings systems are being eliminated as surviving systems undergo radical changes. These changes are needed to overcome traditional coatings technologies' dependence on high solvent content, and thus high volatile organic compound (VOC) content, to facilitate and control the coatings application. The technology changes now taking place in industrial markets such as automotive refinish, maintenance, architectural and OEM applications, are being primarily driven by regulatory pressures but are also coupled with increased customer demands for performance and value. VOCs undergo photochemical decomposition in the environment to a variety of products that lead to the formation of smog, and contribute significantly to overall air pollution. The push to reduce emissions of VOCs has shifted resin development towards a variety of approaches as shown in Table I.

Table I. Approaches to Low VOC Coatings

Approach	Primary Characteristic
1. UV cure	using a polymerizable monomer as the solvent
2. Powder	using heat to induce flow and crosslinking
3. Waterborne	using water as solvent
4. High Solids	low viscosity reactive resins

Approaches to Low VOC Coatings

Currently, the uses of UV or Powder coatings are limited as both require specialized application and cure equipment. They do, however, find use in factory situations where the applications are specifically controlled. UV cure is not applicable to pigmented coatings or for irregular shaped substrates, since in both these cases the UV radiation cannot penetrate to give an evenly cured film. Powder coatings have a major drawback in their requirement for high bake temperatures (>150 °C). These high temperatures limit the number of substrates which can be coated and have the additional disadvantage of increased energy demand and the environmental problems associated with increased fossil fuel combustion to meet the demand.

Waterborne coatings have found wide acceptance as low VOC systems. In these products water acts as the carrier solvent, and film formation is facilitated by addition of coalescing solvents or through the use of low glass transition temperature resins. Consumer architectural latex paints are one successful example of how a waterborne coating can effectively displace organic solvent borne systems without a decrease in product performance or ease of application.

Nevertheless, water still carries with it several drawbacks which limit its use as a universal solution to VOC reduction. First, it is a reactive species which causes hydrolysis of resins and interferes with many common crosslinking reactions. Second, water has a fixed evaporation rate so there is a significant dependence of drying speed and the final quality of coatings applied on curing temperature and relative humidity.

High solids reactive coatings bring relief from many of the difficulties mentioned above. They are the preferred method for producing high quality, weatherable, chemically resistant coatings. In many industries where high performance in terms of durability, solvent/chemical resistance or weatherability is required, high solids polyurethanes are becoming the coatings of choice.

Though high solids, 2-component polyurethanes represent a robust, viable route for significantly reducing emissions while still yielding high performance coatings, formulators are beginning to reach the limit of effectiveness with currently available technology. Further solvent reductions must now be realized by reducing the molecular weight of the paint components, which concurrently reduces their viscosity and in turn their demand for organic solvents. One key effect of this molecular weight reduction is that the resins employed are no longer solid products at 100% concentration so the coating no longer lacquer dries but must dry through chemical reactions.

Based on the limitations described above, the ideal chemistry for a high-solids crosslinking coating would be one in which the:

• rate of reaction is suppressed before application (extended pot-life),
• the rate of reaction is greatly accelerated after application (fast dry times) and
• the method of suppression/acceleration does not generate VOCs.

We have found that high solids coatings that utilize imine-isocyanate based systems over amine-isocyanate/hydroxyl-isocyanate systems meet the regulatory demands of developing less hazardous substances, and the customer demands for product performance. Imine-isocyanate systems significantly reduce the VOCs in a coating when compared with current systems and they do not utilize primary amines, which are generally ecotoxic. The systems developed by us are less ecotoxic than amine-isocyanate based systems because the imines are stable to water hydrolysis and do not generate primary amines when combined with water. This paper will describe our approach to developing imine-isocyanate based systems as environmentally friendly, equally efficacious alternatives to existing coatings systems.

Background

Polyurethane Coatings. Within the coatings industry, the term "urethane" has evolved to describe a broad and often intermixed family of chemistries derived from reactions of the isocyanate group (R-N=C=O). Still, the fundamental chemical reaction taking place in urethane coatings is that of an isocyanate with an alcohol. By employing polyfunctional alcohols (polyols) and polyfunctional isocyanates (polyisocyanates) one obtains crosslinked polyurethanes (Figure 1). This reaction proceeds at a reasonable, well defined rate and can be further controlled by catalysis with tertiary amines, various organo-tin compounds and metal salts *(1)*.

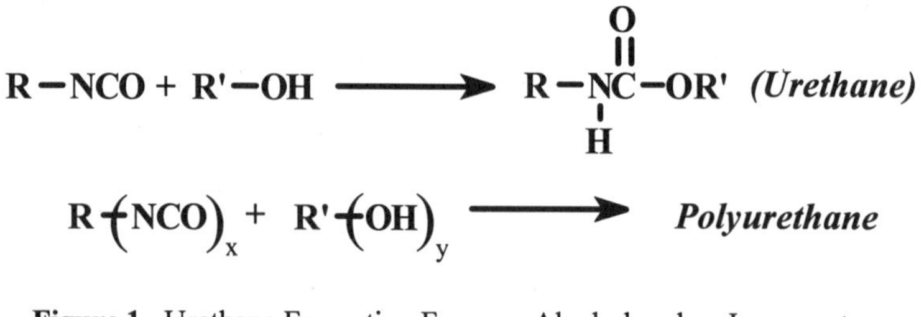

$$R-NCO + R'-OH \longrightarrow R-\underset{H}{\overset{\overset{\displaystyle O}{\|}}{N}}C-OR' \quad \textit{(Urethane)}$$

$$R\text{-}(NCO)_x + R'\text{-}(OH)_y \longrightarrow \textit{Polyurethane}$$

Figure 1. Urethane Formation From an Alcohol and an Isocyanate.

Polyisocyanates. Analysis of the solvent requirements of the individual components in a urethane coating makes apparent that the polyisocyanate portion is the high solids component of the coating (2). Figure 2 shows the viscosity of a typical polyisocyanate solution as a function of percent solids. The commercial polyisocyanate shown here has a neat viscosity of 3,500 mPa•s, which when reduced to a typical application viscosity of 100 mPa•s gives a solution of 84% non-volatile content. (i.e., 16% solvent by weight). This value may be compared to a 50-60% organic solvent demand for typical polyisocyanate coreactants (this will be described in more detail in the next section).

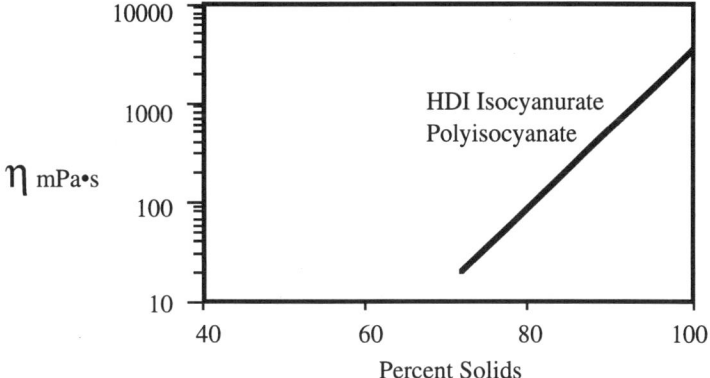

Figure 2. Viscosity of an HDI Isocyanurate Polyisocyanate as a Function of Percent Solids in a 2/1/1 Blend of MEK/MIBK/Exxate 600.

Monoisocyanurate of HDI

Hexamethylene diisocyanate (HDI)-based polyisocyanurates and biurets are being commercially pushed to lower viscosities, with the practical limitations of these products being 1000 and 2000 mPa•s, respectively. Still newer types of polyisocyanate resins are pushing the viscosity envelope below 500 mPa•s (3,4). These materials yield sprayable solution viscosities (100 mPa•s) with non-volatile contents approaching 100%.

For hygiene and practical purposes, it is important to remember that all of the commercial and developmental polyisocyanate adducts used in high solids coatings are free of monomeric diisocyanate and that the products derived from the oligomerization of HDI are at least difunctional.

Hydroxy Functional Coreactants. Acrylic polyols have been favored historically for most applications because of their superior weathering, low cost and low isocyanate demand. These products have traditionally been of relatively high molecular weight (10-20,000 g/mol) with hydroxyl equivalent weights of 1,000 or more. With the large number of hydroxyls present on the polymer, only a moderate degree of reaction is required to get complete network formation. Because these products are solids (without crosslinking), drying of the product was not a major concern. The major drawback of high molecular weight acrylics is their high solvent demand, yielding coatings with only 40-50% solids at application viscosities.

The first step taken toward high solids OH-functional coreactants was the development of lower molecular weight acrylics and polyesters with reduced functionality. These products, with molecular weights typically below 2,500 g/mol, have been successful in providing high quality coatings with solids of 55-65% at application viscosity (for example, see Figure 3).

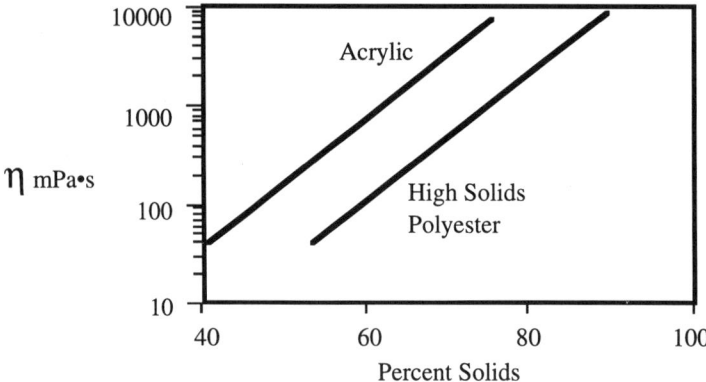

Figure 3. Viscosity of Hydroxy Functional Coreactants as a Function of Percent Solids in a Blend of Methylethyl Ketone/Methyl Isobutyl Ketone/Exxate 600.

Great strides have been made in reducing solvent demand by lowering the molecular weight and changing the molecular architecture of acrylic and polyester polyols, but additional gains will be difficult without sacrificing application properties. For example, further reduction of molecular weight below current levels yields, in the case of acrylics, an unacceptable fraction of monofunctional or non-functional polymer chains *(5)* and in the case of polyesters exceedingly long dry times and/or soft coatings. Clearly new chemistries for and approaches to low viscosity coreactants for polyisocyanates are needed.

Polyurea Coatings. The reaction of isocyanates with amines or polyamines occurs to form ureas and polyureas. Typical primary and secondary aliphatic amines and primary aromatic amines react very rapidly with isocyanates at ambient temperatures (Figure 4). In most cases this reaction is too fast to be useful in coatings applications unless very specialized application equipment and techniques are used. For example, one commercial polyurea system is formed by the reaction of amino terminated polyethers with aromatic polyisocyanates (6). In this system, the reactants are mixed under high pressure at the tip of the spray gun and cure within seconds of application. This limits their widespread use in that the equipment is very expensive and the rapid cure precludes the flow and leveling required to make a high quality finish.

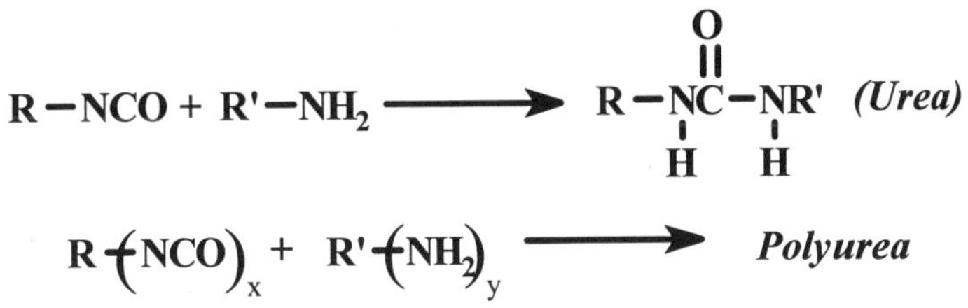

$$R-NCO + R'-NH_2 \longrightarrow R-\underset{H}{\overset{O}{\underset{|}{N}}}\!\!\!\overset{\|}{C}-\underset{H}{\overset{|}{N}}R' \quad (Urea)$$

$$R\text{-}(NCO)_x + R'\text{-}(NH_2)_y \longrightarrow Polyurea$$

Figure 4. Urea Formation From an Amine and an Isocyanate.

Aspartic Acid Esters. Recently there have been advances in hindered amine chemistry which allow amines to be used in high solids coatings without the use of specialized equipment (7-11). These hindered amines are obtained by the Michael Addition of primary amines to dialkyl maleates. The resulting secondary amine, an aspartic acid ester, has reduced reactivity due to steric effects and hydrogen bonding. Like the parent amines, aspartic acid esters react with isocyanates to form ureas (Figure 5).

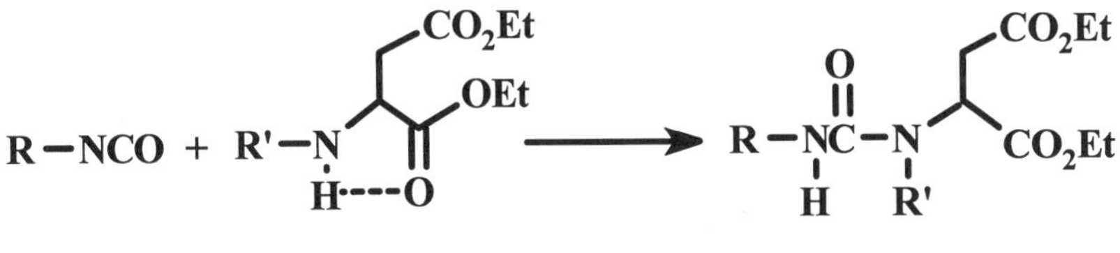

$$R-NCO + R'-\underset{H\text{----}O}{N}\!\!\!\overset{CO_2Et}{\underset{OEt}{\diagup}} \longrightarrow R-\underset{H}{N}\!\!\!\overset{O}{\underset{\|}{C}}-\underset{R'}{N}\!\!\!\overset{CO_2Et}{\underset{CO_2Et}{\diagup}}$$

Figure 5. Reaction of an Aspartic Acid Ester with an Isocyanate.

The reactivity of these amines towards polyisocyanates was found to be dependent on the structure of the parent amine. As shown below in Table II, cycloaliphatic polyaspartic acid esters yield gel times of several hours, with an additional ring methyl group extending the gel time to greater than a day due to steric hindrance. Acyclic structures, such as hexamethylene diamine, are more reactive than their cyclic analogs, but still much slower than typical primary and secondary amines.

Table II. Reactivity of Aspartic Acid Esters Towards Polyisocyanates

$$\text{EtO}_2\text{C-CH(CO}_2\text{Et)-NH-R-NH-CH(CO}_2\text{Et)-CO}_2\text{Et}$$

R	Viscosity *	Gel Time **
4-CH$_3$-C$_6$H$_{10}$-CH$_2$-C$_6$H$_{10}$-4-CH$_3$	1500	>24 h
C$_6$H$_{10}$-CH$_2$-C$_6$H$_{10}$	1200	2-3 h
3,3,5-trimethylcyclohexyl	800	2-3 h
-(CH$_2$)$_6$-	150	<5 min

* 100% Solids, mPa·s at 23 °C
** HDI Polyisocyanurate/Polyaspartic Acid Ester, NCO/NH: 1.0, 65% solids in 1:1 MEK/Aromatic 100

Significant differences in catalysis of these products compared to that of hydroxyl/isocyanate chemistries has also been reported (7). In direct contrast to urethane reactions, the rate of aspartic acid ester reaction with isocyanates was inhibited by the presence of tin (IV) compounds such as dibutyltin dilaurate. Also in direct contrast it was found that carboxylic acids and water were very effective catalysts for the aspartic acid ester reactions.

Though a good understanding of the reaction of aspartic acid esters with isocyanates has been developed and these systems are now moving into commercial use, they still do not represent a perfect partner for isocyanates in two component systems. Specifically, they still react too fast with polyisocyanates in the container after mixing, thus limiting their utility in a fast dry system.

Use of Imines as Isocyanate Coreactants

Imine Formation. Another route to blocked amines is through imine formation. Amines readily react with carbonyl-containing compounds in an equilibrium process to produce imines and water (Figure 6). There are two basic types of imines used in coatings chemistry. The first, a ketimine, is the result of a ketone and a primary amine combining. When an aldehyde is used in place of a ketone, the result is an aldimine. Because imine formation is an equilibrium process, parameters such as reaction temperature, solution pH and reactant concentration play an important role in the effectiveness of amine to imine conversion. The following sections will compare and contrast the preparation and stability of ketimines and aldimines and discuss the unique manner in which these materials react with aliphatic polyisocyanates.

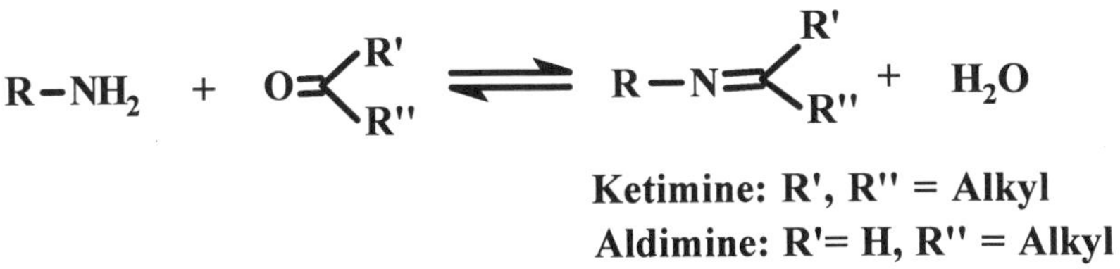

$$R-NH_2 + O=C{\binom{R'}{R''}} \rightleftharpoons R-N=C{\binom{R'}{R''}} + H_2O$$

Ketimine: R', R'' = Alkyl
Aldimine: R'= H, R'' = Alkyl

Figure 6. Imine Formation.

Ketimines in Epoxy Coatings. The use of imine groups as a source of blocked primary amine crosslinking agents has found some acceptance in high solids coatings. For example, polyfunctional ketimine resins have a long history of use in epoxy coatings as moisture activated curing agents. The ketimine functional groups themselves do not readily react with epoxy groups. More typically, exposure to ambient moisture hydrolyzes the ketimine, releasing a primary amine that rapidly reacts with two epoxy groups to crosslink the resin. The drawback to this system is that the hydrolysis step also releases one mole of ketone as a volatile for each amine generated. The VOC impact of this is evident in the hydrolysis of a typical product now in use, the bis-methyl isobutyl ketone (MIBK)-ketimine of ethylene diamine. Hydrolysis of this product results in only a 27% retention of its mass in the final film (Figure 7). The remaining 73% is released as volatiles. So even though ketimines in epoxy systems do possess the reactivity aspects required of high solids, the method of reactivity control generates an unacceptable level of VOC.

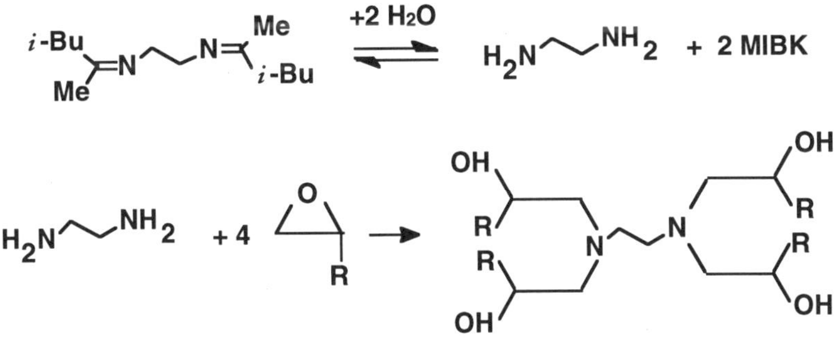

Figure 7. Hydrolysis of the bis-MIBK Ketimine of Ethylene Diamine and Reaction with Epoxy Functional Groups.

Ketimines as Isocyanate Coreactants. It was previously reported that the chemistry of ketimines and aliphatic isocyanates proceeds by a different route *(12)*. With isocyanate-based systems the ketimine no longer has to hydrolyze to react, but instead can react directly. This reaction is extremely slow in the absence of weak acid catalysts or moisture, but proceeds very quickly to a high yield when exposed to ambient moisture.

In the curing film the direct reaction between isocyanate and ketimine is competing with the hydrolysis of the ketimine. The kinetics of this competition are primarily dependent on the structure of the ketimine and the presence of catalytic groups. It was found that between 50 and 75% of the ketimine groups react directly with isocyanates and the remainder hydrolyze before reacting.

As with epoxy coatings, ketimines in isocyanate systems were found to give inhibition of reaction before application with acceleration taking place after application of the coating. However, even with the ability to undergo a direct reaction with an isocyanate, the partial hydrolysis of the ketimines results in too high of a VOC contribution.

Aldimines as Isocyanate Coreactants. The initial promising results with ketimines led us to further investigate the direct reaction between isocyanates and imine groups with emphasis on finding structures with a reduced tendency towards hydrolysis. All attempts at varying the ketone structure resulted in either an increased number of unwanted side reactions or an increased hydrolysis rate. Surprisingly, the use of an aldehyde to form an aldimine showed promise for control of reaction product profile.

Preparation of IPDA Ketimine vs. IPDA Aldimine. The most common method of preparing bis imines is the reaction of diamines with ketones or aldehydes, as shown in Figure 8 *(12)*. For the current study we compared the bis ketimine (**I**) formed from isophorone diamine and methylisobutyl ketone with the corresponding bis aldimine (**II**) derived from isobutyraldehyde. In the preparation of **I**, water must be removed from the reaction mixture to drive it to completion. Amines react with aldehydes at room temperature to produce the desired aldimines quantitatively. Even better, the reaction to form **II** goes to completion without the need to remove the evolved water.

Because the reaction with ketones requires higher temperatures and longer reaction times, the ketimine products develop a yellow color during production. For

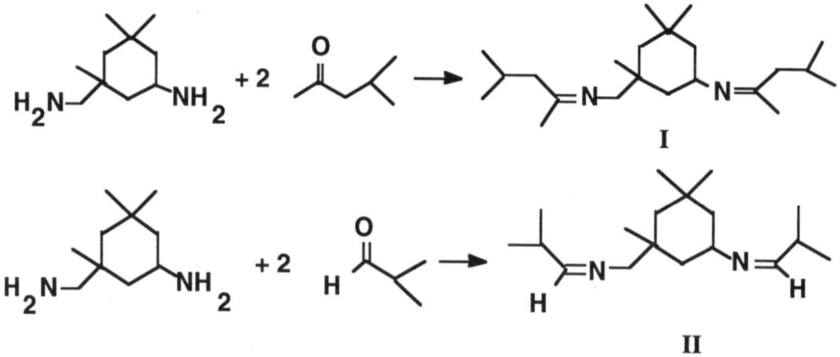

Figure 8. Preparation of Aldimines and Ketimines.

example, **II** has an American Public Health Association (APHA) color of about 20, whereas the ketimine product (**I**) has an APHA color of about 150. The ketimines also contain more impurities, because side reactions occur under these conditions. Some of these impurities are higher molecular weight and lead to higher viscosity products. The viscosity of **II** at 25 °C is 15 mPa•s, whereas the ketimine (**I**) has a viscosity of about 80 mPa•s.

Hydrolytic Stability of Ketimines vs. Aldimines. The hydrolytic stability of the ketimine and aldimine were compared by first dissolving one gram of each compound in five grams of tetrahydrofuran, to which we added 2 equivalents of water for each equivalent of imine. Infrared spectra of the solutions were taken at 0 hours, 3 hours, 1 and 3 days. Evidence for hydrolysis of the imine was taken as the appearance of a peak at about 1720 cm^{-1}, which would be due to carbonyl absorption.

As shown in Figure 9, we found that within 3 hours the ketimine began to produce a peak at 1717 cm^{-1}. This indicated that the ketimine was hydrolyzing to yield the starting ketone and amine. Within 3 days, the carbonyl and imine absorption peaks were almost of the same intensity. In sharp contrast the aldimine did not produce a noticeable carbonyl peak even after 3 days (Figure 10). This is strong evidence that the aldimine is not hydrolyzing to any appreciable amount under ambient conditions.

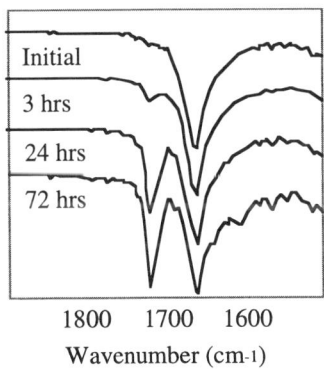

Figure 9. IR Spectra of **I** in THF/Water Over Time.

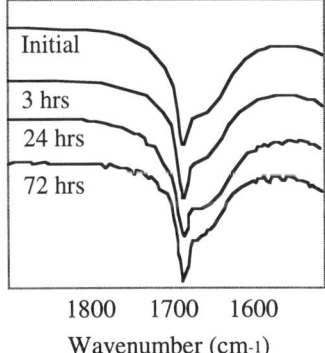

Figure 10. IR spectra of **II** in THF/Water Over Time.

Structure Determination of Reaction Products. In order to better understand the direct reaction between an aldimine and an isocyanate we conducted model reactions with a difunctional aldimine (similar in structure to **II**) and a monofunctional isocyanate (octadecylisocyanate). After heating the mixture at 70-80 °C for 72 hours the reaction products were analyzed by ^{1}H and ^{13}C-NMR. We found that the reaction yielded one predominate product that could best be described as an "enurea". The general structure and postulated reaction pathway are shown in Figure 11.

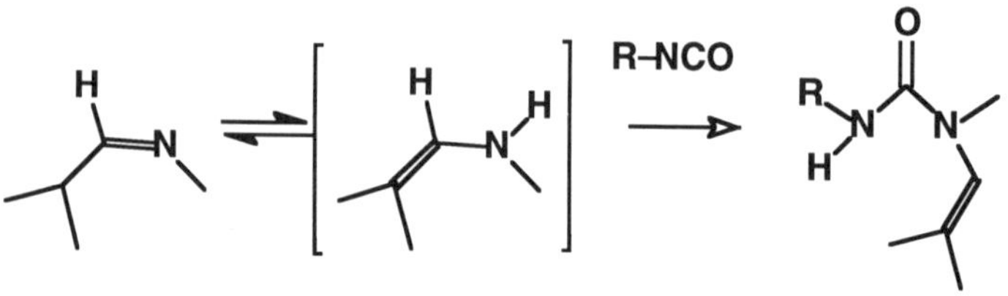

Figure 11. Postulated Mechanism for the Formation of an "Enurea".

Further model studies were carried out and have shown that hydrolysis can indeed take place under ambient application conditions. In practice, the actual retention of solids of the aldimine group in reaction with an isocyanate is dependent on many variables, including:

- structure of the parent amine
- catalysis - i.e. carboxylic acids and alcohols
- solvents used in the formulation
- additives and
- curing conditions.

For example, in the reaction between butyl isocyanate and the isobutyraldehyde aldimine of cyclohexyl amine, it is possible to cleanly form the urea product resulting from hydrolysis when the reaction was carried out at high humidity (Figure 12). However, the same reaction carried out at elevated temperature afforded the previously described "enurea" structure (Figure 13).

Figure 12. IR Spectra of Urea.

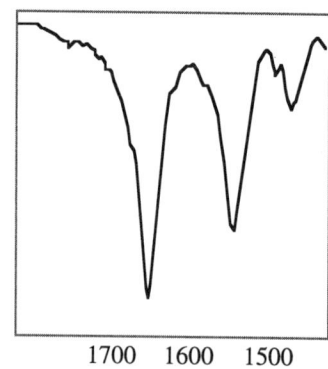

Figure 13. IR Spectra of Enurea.

Performance in a Model Coating Formulation. The differences in the reactivity of **I** and **II** toward isocyanates can be seen in solution viscosity build and drytime measurements of a simplified high solids model system. In this system the imines were combined with a polyisocyanurate of hexamethylene diisocyanate at an NCO/N ratio of 1:1 and reduced with butyl acetate to 75% resin solids.

The ketimine based system had an initial viscosity of 15 mPa•s which increased over a 2 hour period to 35 mPa•s. After 48 hours the reaction mixture had gelled. When the aldimine (**II**) was used the initial viscosity was 10 mPa•s and remained unchanged over a 2 hour period. After 48 hours a viscosity of 15 mPa•s was observed and, under scrupulously dry conditions, gel times of more than 6 weeks after mixing can be obtained! Films cast from both formulations dried to the touch within 1 hour. Both systems showed excellent solvent resistance within 24 hours of application, indicating that complete chemical cure was obtained.

These results demonstrate that both imines can fulfill the reactivity requirements for high solids applications, however comparison of the % solids retention of the two formulations clearly shows the advantage of the aldimine product **II**. Ketimine **I** acts as if it is 75% solids when subjected to testing for VOC content (ASTM D2369-93, 24 hour incubation). This result corresponds to one-half of the ketimine groups reacting directly with the isocyanate and the remainder via hydrolysis. In contrast, the aldimine system fulfills the third requirement for high solids coatings. In this test the aldimine **I** shows a >95% retention of solids without factoring in volatile impurities and the accuracy of the test. This is taken to indicate exclusive direct reaction with the polyisocyanate without hydrolysis of the aldimine. The direct reaction is also supported by the lack of the distinct odor of isobutyraldehyde that would be expected if hydrolysis was taking place.

Conclusions

The direct reaction between isocyanates and aldimines appears to be a significant advancement towards the development of high solids coatings which can meet the increasingly stringent requirements for VOC reduction. Though ketimines were a step in the right direction, they are prone to hydrolysis, thereby detracting from their VOC benefit. The search for a mechanism to suppress the hydrolysis reaction led to the discovery that aldimines not only yielded higher solids but also: required less energy to produce, possessed better color and viscosity, and had a better reactivity profile. This paper demonstrates how safer chemicals can be designed through molecular modifications that reduce toxicity and obviate the need for associated hazardous substances.

Acknowledgments

The authors would like to thank the many people within Bayer Corporation who have contributed to the work discussed in this paper, including, Lanny Venham, A. Donald Meltzer, S. Ming Lee, Gary Allen and Joe Dettorre.

Literature Cited

1. Squiller, E. P.; Rosthauser, J. W.; *Modern Paint and Coatings* **1987**, June , p 88.
2. Jorissen, S. A.; Rumer, R. W.; Wicks, D. A. *Proceedings of the Nineteenth Water-borne, Higher Solids and Powder Coatings Symposium* **1992**, p 182.
3. Wojcik, R. T.; Goldstein, S. L.; Malofsky, A. G.; Barnowski, H. G.; Chandalia, K. B. *Proceedings of the Twentieth Water-Borne, Higher Solids, and Powder Coatings Symposium* **1993**, p 26.
4. Wojcik, R. T. *Proceedings of the American Chemical Society, Division of Polymeric Materials: Science and Engineering* **1994**, *70*, p 116.
5. O'Driscol, *J. Coat. Technol.* **1983**, *55*, Nr. 705, p 57.
6. Schrantz, J.; *Industrial Finishing* **1992,** October, p 14.
7. Wicks, D. A.; Yeske P. E., "Proceedings of the Twentieth Waterborne & Higher Solids and Powder Coatings Symposium" **1993**, p 49.
8. Zwiener, Ch.; Schmalstieg, L.; Sonntag, M.; Nachtkamp, K.; Pedain, J., B; *Farbe und Lack* **1991**, Nr. 12, p 1052.
9. Zwiener, Ch.; Schmalstieg, L.; Sonntag, M.; *European Coatings Journal* **1992** Nr. 10, p 588.
10. Zwiener, Ch., Sonntag, M., Kahl, L.; *Proceedings of the Twentieth FATIPEC Congress* **1990**, p 267.
11. Squiller, E. P.; Wicks, D. A. *Proceedings of the Twentieth Waterborne & Higher Solids and Powder Coatings Symposium* **1993**, p 15.
12. Layer, R. W. *Chem. Rev.* **1963**, *63*, p 489.

Author Index

Arcos, Joseph C., 62
Argus, Mary F., 62
Bodor, Nicholas, 84
Boethling, R. S., 156
DeVito, Stephen C., 16,194
Fung, V., 138
Garrett, Roger L., 2
Jacobson, A. H., 224
Jones, Jeffrey P., 116
Kim, Anne, 172
Lai, David Y., 62
Milne, G. W. A., 138
Nabholz, J. Vincent, 172
Newsome, Larry D., 172
Sieburth, Scott McN., 74
Wang, S., 138
Wicks, Douglas A., 234
Willingham, G. L., 224
Woo, Yin-tak, 62
Yeske, Philip E., 234

Affiliation Index

Bayer Corporation, 234
National Institutes of Health, 138
Rohm and Haas Company, 224
State University of New York—Stony Brook, 74
Tulane University, 62
U.S. Environmental Protection Agency, 2,16,62,156,172,194
University of Florida, 84
University of Rochester, 116

Subject Index

A

Absorption
  gastrointestinal tract, 19–20
  influencing factors, 19,20t
  lung, 20–21
  skin, 20t,21
Acetic acid, 44
Acetoacetates, 48,49f
Activated soft compounds, 94–96
Activation energies, prediction, 123
Active metabolite principle, 96–98
Acute lethality, nitriles, 196–206
ADAPT, toxicity estimation method, 141
Alkenes, 30,32–33
Alkyl phosphate surfactants, biodegradability, 167
4-Alkylphenol(s), 28,29f
Alkylphenol ethoxylates, 165–166
Alkynes, 30,32–33
Allyl alcohols, 28,30,31f
AM1 semiempirical molecular orbital method, use in expert systems for retrometabolic design, 110,112–114
Amphoteric dyes, 186
Anionic dyes, 183–184
Antifoulant, criteria, 225
Aquatic organisms, types of chemical-induced lethality, 172–173
Aquatic toxicity, 173–175
Aquatically safer chemicals, design by modification of chemical structure
  chelation, 179
  dyes, 183–186
  inorganic metallics, 189
  ion pairs, 178

Aquatically safer chemicals, design by modification of chemical structure—*Continued*
  molecular size and weight, 176,178
  narcosis vs. excess toxicity, 179–182
  octanol–water partition coefficient, 175–177f
  organometallics, 189
  prediction of aquatic toxicity, 173–174
  polyanionic monomers, 188–189
  surfactants, 186–188
  water solubility, 176,177f
  zwitterions, 178–179
Aspartic acid esters, 239–240
Associated toxic substances, elimination for need, 50–51
Atenolol, 107,109f
Azo dyes, 45–50

B

Benign by design, concept, 157
Benzene, 44
Benzidines, use of sulfonated diaminobenzanilides as substitutes, 50
Bioactivation
  cytochrome P450 mediated, 116–135
  description, 10,117
  mechanisms
    metabolic activation pathways, 64,66f
    nature of amine–amine-generating groups, 67
    number and nature of aromatic rings, 67
    planarity of molecules, 68–69
    position of amine–amine-generating groups, 65f,67
    reactivity, 64,67
Bioanalogy, description, 75
Bioavailability, definition, 17
BIODEG, prediction of biodegradability, 159–162
Biodegradability, prediction using group contribution method, 159–162
Biodegradable chemical design
  alkyl phosphate surfactants, 167
  alkylphenol ethoxylates, 165–166
  dialkyl quaternaries, 163–165

Biodegradable chemical design—*Continued*
  linear alkylbenzenesulfonates, 162–163
  microbial basis of biodegradation, 157–159
  polypropylene derivatives, 167
  prediction using group contribution method, 159–162
  propylene oxide, 168
  transport of banned substances, 169–170
Biodegradation, microbial basis, 157–159
Bioisosterism, development, 74–75
Biologically safer chemicals, design based on retrometabolic concepts, 84–114
β Blockers, 107,109f

C

Cancer risk reduction through mechanism-based molecular design of chemicals
  aromatic amine dyes
    bioactivation mechanisms, 64,66–69
    molecular design for low carcinogenic potential, 69–71
    structural criteria for carcinogenicity, 63–65f
  background, 62–63
Cannabinoid analogue HU–211, 108,110,111f
Carbon, isosteric replacement with silicon, 74–80
Carbon-based chemicals, similarities and differences with silicon-based chemicals, 76
Carboxylic acids, qualitative structure–activity relationships, 40,41t
Carcinogenicity, measurement, 138–139
CASE, toxicity estimation method, 141–143
Cationic dyes, 184–186
Cetylpyridinium chloride, retrometabolic design, 47
Chelation, role in design of safer chemicals, 179
Chemical carcinogenesis, mechanism studies, 62–63
Chemical delivery system, 89–90
Chemical education, role in design of safer chemicals, 12–14f

INDEX

Chemical-induced lethality in aquatic organisms, types, 172–173
Chemical industry, role in design of safer chemicals, 13–15
Chemical substances, commercial, *See* Commercial chemicals
Chemical toxicity
  adsorption, 18–21
  distribution, 21
  excretion, 22–23
  metabolism, 21–22
  requirements, 17–18
  toxicodynamics, 22
Chlorophenothane substitutes, 108,109f
Coatings, *See* Environmentally friendly high solids coating design
Coatings systems, need for environmentally friendly materials, 233
Commercial chemicals
  design for safety, 107–114
  design using retrometabolic concept, 89
  exposure routes, 85,86f
  testing criteria, 87,89
  toxicity concern, 116–117
  toxicity equation, 87,88f
COMPACT, toxicity estimation method, 144–146f
Computers in toxicology and chemical design, 138–153
  cost, 153
  National Cancer Institute/National Toxicology Program bioassay, 148,151t,f,152t
  toxicity estimation methods, 139–150
CUP101, description, 120
Cytochrome P450 enzymes
  bioactivation rates, 116–135
  catalytic cycle, 120–122

D

DEREK, toxicity estimation method, 144,147
Design for the Environment Program, 5
Design of safer chemicals
  approaches, 17
  aquatically safer chemicals, 172–189

Design of safer chemicals—*Continued*
  aspects of chemical toxicity, 17–23
  biodegradable chemicals, 156–170
  cooperation between U.S. Environmental Protection Agency and industry, 4–5
  criteria, 6–7
  cytochrome P450 mediated bioactivation, 116–135
  definition of concept, 5–6,8f
  description, 16–17,194–195
  external approaches, 7,10
  elimination of need for associated toxic substances, 50–51
  environmentally safe marine antifoulants, 224–232
  foundation, 11–15
  history, 3–4
  identification of equally useful, less toxic chemical substitutes, 47–50
  isosteric replacements, 42–46
  molecular modifications that reduce absorption, 23–24
  obtaining toxicity-related information, 51–54
  retrometabolic design, 47,49f
  role of academia, 3
  structure–activity relationships, 36–41
  toxic mechanisms, 24–37
Dialkyl quaternaries, 163–165
Dibasic esters, substitute for glycol ethers, 51
4,5-Dichloro-2-*n*-octyl-4-isothiazolin-3-one
  biological testing against fouling organisms, 226–227
  comparison to tributyltin oxide, 232
  concentration in fish, 230
  environmental partitioning, 229
  metabolic pathways in aquatic sediment, 229–230
  toxicology in nontarget aquatic organisms, 230
3-(Dimethylamino)propionitrile, 214–216
Domestication of chemistry, definition, 6
Drug substances
  design and development, 85
  development process, 87

Drug substances—*Continued*
federal regulations governing approval, 85,87
metabolism scheme, 85,86f
toxicity, 85,88f
Drug toxicity, equation, 85,88f
Dyes, 182–186

E

Economic credibility, 11
Electron-insulating intercyclic linkages, role in molecular design of low carcinogenic potential, 70–71
Electrophiles, toxic mechanisms, 24–26t
Enurea, formation mechanism, 243–244
Environmental risk, definition, 225
Environmentally friendly high solids coating design
  approaches to low volatile organic compound coatings, 235–236
  imine–isocyanate chemistry
    aspartic acid esters, 239–240
    background, polyurea coatings, 239
    polyurethane coatings, 236–238
    use of imines as isocyanate coreactants, 242–245
  need, 234–235
Environmentally safe marine antifoulants
  biological testing against fouling organisms, 227–229
  characterization of fish metabolites, 230
  comparison of 4,5-dichloro-2-n-octyl-4-isothiazolin-3-one to tributyltin oxide, 232
  concentration in fish, 230
  environmental partitioning, 229
  experimental procedure, 225–226
  metabolic pathway in aquatic sediment, 229–230
  toxicology in nontarget aquatic organisms, 230–231
Equally useful, less toxic chemical substitutes from another class
  acetoacetates, 48,49f
  isothiazolones, 48,50
  process, 47–48
  sulfonated diaminobenzanilides, 50

EXAMS, modeling of environmentally safe marine antifoulants, 224–232
Excess toxicity, comparison to narcosis toxicity, 179–180,181t
Excretion, description, 22–23
Expert systems, retrometabolic drug design, 110,112–114

F

Federal regulations, approval of drug substances, 85,87
Fouling, damage created, 224
Free radicals, 33

G

Glycidyl ethers, qualitative structure–activity relationships, 40,41t
Glycol ethers, 34–37f,51
Green algae, function, 172
Green chemistry
  alternative synthetic pathways, 5
  description, 16
  interpretation of pollution prevention, 3

H

Halogenated hydrocarbons, rate prediction, 126–131
HAZARDEXPERT, toxicity estimation method, 147,149f
Hepatic cytochrome P–450 and related enzymes, disadvantages, 90,93t
n-Hexane substitutes, 36,37f
High solids coatings, See Environmentally friendly high solids coating design
HU–211, 108,110,111f
Hydrocarbons, halogenated, rate prediction, 126–131
Hydrocortisone, 104,105f
Hydrophilic groups, role in molecular design of aromatic amine dyes of low carcinogenic potential, 70t,71
Hydroxy functional coreactants, 238
α-Hydroxy keto and hydroxy substituents of hydrocortisone, 104–107

INDEX

I

Imine–isocyanate chemistry, role in design of environmentally friendly high solids coatings, 234–245
3,3'-Iminodipropionitrile, 214–218
Inactive metabolite approach to soft drug design, 104–109f
Indirect biotoxication, description, 10
Insecticide MTI–800, 46
Intrinsic toxicity, definition, 17
Ion pairs, 178
Isocyanate(s), use of acetoacetates as substitutes, 48,49f
Isocyanate–imine chemistry, role in design of environmentally friendly high solids coatings, 234–245
Isostere
  definition, 42
  structures, 74
Isosteric analogue of cetylpyridinium chloride, 99–100
Isosteric replacement
  carbon with silicon, 74–75,78–80
  organosilanes, 77
  silicon as isostere of carbon, 75
  similarities and differences of silicon- and carbon-based chemicals, 76
  use in design, 42–46
Isothiazolones, 48,50

K

Ketimines, role in design of environmentally friendly high solids coatings, 234–245

L

Linear alkylbenzenesulfonates, 162–163
Long-chain alkyl or bulky N substituents to amino groups, role in molecular design of aromatic amine dyes of low carcinogenic potential, 69,70t
Low volatile organic compound coatings, approaches, 235

Lysine oxidase, inhibition during nitrile-induced osteolathyrism, 209–212

M

Man-made chemicals, use-oriented, 84–85
Marine antifoulants, 224–232
Mechanism-based molecular design of chemicals, cancer risk reduction, 62–71
Mechanistic toxicological research, 12
Metabolic activation pathways, role in bioactivation mechanisms, 64,66f
Metabolism
  description, 21–22
  phase I and II reactions, 117
Metalized acid dyes, 184
Metiamide, 45
Metoprolol, 107,109f
Microbial basis of biodegradation
  chemical structure vs. biodegradability, 158–159
  occurrence, 157
  pathways, 157–158
Molecular design
  aromatic amine dyes of low carcinogenic potential, 69–71
  chemicals, cancer risk reduction, 62–71
Molecular modifications that reduce absorption, 23–24
Molecular size and weight, role in design of safer chemicals, 176,178
Mutagenicity, marker for carcinogenicity, 139

N

Narcosis toxicity
  comparison to excess toxicity, 179–181t
  description, 172–173
National Cancer Institute–National Toxicology Program bioassay, 148,151t,f,152t
Nature of amine–amine-generating groups, role in bioactivation mechanisms, 67
Neurotoxic nitriles, 214–218
Neutral dyes, *See* Nonionic dyes

Nitrile(s)
  acute lethality, 196–206
  description, 196
  rate prediction, 124–126
  toxicity, 124
Nitrile-induced osteolathyrism
  characteristics, 206–207
  design of safer nitriles, 213–214
  mechanism, 207,209–212
  nitrile potency, 207,208t
  occurrence, 206
  structure–activity relationships,
    212–213
p-Nitrosophenoxy radical model, rate
  prediction of cytochrome P450 mediated
  hydrogen atom abstraction, 116–135
Nonionic dyes, 182–183
Number and nature of aromatic rings, role
  in bioactivation mechanisms, 67

O

Octanol–water partition coefficient, role
  in design, 175–177f
Office of Pollution Prevention and Toxics,
  mission, 156–157
Oil-based paints, substitutes, 51
ONCOLOGIC, toxicity estimation method,
  147–150f
Organometallics, 189
Organosilane(s), abiotic degradation, 77
Organosilane fungicides, environmentally
  safe chemicals, 78,79f
Organosilane pyrethroids, environmentally
  safe chemicals, 78–80
Organotin antifoulants, use of
  isothiazolones as substitutes, 48,50
Organotin biocides, ecotoxicological
  problems, 224–225
Osteolathyrism, nitrile induced, 206–214

P

Persistent organic pollutants, resistance
  to biodegradation, 168
Phase I and II reactions of metabolism, 117

Planarity of molecules, role in
  bioactivation mechanisms, 68–69
Pollution prevention
  challenges, 16
  concept of green chemistry, 3
Pollutant Prevention Act of 1990, 5,156
Polyanionic monomers, 188–189
Polyethoxylated nonylphenols, qualitative
  structure–activity relationships, 38
Polyisocyanates, formation, 236–238
Polypropylene derivatives,
  biodegradability, 167
Polyurea coatings, formation, 239
Polyurethane coatings, 236–238
Prevention of pollution, 3,16
Proelectrophiles, description, 179
Propargyl alcohol, 30
Propylene oxide, biodegradability, 168
Prostaglandins, 101–103f

Q

Qualitative structure–activity relationships
  (QSAR)
  carboxylic acids, 40,41t
  glycidyl ethers, 39
  prediction of aquatic toxicity, 174
  polyethoxylated nonylphenols, 38
  1,2,4-triazole-3-thiones, 39–41t

R

Rates of cytochrome P450 mediated
  bioactivation
  bioactivation reactions, 117–120
  enzyme kinetics, 133
  enzymology, 120–122
  halogenated hydrocarbons, 126–131
  model development, 123–125f
  model failure situations, 130,132
  nitriles, 124,126
  pharmacokinetics, 132–133
Reactive toxicity, description, 173
Reactivity, role in bioactivation
  mechanisms, 64,67

INDEX

Retrometabolic drug design
  advantages, 89
  chemical delivery system, 89–91f
  expert systems, 110,112–114
  examples, 47,49f
  safer commercial chemicals, 107–114
  soft drugs, 90–109
Risk reduction for cancer, mechanism-based molecular design of chemicals, 62–71

S

Safer nitriles
  acute lethality data, 196–198
  less acutely toxic nitriles, 205–206
  mechanism, 198–199,202f
  neurotoxic nitriles
    3-(dimethylamino)propionitrile, 214–216
    3,3'-iminodipropionitrile, 214–218
    nitrile-induced osteolathyrism, 199–214
Silafluofen, field activity, 79–80
Silane analogues of 1,1,1-trichloro-2,2-bis(p-chlorophenyl)ethane, 78
Silicon, isosteric replacement of carbon, 74–80
Silicon-based chemicals, similarities and differences with carbon-based chemicals, 76
SIMCA, toxicity estimation method, 141
Soft chemicals and drugs
  activated soft compounds, 94–96
  active metabolite principle, 96–98
  analogues, 98–103f
  definition, 47
  description, 90,92f
  inactive metabolite approach, 101,104–107,109f
  types, 9–94
  See also Retrometabolic drug design
Solubility, effect on biodegradation, 159
Specific toxicity, description, 173
Structural alerts, toxicity estimation method, 143–144

Structure–activity relationships
  description, 36,38
  prediction of aquatic toxicity, 173–174
  QSAR, 40,41t
  qualitative relationships, 38–40,41t
Substituents
  on ring, role in bioactivation mechanisms, 68
  ortho to amine–amine-generating groups, role in molecular design of aromatic amine dyes of low carcinogenic potential, 69,70t
Sulfonated diaminobenzanilides, 50
Supercritical CO_2, substitute for volatile organic compounds, 51
Surfactants, 186–188

T

33–50 Program, description, 5
Textile processing, opportunities for pollution prevention by molecular design, 167
Toluene, 34,35f
TOPKAT, toxicity estimation method, 139–140,142f
Toxic mechanisms
  bioactivation to electrophiles
    alkenes and alkynes, 30,32–33
    4-alkylphenols, 28,29f
    allyl alcohols, 28,30,31f
    mechanism, 28
    propargyl alcohol, 30
  electrophiles, 24–27
  free radicals, 24,33
  glycol ethers, 34–36,37f
  n-hexane substitutes, 36,37f
  process, 33
  toluene, 34,35f
Toxic Release Inventory Program, 5
Toxic Substances Control Act, 173–174
Toxicity estimation methods
  ADAPT, 141
  CASE, 141–143
  COMPACT, 144,145–146f
  DEREK, 144,147

Toxicity estimation methods—*Continued*
  HAZARDEXPERT, 147,149f
  ONCOLOGIC, 147–148,149–150f
  SIMCA, 141
  structural alerts, 143–144
  TOPKAT, 139–140,142f
Toxicity of chemical compounds, measurement, 138
Toxicity testing in animals, cost, 139
Toxicodynamics, description, 22
Toxicological chemistry, development, 13,14f
Toxicological considerations for chemists, 16–55
Toxicology, use of computers, 138–153
1,2,4-Triazole-3-thiones, QSAR, 39–41t
Tributyltin oxide, comparison to 4,5-dichloro-2-*n*-octyl-4-iso-thiazolin-3-one, 232

V

Volatile organic compound(s)
  emission reductions, 234–235
  use of supercritical CO_2 as substitute, 51
Volatile organic compound coatings, 235

W

Water-based paints, substitutes for oil-based paints, 51
Water solubility, role in design of safer chemicals, 176,177f

Z

Zwitterions, 178–179